Is Water H$_2$O?

BOSTON STUDIES IN THE PHILOSOPHY OF SCIENCE

VOLUME 293

For further volumes:
http://www.springer.com/series/5710

Hasok Chang

Is Water H$_2$O?

Evidence, Realism and Pluralism

Springer

Hasok Chang
Department of History and Philosophy of Science
Free School Lane
University of Cambridge
Cambridge, England CB2 3RH
United Kingdom

ISSN 0068-0346
ISBN 978-94-007-3931-4   ISBN 978-94-007-3932-1 (eBook)
DOI 10.1007/978-94-007-3932-1
Springer Dordrecht Heidelberg New York London

Library of Congress Control Number: 2012934281

Printed on acid-free paper

Springer is part of Springer Science+Business Media (www.springer.com)

To Seung-Joon Ahn (1966–1991),

Alter Ego

Acknowledgments

So many people have helped in the making of this book and its author, and it will be impossible to remember them all here. I would like to thank once again the very long list of people acknowledged in my first book, without naming all of them. That list begins with my parents, whose love, support and trust continue to be the bedrock of my life, and ends with my wife Gretchen Siglar, who has lovingly shared all the highs and the lows of the long gestation of this book as parts of our life together. In the middle are my brother and sister and their families, the best guides and companions in life that anyone could ask for. I must also mention my old teachers who are still mentoring me through the academic journey, especially Nancy Cartwright, Gerald Holton, Peter Galison, John Dupré, and Jim Woodward.

To my old colleagues and students at University College London, I owe much gratitude for my 16 wonderful years there. Even if they didn't help me specifically on the work going into this book, the intellectual stimulation and conviviality they provided have been essential. It is impossible to name everyone. For the closest collaboration and the most generous help in the latter half of my time there, I especially thank my colleagues Joe Cain, Daren Caruana, Andrea Sella, Michela Massimi, Steve Miller, Brian Balmer, Philip Dawid, Caroline Essex, Beck Hurst, and Helen Wickham. At the institutional level, I have been most grateful for the support and encouragement from Richard Catlow and Malcolm Grant. And the greatest pleasure of academic life has been to see students develop into colleagues; the most notable cases include Grant Fisher, Sabina Leonelli, Georgette Taylor, Catherine Jackson, and Chiara Ambrosio, who have each helped me write this book in numerous ways that they may not even realize. Similar pleasure comes when students become published authors, and in that context I thank especially the participants in the undergraduate research projects on chlorine and electricity.

I would like to thank my new colleagues and students at Cambridge, especially at the Department of History and Philosophy of Science and at Clare Hall, for providing a congenial and stimulating environment for the completion of this project. I look so much forward to the coming years. I would also like to record my thanks to Lisbet Rausing for establishing the chair that I now occupy, and appreciation for the work of my predecessors Peter Lipton, Michael Redhead and Mary Hesse.

Various organized research groups have been very important in helping me learn and think, and I thank my colleagues in those groups, and the organizations that made them possible. These include the Evidence, Inference and Enquiry project at UCL (funded by the Leverhulme Trust and the ESRC), the Analysis and Synthesis project (funded by the Leverhulme Trust), the Pragmatism, Pluralism and Phenomenology study group, AD HOC (supported in recent years by SHAC), the Society for Philosophy of Science in Practice (SPSP), and the UK and international networks for Integrated HPS.

Many people have been instrumental in welcoming me to the world of chemistry and its history and philosophy, and guiding me through what was very unfamiliar territory to start with. In addition to those already mentioned for other reasons, I would like to thank Rosie Coates, Jenny Rampling, Ursula Klein, David Knight, Robert Anderson, Peter Wothers, Bill Brock, Bernadette Bensaude-Vincent, Mary Jo Nye, Piyo Rattansi, Kat Austen, John Perkins, Mimi Kim, Seymour Mauskopf, Robin Hendry, Joachim Schummer, John McEvoy, Matthew Eddy, Fred Parrett, and Anna Simmons.

For countless valuable suggestions and much-needed practical help, I would like to thank Chiara Ambrosio and Shahar Avin, consummate research assistants; Shahar's duties included commenting on the entire penultimate draft of this book. I also thank: members of the All-London HPS seminar group, who most helpfully read and discussed with me earlier drafts of the first four chapters during the spring of 2010; two anonymous referees for Springer, for a great number of insightful and helpful critical suggestions; and the anonymous referee for Oxford University Press, to which the proposal for this book was also submitted.

Writing this book has been a humbling experience. Within its subject matter are so many things that I have reluctantly come to accept that I will never be able to master to my own satisfaction. I thought I would be able to manage by standing on the shoulders of giants, but that proved impossible when there were so many giants and such high shoulders. There are many contemporary authors whose work I wished to study better or incorporate more convincingly into my discussions, including Alan Rocke, Alan Chalmers, Ursula Klein, Howard Stein, William Wimsatt, Sandy Mitchell, Gerald Doppelt, and Deborah Mayo. I would also like to apologize to various people who offered me valuable suggestions that I have not been able to take up sufficiently. Erik Curiel is at the top of that list, but I should also mention Christian Hennig, Alex Bellamy, Jinho Kang, and Marita Hübner.

For all the help, encouragement and suggestions that I was able to absorb and use, I would like to thank the following people, in addition to those already mentioned. There are senior scholars whose encouragement has worked like a much-needed shot in the arm as this project evolved: Alan Chalmers, John Heilbron, Paul Hoyningen-Huene, Philip Kitcher, Helen Longino, Roberto Torretti, Ken Waters, Alison Wylie, and also Harvey Brown, Jed Buchwald, Jeremy Butterfield, Rom Harré, Don Howard, Frank James, and John Norton. Similarly, friends and colleagues whose enthusiasm and insights kept me going include Douglas Allchin, Hanne Andersen, Theodore Arabatzis, Katherine Brading, Julia Bursten, Cathryn Carson, Elizabeth Cavicchi, Jason Davies, Silvia De Bianchi, Sophia Efstathiou,

Melinda Fagan, Uljana Feest, Graeme Gooday, Jeff Hughes, Katie Kendig, J.B. Kennedy, Ian Kidd, Dongwon Kim, Sung Ho Kim, Vasso Kindi, Lucia Lewowicz, Felicia McCarren, Josipa Petrunic, Greg Radick, Léna Soler (and the PratiScienS), Katie Tabb, Junhee Won, and Sang Wook Yi. For their help with various chapters, including invitations to present various materials, I would like to thank Seamus Bradley, Sharyn Clough, Patrick Coffey, Marilena Di Bucchianico, Jonathan Everett, Roman Frigg, Susannah Gibson, Ron Giere, Florence Grant, Lina Hakim, Marta Halina, Jane Heal, Nick Huggett, Minsoo Kang, Jon Kaplan, Lena Kästner, Ismo Koponen, Antigone Nounou, Paolo Palmieri, Lydia Patton, Dean Peters, Stathis Psillos, Andrew Rabeneck, Miklos Redei, Felix Rietmann, Tad Schmaltz, Kyle Stanford, Jacob Stegenga, Thomas Sturm, Iain Watts, Michael Weatherburn, and Brad Wray. I also thank audiences at various places including Berkeley, Caltech, Cambridge, UIC, Duke, Durham, Exeter, Leeds, Les Treilles, LSE, Manchester, Minnesota, Missouri (St. Louis), MPI (Berlin), Oxford, Reading, the Royal Society, San Diego, SNU, Pittsburgh, Stanford, and the BSHS, BSPS, HSS, PSA, SPSP and HPS conferences.

Last but not least, I would like to thank the wonderful team at Springer who called forth this book from the realm of possibility to that of reality. Charles Erkelens and Ties Nijssen had enough faith in my work to commission and accept this project. Lucy Fleet saw through the rest of the process with admirable patience and efficiency. I also thank Corina van der Giessen, Sundaramoorthy Karthigayan, and their colleagues for all their painstaking work in the production process.

This book is dedicated to the unfading memory of Seung-Joon Ahn, my first and best friend, who would have understood and appreciated this work like no one else.

Contents

Introduction

Anyone with even the slightest acquaintance with modern science knows that water is H_2O. Yet it was a very difficult thing for scientists to learn. In this book I intend to show how contingent the series of decisions were that led people from the traditional assumption that water was an element to the consensus that it was a compound with the chemical formula H_2O, which was not reached till the late nineteenth century. Through this story of the changing conceptions of water, I also wish to advance the debate on some major philosophical issues, including realism and pluralism. I have deliberately chosen as the subject of my study one of the most familiar substances in human life and one of the most basic scientific facts about that substance. My aim is to make us all aware of the challenges involved in building scientific knowledge, no matter how simple or taken for granted. Without such awareness, we can reach neither a true appreciation of the achievements of science nor a properly critical attitude regarding the claims of science.

Over half of the book consists of three chapters containing a philosophical history of water from the middle of the eighteenth century to the late nineteenth century. I begin with a re-telling of the Chemical Revolution, in which water came to be recognized as a compound for the first time in Western science; I will attempt, and fail, to remove a lingering suspicion that there never was such a conclusive reason for rejecting the infamous phlogiston theory. Next I examine the early history of electrochemistry, in which electricity decomposed water into hydrogen and oxygen as expected, but a serious puzzle was raised about why the two gases came out from distant places while they were presumed to originate from the same water molecule in each case. This is followed by a slice of the early history of chemical atomism, in which chemists took more than half a century to agree on changing John Dalton's original formula for water from HO to H_2O. In these studies I intend to make some original historiographical contributions, as well as craft new philosophical ideas fit for the purpose of framing the historical accounts.

These concrete chapters will be followed by two that are more abstract, which systematically develop the philosophical ideas generated through the historical investigations. One general question plagues me through all the historical discussion: did the scientists have sufficient evidence to justify the verdicts that they reached?

A careful consideration of this question will lead me to formulate a fully contextual and practice-based view of evidence in scientific inquiry. This inevitably leads into the realism question: if scientific knowledge is contingent, can we still preserve the notion of scientific truth and its pursuit? Contingency also implies choice: past scientists could reasonably have made choices that would have led to systems of science that are different from what we have today. Rather than try to avoid this implication, I embrace it and develop it into a full-blown doctrine of pluralism in science.

My questioning of the simple and unique truth of the statement "Water is H_2O" will raise eyebrows and disturb commonplace assumptions, and that is fully intended. Independently of the details of my various arguments, it will be beneficial for people to realize that it is not crazy to subject the most fundamental truths of modern science to critical scrutiny, and to contemplate the possibility of scientific systems which deny or do without them. After all, many of the great and rational thinkers whose political, philosophical and scientific writings we still study with reverence did not have any idea that water was H_2O: Newton, Voltaire, Hume, Franklin, Goethe and Kant, just to mention a few out of a myriad. In any case, *very* modern science no longer subscribes to the notion that water is simply H_2O.[1] Not only does water contain rarer isotopes such as deuterium, but its familiar chemical and physical properties depend essentially on the presence of various ions, and on the continual connections and re-connections between neighboring molecules which belie the single-molecule formula of H_2O. If we had a simple heap of H_2O molecules, it would not be recognizable as water. Of course, the "H_2O" view still contains an important element of truth about the constitution of water, and continues to have heuristic utility. But it would be wrong to take it as an eternal and unqualified truth; rather, it was merely one important resting-point in the continuing progressive saga of science. This illustrates a general point: there is no benefit to be gained from a dogmatic adherence to a simple-minded scientific truth that science itself has already modified.

This book is a continuation of the "complementary science" project, which I launched with my previous book, *Inventing Temperature*. The studies contained here have further fulfilled the promise of complementary science, which is to use history and philosophy of science to address scientific questions that science itself neglects. In the earlier work I outlined the character of knowledge generated by complementary science in three categories: recovery, critical awareness, and new developments.[2] My investigations of the history of water have generated contributions in each of those categories.

What I offer here is a study in history, philosophy and science simultaneously. Any real originality in my work probably lies in the way I weave these three strands together, not within any single thread. Still, I hope that the first three chapters

---

[1] For scientific details see Eisenberg and Kauzmann (2005). Hendry (2008) gives a succinct summary.

[2] Chang (2004), 240–247.

constitute a modest original contribution to the history of chemistry in the eighteenth and nineteenth centuries, that the last two chapters provide at least a refreshing point of view in the philosophy of science, and that here and there my discussions raise interesting and unorthodox scientific ideas. The broad disciplinary reach and the unusual directions of my thought have meant that I have not been able to bring most aspects of the work anywhere near perfection. But I am emboldened and comforted by the words of an anonymous referee who reviewed the original proposal for this book, which I take the liberty to quote here: "in any brave and novel project… perfection can only be a pernicious dream."

The following is a very brief synopsis of the five main chapters of the book. Chapter 1, opening our philosophical history of water, is about the Chemical Revolution of the late eighteenth century. This is a very familiar topic in history and philosophy of science, but my re-examination of it will show that there never was sufficiently strong evidence at the time to warrant the triumph of Antoine-Laurent Lavoisier's oxygen theory (with water as a compound of hydrogen and oxygen) over the phlogiston theory (with water as an element). Phlogiston-based chemistry was actually a highly cogent system of knowledge, grounded in very concrete laboratory operations such as the calcination and reduction of metals. The concept of phlogiston provided some important unifying explanations, and played an important heuristic role in many empirical discoveries, including that of oxygen itself. Lavoisier's chemistry had many difficulties, both as recognized by his contemporaries and from a modern (whiggish) point of view. The very name of "oxygen" (acid-generator) embodies a mistaken theory of acidity, and Lavoisier's theory of combustion rested crucially on the concept of caloric, an imponderable fluid just like phlogiston.

All in all, I argue, there were no conclusive grounds of empirical evidence, simplicity or progressiveness that supported the complete elimination of the phlogiston theory. Rather, there was a genuine methodological incommensurability between the two systems of chemistry. Joseph Priestley was not irrational or unreasonable in his resistance to Lavoisierian chemistry, nor was he alone. So I conclude that phlogiston was killed prematurely; that is a shocking claim, and its implications must be considered seriously. I argue that the concept of phlogiston should have been kept on; it was not, so we might contemplate reviving it. But a look back at the subsequent history of chemistry reveals that phlogiston was in effect brought back, under different names. Lavoisier's chemistry never explained why chemical reactions happened, and phlogiston was later seen to have held the conceptual space that chemical potential energy would fill. On the other hand, phlogiston was even at the time commonly identified with electricity, and could easily have been kept and developed into the concept of free electrons. Eminent chemists such as William Odling, Justus Liebig and G.N. Lewis have recognized and expressed these phlogistic connections.

If Lavoisier was right about water, it should also have been possible to decompose water into hydrogen and oxygen. Chapter 2 begins by noting the great excitement following the invention of Alessandro Volta's "pile" (battery), which allowed the electrolysis of water in the year 1800. What more could one ask, as proof of the compound nature of water? But there was a problem, which was already recognized in the very first paper on the subject, by William Nicholson: if electrolysis broke down each molecule of water into hydrogen and oxygen, how could it be that the two gases emerged separately, at positive and negative electrodes separated by a macroscopic distance from each other? If this problem was not solved, the electrolysis of water threatened to become a piece of evidence *against* Lavoisier's theory. Indeed, Johann Wilhelm Ritter advanced an anti-Lavoisierian interpretation, according to which electrolysis was *synthesis*: at one electrode, the combination of water with negative electricity forms hydrogen; at the other electrode, positive electricity and water make oxygen; water is an element, and hydrogen and oxygen are compounds.

Ritter's view was not so much refuted as repelled by the mainstream of chemistry, by this time heavily Lavoisierian. There was never a convincing solution to the distance problem until the end of the nineteenth century, when Svante Arrhenius's theory of free ionic dissociation was proposed and accepted. Meanwhile chemists and physicists consoled themselves with hypothetical mechanisms, such as an invisible transfer of oxygen or hydrogen through the water over to the other side, or a chain of partner-swapping water molecules linking the two electrodes. Who advanced and advocated which views, on what basis? Why was Ritter's view rejected, and was there sufficient evidence supporting that rejection? There is not a great deal of modern literature on this episode. From an examination of some primary sources and older secondary sources, I weave together an account of the development of various competing views. I also note how electrochemistry forged ahead as a productive research science without a clear agreement on the fundamental mechanism of electrolysis (and of the battery).

For those accepting that water was made up of hydrogen and oxygen, the advent of chemical atomic theory raised a further question: how many atoms of each element combined to make water? This is the subject of Chap. 3. John Dalton, from his original 1808 publication onward, candidly acknowledged that he had no way of answering such questions with certainty. There is a fundamental circularity between atomic weights and molecular formulas. In the case of water, what was known from experiments was that hydrogen and oxygen always combined in a 1:8 ratio by bulk weight (in approximate modern numbers). From that we can deduce that the ratio of atomic weights is 1:16, if we know that water is H_2O; or we can deduce that the molecular formula of water is H_2O, if we know that the atomic weight ratio is 1:16. But we need to know one in order to know the other—and to begin with, we know neither. Dalton applied his "rules of greatest simplicity" to break the circularity: since water was the only chemical compound of hydrogen and oxygen that he knew, he assumed that it was the simplest possible atomic combination: HO. Amedeo Avogadro almost immediately proposed a system familiar to the modern eye: two volumes of H_2 and one volume of O_2 make two volumes of H_2O. Interestingly, Avogadro's ideas were rejected by Dalton and most other chemists as ad hoc, speculative and implausible, and not generally adopted until half a century later.

Retracing the history of early chemical atomism, I discern at least five different systems of atomic–molecular chemistry in operation in the first half-century. Each system had its own distinct aims, and its own list of successes and failures, too. It was only as a result of some complicated developments and interactions of these systems that the consensus on the H_2O formula slowly emerged. It was not simply a matter of reviving and publicizing Avogadro's hypothesis in a clearer and more convincing form. Many clues had to be fitted together, and some of the decisive clues arose from very subtle developments in organic chemistry in the 1840s and the 1850s. When the consensus on molecular formulas and atomic weights (including H_2O for water) did come, it was not taken by everyone in a realist manner; many leading chemists still doubted the existence of physical atoms, and had reservations about taking the models of structural chemistry literally. And the synthesis of systems leading to this consensus also left some important questions unanswered, which were taken up by the newly emerging field of physical chemistry.

Regarding each of the episodes treated in the first three chapters, I arrive at the judgment that there was no system that deserved a monopolistic dominance, and that not having one dominant system in each situation did not hamper, or would not have hampered, the progress of science. There can be, and have been, successful systems of science which do not affirm the truth of the statement that water is H_2O. What does this judgment imply about the traditional conception of the pursuit of truth in science? Addressing this question in Chap. 4, I advance a novel doctrine called active scientific realism, which affirms that science should strive to maximize our contact with reality in order to learn as much as we can. "Reality" is taken to mean whatever is not subject to one's own will; reality offers resistance to our ill-conceived schemes, as the pragmatists put it. Nearly all sides in the scientific realism debate should be able to subscribe to active realism. But there is a more controversial side to it, too. Active realism recommends that we should pursue *all* systems of knowledge that can provide us an informative contact with reality; if there are mutually incommensurable paradigms, we should retain all of them at once. But will that not interfere with the pursuit of the one truth about nature? I maintain that we need to come away from such an inoperable notion of truth. When we come to consider what "truth" means in practice, the concept splinters into several different ones, including one that is internal to a given system and nearly synonymous with "success". Realism should be a commitment to promote realistic ways of learning from reality, not a vain and hubristic attempt to prove that we are in possession of the unique truth about nature.

My discussion of evidence and realism leads to a general pluralism about science, which I explicate and advocate fully in Chap. 5. When we take a fully contextual view of evidence, we will come to see that any serious scientific topic is bound to admit more than one rationally justified treatment. To the extent that scientists have a tendency to agree on one theory (or system) at a given time, we need to be aware of the possibility that there might be worthwhile alternatives that are rejected without sufficient reasons. This tendency is amply exhibited in the episodes discussed in Chaps. 1 and 2, and there are many other apt cases in the history of science. The judgment that a system of knowledge was rejected without sufficient epistemic warrant is a weighty one to make. First of all, it involves a claim that it would have been better to let it survive. Secondly, judgment comes with a demand for action: if I think, for instance, that phlogiston chemistry was killed off prematurely, what am I going to *do* about it? If there is lost potential there, it should be recovered and developed. This is pluralism in practice — not the armchair pluralism of declaring "Let a hundred flowers bloom", but an active pluralism of actually cultivating the 99 neglected flowers.

But why is it better to be pluralistic? Why keep multiple systems of knowledge alive? The immediate reason for this is the sense that we are not likely to arrive at the one perfect theory or viewpoint that will satisfy all our needs. Call it pessimism, but I do not think it is unwarranted pessimism. I would rather think of it as reasonable humility about human capabilities. If we are not likely to find the one perfect system, it makes sense to keep multiple ones, which will each have different strengths. Different benefits, practical and intellectual, will spring from different systems of knowledge. It is also important to note that the co-existence of multiple systems can facilitate productive interactions between them through integration, co-optation and competition. These benefits of interaction are just as important as the more widely recognized benefits of toleration; both are essential planks in the program of pluralism that I advocate. It is important to distinguish pluralism from relativism. Relativism involves an idle permissiveness and renunciation of judgment. Pluralism does not renounce judgment, yet maintains that it is better to foster a multitude of worthwhile systems, rather than only one. Pluralism as I conceive it actively engages in the work of proliferation; it is about knowledge-building, not just knowledge-evaluation. In that sense, pluralism emerges as the underlying spirit behind the project of complementary science.

This book has an unusual structure, which deserves some explanation at the outset. Each chapter has three main sections. Section 1 gives an engaging surface-level introduction and summary, intended to be accessible to non-specialists; it is at the level of depth and detail that I may be drawn to give in a serious sociable conversation with interested friends. Section 2 contains a full exposition of my position without constraints; it says what I want to say the way I want to say it, in a linear, focused and systematic way, assuming a fair amount of background. Then Section 3 follows up with esoteric details and anticipated objections that would interest specialists on particular topics; it is a mix of in-depth discussion, self-defence, apologies, qualifications, and projections for future work; some of it will be shallow and sketchy, merely registering an awareness of certain issues and questions and encouraging future work by myself and others.

For readers who first want to get a sense of what this book is all about, or those who do not think they can invest much time reading it, or those who just want the big story without esoteric details, I recommend the surface approach: read this Introduction, and then Section 1 of all the chapters. If that intrigues you sufficiently, or if you are already determined to find out in full what it is that I have learned and think you would enjoy learning, too, then you can take the full-content approach by reading Sections 1 and 2 of all the chapters. If you are a philosopher who can't stand historical details, then you can take the surface approach in Chapters 1, 2, and 3, and follow that with a full reading of Chapters 4 and 5. But I make a gentle request: be open to the possibility that you may get intrigued about the history through reading

Section 1 of the early chapters. As I tell my students: history will get you eventually, if your eyes are open.

If you have specialist interest in any of the topics discussed in a given chapter, you will probably want to read at least some of Section 3. If you easily agree with what I say in Section 2, some portions of Section 3 are probably not necessary. But if there are points on which you have objections or doubts, then relevant bits of Section 3 will hopefully satisfy your demand. Even if you agree with me in Section 2, you may want to read these bits of Section 3 if you are interested in defending the position against doubters and opponents. Other parts of Section 3 are not adversarial; they simply delve into more depth and detail on particular topics, in ways that would have interrupted the flow of thought in Section 2.

The primary intended audiences for this book are academic communities in the history and philosophy of science, for research and for all levels of learning. I think it would also be of interest to various research scientists, science educators and science students, especially in the areas of chemistry and physics. There is also some cross-over potential to popular science, especially given that Section 1 of each chapter is intended to be accessible to wider audiences. Ultimately, disciplinary and professional boundaries are not important to me. As I wrote in the introduction to my first book: if you can glimpse through my words any of the fascination that has forced me write them, then this book is for you.

References

Chang, Hasok. 2004. *Inventing temperature: Measurement and scientific progress.* New York: Oxford University Press.

Eisenberg, D., and W. Kauzmann. 2005. *The structure and properties of water.* Oxford: Oxford University Press.

Hendry, Robin Findlay. 2008. Chemistry. In *The Routledge companion to the philosophy of science*, ed. Stathis Psillos and Martin Curd, 520–530. London: Routledge.

Chapter 1
Water and the Chemical Revolution

Abstract It was through the Chemical Revolution of the late eighteenth century that water first came to be recognized as a compound, having been considered an element since ancient times. In this chapter I offer a revisionist account of that momentous event. A systematic appraisal shows that the old phlogistonist system of chemistry was not clearly inferior to Lavoisier's oxygenist system of chemistry. Lavoisier's system actually suffered from significant empirical and theoretical problems already recognized at the time, and there was significant methodological incommensurability between the two systems, though only mild semantic incommensurability. Aside from the effective and ruthless campaigning by the Lavoisierians, the demise of phlogiston (which was not as sudden or complete as often imagined) was most of all due to the advent of *compositionism* as a dominant trend in chemistry, into which phlogistonist practices did not easily fit. With the demise of phlogiston, many valuable elements of knowledge were lost; in effect, these were recovered and developed later with the help of different concepts (e.g., potential energy and electrons), but I argue that it would have been better for science if the phlogistonist system had been allowed to continue its work. This conclusion also anticipates the more general argument for pluralism in science, to be given fully in Chap. 5. In order to give more precision in the articulation and defence of these ideas, I introduce and use the notion of *system of practice* as a unit of analysis.

1.1 The Premature Death of Phlogiston

In the middle of the eighteenth century people still considered water as an element. For Europeans this idea dated at least back to the ancient Greeks at the time of Aristotle, according to whom water was one of the four basic elements (along with earth, air and fire) which constituted all the substances making up the terrestrial world. Thales had even postulated that water was *the* element from which every-thing was made. We now know that water is not an element, but a compound made

H. Chang, *Is Water H₂O?: Evidence, Realism and Pluralism*, Boston Studies in the Philosophy of Science 293, DOI 10.1007/978-94-007-3932-1_1, © Springer Science+Business Media B.V. 2012

up of oxygen and hydrogen. This chapter tells the story of how we came to know that, as a result of the Chemical Revolution just over 200 years ago. It is a story that has been told very many times, and knowledgeable readers may wonder why I should try to tell it again. The reason is simple: it is all too often told all wrong— "wrong" in several senses: incorrect about the historical circumstances, ignorant of the relevant scientific arguments, judgmental on the basis of deep misunderstandings, and philosophically naïve and simplistic. The best available insights are usually buried in specialist works that are neglected even by the majority of professional historians and philosophers of science.

I hope that my re-telling of this story will enhance your interest in the Chemical Revolution as an exciting topic of historical, philosophical and scientific thinking. As will be the case in each chapter of this book, the first section (Sect. 1.1) will provide a brief surface-level presentation of the main ideas, intended to be accessible to readers without much background in the relevant philosophical, historical and scientific areas. My hope is that even non-specialist readers will become intrigued enough to go on to read the second section (Sect. 1.2), and even some of the third section (Sect. 1.3).

1.1.1   Joseph Priestley

Our story begins with Joseph Priestley (1733–1804), a paragon of eighteenth-century amateur science.[1] A dissenting preacher and political consultant, Priestley flourished in the benign exclusion imposed by the Anglican orthodoxy. He was one of those British men of science who never went near a university, either for learning or teaching. His scientific research was done at home, initially in the "warm mice-ridden Yorkshire cottage kitchens" (Crowther 1962, 218). His great work in chemistry began when he moved to Leeds in 1767, where he had the good fortune of "inhabiting a house adjoining to a public brewery" (Priestley 1970, 94). Experimenting with the "fixed air" that was found collecting in the fermenting vats (what we now call carbon dioxide), he became a celebrity all over Europe when he found a way of making artificially carbonated water. This work also marked the beginning of his long-term research programme in "pneumatic chemistry", the chemistry of gases—or "airs", as he and his contemporaries more commonly conceived them.

Priestley was the most prolific discoverer and manufacturer of new airs. Not long before his work, most people had considered air to be a pure element as much as water, though some isolated observations of certain gases different from ordinary air had been reported before. After his work, there was no doubt left that ordinary air had at least two components, and that different types of air could be produced by various chemical reactions. Priestley's *Experiments and Observations on Different*

---

[1] For further details on Priestley's life and work, see Chang (2012a) and references therein. The most authoritative and extensive biography is Schofield (1997, 2004).

Kinds of Air (1774, 1790) is a sheer delight for those who share a sense of fascination about all the diverse phenomena of nature and a childlike wonder at our own ability to call them forth.

Priestley was the first person to make and bottle what we now call oxygen and tell the wide world about it. In Carl Djerassi and Roald Hoffmann's amusing play *Oxygen* (2001), it is left uncertain who should win the first "Retro-Nobel Prize" in Chemistry for the discovery of oxygen. But those authors, or anyone else well enough informed, would not deny Priestley's priority over Carl Wilhelm Scheele (1742–1786) in publication, and over Antoine-Laurent Lavoisier (1743–1794) in deed. Priestley's excitement is palpable as he reports, in his letter of 15 March 1775 to James Pringle, the President of the Royal Society of London: "the most remarkable of all the kinds of air that I have produced . . . is one that is five or six times better than common air, for the purpose of respiration, inflammation, and, I believe, every other use of common atmospherical air." First he tested this new air by burning things in it. And then, "to complete the proof of the superior quality of this air, I introduced a mouse into it; and in a quantity in which, had it been in common air, it would have died in about a quarter of an hour, it lived, at two different times, a whole hour, and was taken out quite vigorous." (Priestley 1775, 387–388) After that he found the courage to breathe the new air himself. "The feeling of it in my lungs," Priestley reported, "was not sensibly different from that of common air, but I fancied that my breast felt peculiarly light and easy for some time afterwards. Who can tell but that in time, this pure air may become a fashionable article in luxury. Hitherto only two mice and myself have had the privilege of breathing it."

On the new site of the Mill Hill Chapel in Leeds, for which he preached for several years during his scientific heyday, a blue plaque proudly proclaims (Fig. 1.1): "Joseph Priestley, discoverer of oxygen, was minister here 1767–1773." Such commemoration would have annoyed Priestley, for he did not call his new gas "oxygen". He called it "dephlogisticated air", and that was not just a matter of words. By that phrase he really meant common air cleansed of the "phlogiston" that is normally mixed up in it. What was phlogiston? In short, it was the principle of inflammability; "principle" here did not mean a fundamental rule, but rather a fundamental substance that combined with other substances and gave them its characteristic properties. Phlogiston was the principle that imparted combustibility to combustibles. A combustible substance was rich in phlogiston, and when it burned it released its phlogiston, which then manifested itself in the flame that came out.

Certain experiments seemed to indicate that metals, too, were rich in phlogiston, and that it was phlogiston that gave them the characteristic metallic properties, such as their shiny luster, their malleability and ductility, and their electrical conductivity (and their inflammability actually, under the right circumstances). When a metal was deprived of phlogiston, it lost its key metallic properties and became an earthy substance called "calx" (which we would now identify as rust or oxide). All of this sounds much too fanciful to our modern ear. Let us see if we can make the phlogiston-theorist to see some common sense. If calx is really metal that has lost its phlogiston, then you should be able to turn it back into metal by giving some phlogiston to it. Can you do that? "Sure", says the phlogistonist. That is what smelters have been doing for

Fig. 1.1 Memorial plaque to Priestley at the Mill Hill Chapel, Leeds (photo by Hasok Chang)

thousands of years. Take a metallic ore, which often contains calx rather than pure metal, and mix it up with a phlogiston-rich substance, say, charcoal; heat the mixture strongly, to effect a transfer of phlogiston from the charcoal to the calx. And there it comes, the shiny metal! Similar work on the interconversion of sulphur and sulphuric acid by the German physician and chemist Georg Ernst Stahl (1659–1734) was one of the founding experiments of phlogistonist chemistry. This work elicited the admiration of Immanuel Kant, who chose it as one of the three prime illustrations of how empirical science began to grapple with nature in a principled way, in his *Critique of Pure Reason*: when "Stahl changed metals into calx and then changed the latter back into metal by first removing something and then putting it back again, a light dawned on all those who study nature."[2]

Similarly, Priestley had produced his oxygen (dephlogisticated air) through a process in which he thought mercury calx was "revivified" into its metallic form by absorbing phlogiston from air. As a result, the air would have been "de-phlogisticated".

---

[2] Kant [1787] (1998), 108–109. This passage occurs in the Preface to the second edition (B xii–xiii). My attention was drawn to this passage by its quotation in Bensaude-Vincent and Simon (2008), 87; they quote from the classic translation by Norman Kemp Smith, who committed an interesting anachronism by rendering the German "Kalk" as "oxide". Kant's other examples are Galileo on inclined planes and Torricelli on barometric pressure.

Such an air should be an exceptionally good supporter of combustion, since it would re-absorb phlogiston very eagerly. And so it was, when Priestley tried the experiment. He also reasoned, because respiration was a process in which the phlogiston produced by the workings of the body was removed from the lungs, dephlogisticated air should be particularly good to breathe. And so it was.

One can see why Georges Cuvier quipped that Priestley was a father of modern chemistry, but "a father who never wanted to acknowledge his daughter."[3] Priestley is often seen as a tragic figure, possessed of consummate experimental skill and full of good scientific intentions, but blinded by a dogmatic adherence to an outmoded way of thinking. His chemical misfortune was compounded by political injustice, when a reactionary mob ransacked his house and laboratory in Birmingham in 1791 for his support of the French Revolution, on the second anniversary of the storming of the Bastille. After that he tried a life in London, but in the end only found refuge in America. It was sad but inevitably right, it is said, that his work was swept aside by the tide of scientific progress brought in by Antoine-Laurent Lavoisier, the young urbane Parisian with brilliance and ambition in equal measure. Lavoisier had a different way of accounting for Priestley's experiments and observations. Combustion was combination with oxygen, as was calcination (turning metal into calx). Where Priestley saw de-phlogistication, Lavoisier saw oxidation. Having seen the light shown by Lavoisier, chemists have never looked back to phlogiston…

Even Thomas Kuhn, who famously refused to say that the losing side in a scientific revolution was simply wrong, was surprisingly downbeat about Priestley. Although he denied that Priestley's resistance to Lavoisierian chemistry was ever "illogical or unscientific", Kuhn thought he was "unreasonable" to resist as long as he did; the historian "may wish to say that the man who continues to resist after his whole profession has been converted has *ipso facto* ceased to be a scientist" (Kuhn 1970, 159). The tale of stubborn old Priestley blinded by phlogistic dogma has captured many people's imagination, but it is a misleading story on so many levels. By the time you finish reading this chapter, I hope I will have given you some considerable doubt about that version of events. The best way to start seeing the problem is to ask: what was really so wrong with Priestley's stance? To the end of his life he continued his chemical research and published well-informed and closely reasoned defences of phlogiston (Priestley [1796] 1969, 1803). He continued to find the phlogiston theory a sensible and fruitful framework for understanding new phenomena (such as electrolysis, as I will discuss further in Chap. 2), and there were still no phenomena which in his view clearly refuted the phlogiston theory. And there are similar stories to be told about each of the other brilliant and dedicated scientists who refused to let go of the phlogiston theory, some of whom we shall meet shortly.

---

[3] Quoted in the *Encyclopaedia Britannica*, 9th edition, vol. 5 (1876), 462.

1.1.2   Water

A decisive moment in the competition between the oxygen theory and the phlogiston theory was Lavoisier's argument that water was not an element at all, but a compound of oxygen and hydrogen. The irony of the situation is worth noting. No one knew better than Priestley how to make oxygen (dephlogisticated air), and hydrogen (called inflammable air) had been discovered and studied in 1766 by his compatriot and fellow phlogistonist Henry Cavendish (1731–1810), by dropping pieces of metal into acids (see Jungnickel and McCormmach 1999, 202ff). Cavendish also discovered how to explode those two airs together to make water, an experiment which Priestley repeated successfully. It was Priestley, and Cavendish (via his friend Charles Blagden), who taught Lavoisier how to do these things. Still, the common story goes, it was Lavoisier who came up with the correct interpretation of what these gases were and what happened when they reacted with each other.

   In fact, there is no better case than the composition of water for illustrating the *surprising* cogency of the phlogiston theory. Cavendish and Priestley both thought for a time that hydrogen, or "inflammable air" as they called it, was pure phlogiston, driven from the metal by the action of acid (the metal thereby turned into calx, and dissolved into the acid to form a salt[4]). If a calx was put into the acid, it dissolved without the production of inflammable air, because the calx did not contain phlogiston. A more considered version of this view served to explain the formation of water. The later Cavendish–Priestley view was that inflammable air was "phlogisticated water", that is, water containing an excess of phlogiston. As for oxygen, or dephlogisticated air, that was "dephlogisticated water". When phlogisticated water and dephlogisticated water combined with each other, the excess and deficit of phlogiston cancelled out and plain water was produced. To summarize: there were (at least) two competing views on the formation of water, both of which were cogent and self-consistent:

(1)   Hydrogen + Oxygen → Water
(2)   Phlogisticated water + Dephlogisticated water → Water

   The second account here is not a fairy-tale that phlogiston theorists simply manufactured in order to avoid the refutation of their theory by the facts about the composition of water. Cavendish and Priestley had good reasons to think that water was an essential constituent of gases. At the most basic level this idea is suggested by the fact that vapors are produced from liquids. Might the situation not be similar with other kinds of gases, too? Lavoisier ([1789] 1965, ch. 1) also readily agreed that gases were produced from water and other liquids, by the addition of caloric (the matter of heat). Priestley (1788, 154) was entirely comfortable with "the supposition of water entering into the constitution of all the kinds of air, and being, as it were, their proper basis, without which no aëriform substance can subsist". He noted

---

[4] "Salt" here designates a whole class of chemical substances, including common salt and many others.

(p. 152) that he did not know of any ways of producing inflammable air without water, and surmised that "the same may be true of every other kind of air, since water is used in the production of them all."[5] Within this framework of thinking, it made perfectly good sense to think that phlogistication or dephlogistication would affect the process of aerification of water, resulting in different types of airs.

Historically well-informed philosophers have struggled to say what exactly was wrong with Priestley's stance. We must resist the impulse to say "we *know* he was wrong, because phlogiston simply doesn't exist", since that only begs the question of how we know *that*. The complaint that it was not possible to isolate phlogiston in its pure form has no force. If we always required such material isolatability, science would look very different, as we would have to renounce a whole range of concepts from quarks to energy. And at the core of Lavoisier's own theory was *caloric*, the matter of heat, which was not isolatable in its pure form, either. It also won't do to say that phlogiston was an illegitimate scientific concept because it was unobservable. However one may define "observable", science to this day is full of unobservable entities that are postulated because of theoretical necessities (dark matter and super-strings come to mind). And it is not clear that phlogiston was not observable; to the phlogistonists, phlogiston was not only observable (in the flame that comes out of combustion, for example), but even directly manipulable (when it was transferred from one substance to another, as in smelting or in the production of inflammable air by the solution of metals in acids). This feeling was evident within the Lunar Society, that remarkable association of scientific people around Birmingham, which included Priestley.[6] In 1782, as the Chemical Revolution began to move into its last phase, Matthew Boulton, James Watt's business partner, wrote to the potter Josiah Wedgwood, marveling about Priestley's new experiment in which a calx turned into metal by "imbibing" inflammable air (hydrogen), which Priestley then considered pure phlogiston: "We have long talked of phlogiston without knowing what we talked about, but now Dr Priestley hath brought ye matter to light. We can pour that Element out of one Vessell into another, can tell how much of it by accurate measure is neces-sary to reduce a Calx to a Metal ...." (Boulton quoted in Musgrave 1976, 200).

A related complaint against phlogiston concerns weight. The crude version of the complaint is that phlogiston is an "imponderable" substance (meaning a weight-less substance, not an unthinkable one), and therefore it should not be accepted in science. But isn't today's physics quite sanguine about weightless particles, such as photons? And in the days of phlogiston, other imponderables were postulated with impunity, such as the electric fluid(s), and not least of all Lavoisier's caloric. Another version of the complaint focuses on the weight gain in calcination: a metal gains weight in becoming a calx, which would not happen if it were losing something, namely phlogiston. The weight gain is explained very nicely by the oxygen theory,

---

[5] Priestley even gave an experimental estimate that "fixed air consists of about half its weight of water", from his experiment of generating fixed air by passing steam over *terra ponderosa aerata* in a red heat, noting that heating alone did not yield fixed air.

[6] About the Lunar Society, see Schofield (1963) and Uglow (2002).

in which calx is metal combined with oxygen, hence obviously heavier than the metal by itself. However, this does not quite work as a refutation of the phlogiston theory, as there are ways of accounting for the weight gain. It was not necessary to resort to the much-ridiculed idea that phlogiston had "levity".[7] A much more serious explanation, advanced by Priestley and also by Richard Kirwan (c.1733–1812), was that in calcination the metal combined with water, while losing phlogiston. When the calx was reduced back to metal, it gave out the water and absorbed phlogiston. If there was no external source of phlogiston, the calx sometimes took it from the water that it was emitting, which means that what it emitted was dephlogisticated water (oxygen, in Lavoisier's terms)!

1.1.3   The Trouble with Lavoisier

So we can begin to see how it was that Priestley could hold on to his phlogiston theory, in quite a rational way. What is actually more difficult to see is why nearly everyone else should have signed up to Lavoisier's position, or stuck to it for any length of time. In order to free up our thinking from old clichés about the Chemical Revolution, I must begin by pointing out *just how wrong Lavoisier was, if we judge him from the view of modern chemistry and physics.*[8] As John McEvoy says (1997, 22–23), it is "a simple fact" that already "by the end of the eighteenth century, almost every major theoretical claim that Lavoisier made about the nature and function of oxygen was found wanting." Similarly, Robert Siegfried (1988, 35) states that "the central assumptions that had guided his work so fruitfully were proved empirically false by about 1815". Let us take a closer look. Three major pillars of Lavoisier's system of chemistry marked clear departures from previous chemistry: the theory of acids, the theory of combustion, and the caloric theory. All three are clearly wrong, from the viewpoint of modern chemistry, or even from the viewpoint of nineteenth-century chemistry. (I will be discussing more lasting parts of Lavoisier's achievements later.)

Even the most robust Lavoisier-enthusiasts will easily concede that his theory of acids was mistaken. Lavoisier said that all acids contained oxygen, but the Lavoisierians knew as well as anyone that there was no evidence of oxygen in certain acids, including muriatic acid (in modern terms, hydrochloric acid, HCl) and prussic acid (hydrocyanic acid, HCN). As the twentieth-century Oxford chemist Harold Hartley put it (1971, 110), "the rigid acceptance of this doctrine" was "responsible for so much make-believe in chemists' minds", including their failure to recognize chlorine as an

---

[7] Guyton de Morveau, before he "converted" to the Lavoisierian cause, gave a very reasonable version of this idea, that phlogiston had *relative* levity; since it must be lighter than air, its addition to an object would reduce the object's *apparent* weight. See passage included in Crosland (1971), 135, from *Dissertation sur le phlogistique* (1770), and also the exposition in Poirier (1996), 62.

[8] For an extended discussion in this vein, see Chang (2009b).

element for 20 years after its isolation by Scheele.[9] Was this theory of acids just an unfortunate non-essential adjunct to the rest of Lavoisier's system, which could safely be discarded?[10] At least Lavoisier himself didn't think so, as we can glimpse from the way he named his beloved "oxygen", meaning "acid-generator".[11]

Even more central to Lavoisier's "antiphlogistic" system was his theory of combustion. Surely this undeniably essential bit of Lavoisier's system was correct, and is still preserved in modern chemistry? To grant that would be to participate in an amnesia that pro-Lavoisierian historiography has carefully orchestrated. It is in fact quite incredible that any modern person can think: "Combustion is combination with oxygen, that's what causes the emission of heat and light, and Lavoisier discovered all that." What on earth does oxygen have to do with heat and light? Lavoisier's explanation of the production of heat in combustion was that it was the freeing of caloric fluid from oxygen gas, the caloric that was responsible for the gaseous state of the oxygen in the first place. But it was widely recognized by Lavoisier's contemporaries (and even Lavoisier himself) that there were serious difficulties with this story. Thomas Thomson (1773–1852), the leading Scottish chemist of the period just after the Chemical Revolution, gave a calm and devastating summary of well-known difficulties in his *System of Chemistry* first published in 1802 (vol. 1, 354–358). These difficulties included cases of combustion without oxygen in the gaseous state (such as the explosion of gunpowder), and combustion without any oxygen at all.[12] Thomson judged that "upon the whole, it cannot be denied that Lavoisier's theory does not afford a sufficient explanation of combustion." (p. 358) Thomson was not advocating a return to phlogiston, but he wanted chemistry to move on beyond Lavoisier, less than 15 years after the culmination of the Chemical Revolution.

There was also growing discontent with Lavoisier's caloric theory of heat in more general terms, particularly in London, where around 1800 there was a remarkable concentration of advocates of the notion that heat was a form of motion, including Count Rumford, Humphry Davy, Thomas Young and Henry Cavendish. It is important to remember that caloric in Lavoisier's system was not merely a device for explaining the release of heat in combustion; rather, it was an essential element in his cosmology, for example in explaining the three states of matter. Lavoisier ([1789] 1965, 175) clearly considered caloric a cornerstone of his chemical system, putting it (along with light) at the top of his list of chemical elements (see Fig. 1.2), and devoting

---

[9] For details on the story of chlorine and its elementary nature, see Ashbee (2007) and Gray et al. (2007), which are the first two chapters in Chang and Jackson (2007). See also Brooke (1980) and Crosland (1980).

[10] It was certainly possible to accept the rest of Lavoisier's theory without accepting the theory of acids. Even Claude-Louis Berthollet, one of Lavoisier's closest associates, seems to have done so, as Homer Le Grand (1975, 69) argues. Ramón Gago (1988) states that this was a widespread attitude in Spain.

[11] William H. Brock (1992, 125) goes as far as to suggest: "Once Lavoisier had the concept of a gas, it was the issue of acidity, not combustion, that led him to oxygen—as its very name implies."

[12] See also Mauskopf (1988), 93–118.

	Noms nouveaux.	*Noms anciens correspondans.*
	Lumière.........	Lumière.
Substances simples qui appartiennent aux trois règnes & qu'on peut regarder comme les élémens des corps.	Calorique........	Chaleur. Principe de la chaleur. Fluide igné. Feu. Matière du feu & de la chaleur.
	Oxygène.........	Air déphlogistiqué. Air empiréal. Air vital. Base de l'air vital.
	Azote...........	Gaz phlogistiqué. Mofète. Base de la mofete.
	Hydrogène.	Gaz inflammable. Base du gaz inflammable.
Substances simples non métalliques oxidables & acidifiables.	Soufre...........	Soufre.
	Phosphore........	Phosphore.
	Carbone..........	Charbon pur.
	Radical muriatique.	Inconnu.
	Radical fluorique .	Inconnu.
	Radical boracique,.	Inconnu.
Substances simples métalliques oxidables & acidifiables.	Antimoine........	Antimoine.
	Argent...........	Argent.
	Arsenic..........	Arsenic.
	Bismuth..........	Bismuth.
	Cobolt.	Cobolt.
	Cuivre..........	Cuivre.
	Etain............	Etain.
	Fer.	Fer.
	Manganèse.	Manganèse.
	Mercure.	Mercure.
	Molybdène.......	Molybdène.
	Nickel..........	Nickel.
	Or.	Or.
	Platine..........	Platine.
	Plomb...........	Plomb.
	Tungstène........	Tungstene.
	Zinc..	Zinc.
Substances simples salifiables terreuses.	Chaux..........	Terre calcaire, chaux.
	Magnésie........	Magnésie, base du sel d'Epsom.
	Baryte..........	Barote, terre pesante.
	Alumine.........	Argile, terre de l'alun, base de l'alun.
	Silice...........	Terre siliceuse, terre vitrifiable.

Fig. 1.2 Lavoisier's table of simple substances; the *left-hand column* gives Lavoisier's neologisms, and the *right-hand column* gives the corresponding old terms (from the original French edition of 1789, p. 192)

the entire first chapter of his definitive textbook of new chemistry to the elucidation of the nature and role of caloric. It was a beautiful and sensible system, but only as much as the phlogiston theory was. Both theories are equally wrong, from the modern point of view.

Recognizing that the debate was not a simple matter of truth and falsity, various philosophers and historians of science have attempted to meet the challenge of explaining why the vast majority of chemists went over to Lavoisier's side. Alan Musgrave's classic paper on the subject discusses and dismisses several common attempts (1976, 182–186). It is not the case that Lavoisier's new chemistry was established by induction from observations. It is equally wrong to say that the phlogiston theory was simply falsified by facts like the increase of weight in metals when they are oxidized (or, calcined). What may be more appealing but equally fallacious is the notion that Lavoisier's theory won because it was inherently simpler than the phlogiston theory. The crudest version of this idea, which Musgrave calls "simplicism" (or conventionalism) in his critique, says that the phlogiston theory unnecessarily complicated things by postulating an unobservable substance, phlogiston; this argument ignores the fact that Lavoisier had to postulate an equally unobservable substance, caloric.

After dismissing these explanations, Musgrave proposes that the crucial factor was that the oxygen research programme was more progressive, with "progress" as defined in Imre Lakatos's philosophy. Musgrave argues that it was rational for chemists to abandon the phlogiston theory because after a certain point it stopped making successful new predictions; it only continued to make *ad hoc* hypotheses—excuses made up in order to protect a failing theory. Unfortunately, I do not think this Lakatosian explanation works, as I will explain in more detail in Sect. 1.3.1. Musgrave states: "Between 1770 and 1785 the oxygen programme clearly demonstrated its superiority to phlogistonism: it developed coherently and each new version was theoretically and empirically progressive, whereas after 1770 the phlogiston programme did neither." (p. 205) I find this claim difficult to substantiate, attractive as it is. The idea that the progress of the phlogiston program ended in 1770 (while the oxygen program continued in its smooth progress) is contradicted by Musgrave's own statements earlier in the paper (p. 199): "While Lavoisier was failing, Priestley was having great success with the 1766 version of phlogistonism. . . . the most impressive experiment of all came in early 1783." This was the confirmation of the phlogistonist prediction that calxes would be reduced to metals by heating in inflammable air. In order to sustain Musgrave's thesis convincingly, we need to find successful novel predictions that Lavoisier's research program made after the phlogiston program stopped making any, and that is 1783, not 1770. Where are these predictions? Are we thinking of the prediction that the oxidation of inflammable air (hydrogen) would produce an acid?[13] Or the prediction that muriatic (hydrochloric) acid would be decomposed into oxygen and the "muriatic radical" (see Fig. 1.2, second group)?

After the failure of all these attempts to argue for the rationality of the consensus in favour of Lavoisier, the story gets more involved. Two common lines of thought are often pursued: defend the presumed rationality of the Chemical Revolution by an appeal to a complicated version of simplicism, or shift attention away from the traditional philosophical concern with rationality by seeking social explanations. I will consider these lines of thought in Sects. 1.3.1 and 1.3.2, and why I do not follow

---

[13] Again, Musgrave (1976, 199–200) himself points out this failed Lavoisierian prediction, and how long Lavoisier had struggled with it.

them will already be clear by then from the statement of my own position given in the second section of this chapter (Sect. 1.2). For now, however, let me pursue the implication of the conclusion that I am going to reach after the full deliberation: *the phlogiston theory was prematurely killed.*

1.1.4 Could Water Be an Element?

The burden of judgment is a responsibility to act: if I really believe that the phlogiston theory was discarded prematurely, I must consider what could have been achieved by keeping it. My main aim is not the counterfactual history of "what if" (see Sect. 1.3.5 for more on this issue). I am ultimately advocating a more activist type of scholarship, in which we actually open up the possibility of reviving the unjustly discarded line of thinking, and see what comes of it. What I seek is a complete view: I want to know what contributions to scientific knowledge the phlogiston theory did make, what contributions it could have made if it had been kept longer, and what contributions it could still make if it were revived. If all of those categories of contributions have been lost or missed because of the premature abandonment of phlogiston theory, then we should recover, imagine and create them. If you would object that such an enterprise is neither history nor philosophy nor real science, so be it: I call it "complementary science", as I explained in my previous book (Chang 2004, esp. ch. 6). I aim to give a novel function to history and philosophy of science, without denying its traditional functions. And of course I am not proposing to do away with oxygen or the tradition of modern chemistry descended from Lavoisier, even if such a thing would be possible. No, the whole enterprise is a pluralistic one, as I will be explaining further in Sect. 1.2.4, and in Chap. 5 in more general terms.

There are various questions we need to ask. (1) Was there any knowledge that scientists lost when they rejected the phlogistonist system? (From here on I will say "system" instead of "theory", in order to emphasize that there is more than just theory involved.[14]) In other words, was there something good that the phlogistonist system did that the oxygenist system could not do? Kuhn thought that there was typically such knowledge-loss when a scientific revolution happens; this has been dubbed "Kuhn loss" in his honour.[15] (2) Was there any knowledge that could have come from keeping the phlogistonist system, whose development was delayed or prevented because of its demise? (3) Was there a beneficial effect of having both phlogistonist and oxygenist systems present, in terms of what was produced by interactions between them? (4) Would there have been further beneficial

---

[14] This seems consistent with the usage of the time. There were various textbooks of chemistry that were called "Systems", for example Thomson (1802), which I have already mentioned. Other examples included the *Système des connaissances chimique* (1801) by Lavoisier's colleague Antoine Fourcroy. I will give a more thorough definition of a "system of practice" in Sect. 1.2.1.1.

[15] This is a term coined by Heinz Post, according to Ioannis Votsis.

interactions between the oxygenist system and the phlogistonist system, if the latter had been maintained?

I will attempt to answer these questions in some detail in Sect. 1.2.4, but here are some preliminary thoughts to stimulate your imagination. Although a lot of the explanations of chemical phenomena given in the phlogistonist system were successfully co-opted into the oxygenist system, that was not always possible. For example, the phlogiston theory gave a nice explanation of why all the metals had similar properties, and very little about that could be said in Lavoisier's theory. The identification of phlogiston with (negative) electricity was a promising avenue of thought shut down by the demise of phlogiston. While phlogiston theorists are often faulted for not having paid sufficient attention to the balancing of weights in chemical equations, phlogiston did serve as a useful reminder that not everything in chemistry could be explained by weights. The eminent English chemist William Odling opined in 1871 that phlogiston had been a clear precursor of chemical potential energy. These ideas should at least give us a pause, and plenty of food for thought.

My idiosyncratic view of the Chemical Revolution will make many people uncomfortable. If the rejection of phlogiston theory was premature and unwarranted, so was the rejection of the idea that water is an element. But surely water is not an element? If you believe something so outlandish, the whole weight of modern science will crush you. But could there really not be a sensible system of science in which water is an element? The easy answer is: "Yes, and it was called the phlogistonist system"—or, really, every scientific system before the Chemical Revolution itself. The tougher question, however, is whether there is a system of science based on elementary water that we, here and now, could plausibly believe, or at least work with and profit from. In considering this question we must, again, resist the temptation to say "of course *we* now *know* that water is a compound of hydrogen and oxygen; any theory that says otherwise is simply false, and not worth considering". We only think like this because we are prisoners of a scientific worldview which is founded on premises such as the compound nature of water. The truly pluralistic challenge is to ask whether we could step outside that worldview and find another one that is not premised on the compound nature of water; and, if there is one, whether there would be any benefit in developing it. As Léna Soler puts it (2008, 230), following Ian Hacking (2000), the general question is whether well-established scientific results are inevitable or contingent: is it "possible for there to be a science that is . . . as successful and progressive as ours but radically different in content"?

A terminological insight, trivial in itself, will make this prospect seem infinitely more plausible. When we talk about "elements" in modern science, for example when we say that oxygen is a chemical element, we do not mean ultimately simple bodies that cannot be broken down any further, or even bodies that have not been broken down yet. Such simple-minded views of chemical elements had to be rejected in later science—just like the view that water was an element and could not be broken down any further. According to modern science the atoms and molecules of oxygen, hydrogen, or water—each and every chemical substance—is made up of a number of other, simpler particles, such as neutrons, electrons, and protons. If we take modern physics really seriously, even an electron is not a "simple" particle; it is not a "particle" at all in its ordinary sense, but a packet of energy exhibiting wave–particle duality, or not even that, but some fluctuating state of the quantum field. Never mind that—the point is that "Water is an element" is really only about as wrong as "Oxygen is an element". So there is no reason to feel so desperately apologetic in considering the potential merits of a theory that includes elementary water, any more than in considering the merits of a theory that includes elementary oxygen. The real question is what "element" means in a system of chemistry, in terms of its functions. When Priestley and Cavendish maintained that water was an element, they clearly meant it as a substance that could be *modified*, by the addition or subtraction of phlogiston, for example. We can certainly consider whether there are potential insights we can gain from that way of thinking.

Let us take the case of water in the Chemical Revolution as a reminder of the poverty of single vision. The tale of phlogiston illustrates how actual and potential knowledge can be lost if we insist on having only one way to pursue it.

1.2   Why Phlogiston Should Have Lived

Having raised some doubts and questions concerning the outcome of the Chemical Revolution, I will now give a full and free exposition of my views, as I will do in Section 2 of each chapter. Here I assume a fair amount of background knowledge (some of which was provided in Sect. 1.1), and discuss matters with little regard to disciplinary boundaries or worries about objections that might come from various specialists; afterwards, Sect. 1.3 will deal with anticipated objections and try to position my views with respect to existing literature. I will begin, in Sect. 1.2.1, with a systematic appraisal of the evidential situation in the choice between oxygen and phlogiston. The verdict of that appraisal will be along the lines already indicated in Sect. 1.1: there were no conclusive reasons of evidence to reject the phlogiston theory in favor of Lavoisier's theory. This raises a question of historical explanation: why, then, did chemists make that unjustified choice? Section 1.2.2 will argue that chemists at the time did not actually reach a simple, rapid and universal agreement in favor of Lavoisier. Still, admitting that a clear majority did eventually reject phlogiston, Sect. 1.2.3 will attempt to explain why that collective decision was reached, by reference to a larger and longer-term trend underlying the Chemical Revolution, namely the advent of "compositionism". Providing an explanation does not mean advocating what one has explained. After all the evaluations and explanations, Sect. 1.2.4 will take a better-informed look at the fate of phlogiston, and ask what benefits might have followed from retaining it in chemistry.

1.2.1   Phlogiston vs. Oxygen

Was there sufficient scientific justification for the rejection of phlogiston-based chemistry in favor of Lavoisier's oxygen-based chemistry? Was there enough evidence to support that decision?[16] My sense is that there has not yet been a truly systematic assessment of this matter. Now, that is a grandiose claim to make about a subject that has been studied so thoroughly, so I need to explain what I have in mind more carefully. First of all, much of the existing historical literature on the Chemical Revolution is not concerned with the questions of justification. And most of the philosophical analyses have been carried out from particular perspectives, each of which neglects or obscures certain aspects of the situation, as I will explain further as I go along. So I will begin by outlining a more commodious framework of analysis.

1.2.1.1 Evaluating Systems of Practice

At least in the Anglophone traditions, philosophical analyses of science have been unduly limited by the common habit of viewing science as a collection of propositions, focusing on the truth-value of those propositions and the logical relationships between them. The premier subject of discussion in philosophy of science has been *theories* as organized bodies of propositions. This has led to the neglect of experimentation and other non-verbal and non-propositional dimensions of science in philosophical analyses. Many historians, sociologists and philosophers have pointed out this problem, but so far no clear alternative philosophical framework has been agreed upon to provide a language for fuller analyses of scientific practice. A serious study of scientific practice must be concerned with what it is that we actually *do* in scientific work. This requires a change of focus from propositions to actions. I begin with the recognition that all scientific work, including pure theorizing, consists of actions—physical, mental, and "paper-and-pencil" operations, to put it in Percy Bridgman's terms (1959, 3). Of course, all verbal descriptions we make of scientific work must be put into propositions, but we must avoid the mistake of only paying attention to the propositional aspects of the scientific actions.

I propose to frame my analyses in terms of "systems of (scientific) practice" that are made up of "epistemic activities".[17] (I will also use the phrase "system of knowledge" interchangeably with "system of practice", especially where it seems there is no danger of forgetting that knowledge is rooted in practice.[18]) An *epistemic activity* is a more-or-less coherent set of mental or physical operations that are intended to contribute to the production or improvement of knowledge in a particular way, in

---

[16] What exactly "evidence" means is, of course, a serious philosophical question, which I will address as I go on. For the time being I rely on an intuitive notion that seems sufficiently widespread and coherent, of epistemic justification that a theory receives from observations or facts.

[17] I have made preliminary attempts to spell out these notions in Chang (2011a, d), from which I draw here. Some further details are provided in those papers.

[18] In my relatively informal discussion in Sect. 1.1, there was already a felt need to consider more than just theories, and I began to use the term "system", following the contemporary chemists who often spoke of the "phlogistic system" and the "antiphlogistic system" (or the "French system of chemistry").

accordance with some discernible rules (though the rules may be unarticulated). An important part of my proposal is to keep in mind the aims that scientists are trying to achieve in each situation. The presence of an identifiable aim (even if not articulated explicitly by the actors themselves) is what distinguishes activities from mere physical happenings involving human bodies, and the coherence of an activity is defined by how well the activity succeeds in achieving its aim. Common types of epistemic activities include measurement, prediction, and hypothesis-testing. Some epistemic activities are primarily mental. There is such a thing as theoretical practice, and in chemistry it consists of activities such as classification, equation-balancing, and the modeling of molecular structures. In reality, most epistemic activities are both mental and physical at once. When we start thinking of scientific work as a collection of activities, an immediately obvious thing is the sheer variety in the types of epistemic activities that scientists engage in. Here is a partial list of types of epistemic activities: describing, predicting, explaining, hypothesizing, testing, observing, detecting, measuring, classifying, representing, modeling, simulating, synthesizing, analyzing, abstracting, idealizing.

Epistemic activities normally do not, and should not, occur in isolation. Rather, each one tends to be practiced in relation to others, constituting a whole system. A *system of practice* is formed by a coherent set of epistemic activities performed with a view to achieve certain aims. For instance, Lavoisier created a system of chemistry whose activities included collecting gases produced by chemical reactions, measuring the weights of the ingredients and products of reactions, combusting organic substances for analytical purposes, and classifying compounds according to their compositions. The overall aims of this system included determining the composition of various substances, and explaining chemical reactions in terms of the composition of the substances. Similarly as with the coherence of each activity, it is the overall aims of a system of practice that define what it means for the system to be coherent. The coherence of a system goes beyond mere consistency between the propositions involved in its activities; rather, coherence consists in various activities coming together in an effective way toward the achievement of the aims of the system. Coherence comes in degrees and different shapes, and it is necessarily a less precise concept than consistency, which comes well-defined through logical axioms.

It may seem difficult to make a sharp distinction between an epistemic activity and a system of practice, and this is intentional. The distinction is only relative and context-dependent. In each situation in which we study a body of scientific practice, I am proposing to call the overall object a *system*; when we want to study more closely more specific aspects of that system, we can analyze it into different subordinate *activities*. What we take as a whole system in a given situation may be seen as a constituent activity of a larger system, and what we see as a constituent activity in a given situation may in a different situation be analyzed as a whole system made up from other activities. (When I initially introduced the notion of epistemic activity two paragraphs ago, I used the expedient of saying that an activity was made up of operations; the terminology of "operation" is strictly speaking unnecessary, though it is convenient when we want to keep *three* levels of description in view all at once.)

In this way, my framework is applicable at all levels, and can be zoomed in and out to suit any level that we want to focus on. At each focus-point, we call the overall practice "system" and its constituents "activities" (and their constituents "operations"), without intending to stick those categorical labels to anything on a permanent basis.

I do not propose to go deeply into the metaphysics of action, but it will be useful to stress the non-reductive nature of the relationship between the levels of description here. The structure of actions and processes is not atomistic in a reductive way, unlike the structure of things and statements.[19] Each epistemic activity can itself be analyzed as a system of activities, but the "component" activities are not necessarily simpler than the "whole" activity in an absolute sense, and the analysis can go on indefinitely. For example, take the combustion-analysis of a chemical substance. This can be analyzed as consisting of various other activities: burning the target substance; absorbing the combustion-products using other chemicals; weighing with a balance; making percentage-calculations; etc. And those component activities in themselves consist of other activities; for example, the activity of weighing with a balance consists in placing samples and weights on balance-pans, reading the number off the scale, etc. Now it may seem that we are getting to simpler and simpler activities as we continue in our analysis of actions, hopefully to reach a rock-bottom of atomic operations. But we are forgetting something less convenient. The activity of weighing with a balance also includes a certification-activity, to support our assumption that what we are handling are the correct standard weights; without that, the whole activity of weighing with a balance becomes incoherent. This certification-activity may consist in ordering the weights from a reliable supplier, or comparing them to a more trusted set of weights, or checking them against certain natural phenomena (e.g., the weight of a certain volume of water at a certain temperature). Whichever option we go with, it is clear that this component activity is not simpler in any clear sense than the main activity of weighing with a balance. The relation between various epistemic activities is ultimately non-reductive and reticular, although in many situations we can gain useful insights from analyzing an activity into its apparent components. There is no lowest level of description, and no clear end to the process of activity-analysis. Rather, the analysis should be carried out wherever, and as far as, it is productive.

Before I begin to apply the concepts of epistemic activity and system of practice to the case at hand, a few more general remarks will help clarify the framework further. First, the focus on activities and their aims throws new light on the old philosophical preoccupation with propositions and their truth. Of course propositions are important in science, including observation statements, empirical laws and theoretical principles. However, what is really important is to understand how they function within various epistemic activities. Evaluating the correctness of statements

---

[19] There are, of course, some reasons to be skeptical about the atomistic–reductive structure of things and statements, too, but that is a story for another day.

is surely one of the most important epistemic activities; however, the standards of correctness can easily vary from system to system (see Chap. 4, Sect. 4.3.1 for further reflections on this issue). If we are talking about "Truth with a capital *T*" that is not at all dependent on the system of practice one works in, it is doubtful that there are any actual scientific activities that are concerned with it. The success of each activity or system needs to be judged first of all in terms of how well it achieves the aims that it sets for itself; in addition, we may make judgments on the value of the aims themselves. But there will be precious few occasions on which "Truth" is an operable aim or standard of judgment in science.[20]

Finally, it will be instructive to draw a comparison-and-contrast between systems of practice and some other notions already current in the literature of the history and philosophy of science. The most obvious comparison is the Kuhnian concept of a paradigm. Despite some clear similarities, I think there are also some significant differences, sufficient to warrant the articulation of a new concept. As Kuhn freely admitted, he used the term "paradigm" in two main senses.[21] The first, of an "exemplar", does not match my sense of "system". The second sense, the "disciplinary matrix", is akin to my concept, but for two main reasons I do not find it helpful. First, I think we need a concept that is more definite and orderly than the Kuhnian disciplinary matrix, which incorporates all kinds of elements ranging from fundamental metaphysical principles to institutional structures, with no definite indication of how the whole thing holds together. Hacking's (1992, 44–50) characterization of laboratory practice also has a similar difficulty, specifying 15 different kinds of elements that enter into experimental practice without elaborating how they combine and interact with each other. In addition, Kuhn's paradigm concept is too closely tied to his insistence that a paradigm does and should enjoy a monopoly over an entire scientific discipline in normal phases of science; as explained further in Chap. 5, I have both descriptive and normative objections to that presumption of monopoly. Another interesting comparison is with John Pickstone's "ways of knowing", which I think are *types* of systems of practice (or, system-types). A *system-type* is a class consisting of various systems all of which share some core activities. A system-type is not an actual system, not even a very general one, but an incomplete characterization of a system with blanks consciously left to be filled in (and similarly for activity-type, a notion I used briefly above). Depending on how the blanks are filled, we have different instantiations of the type. A system-type defines a class of systems consisting of these instantiations. I will have occasion to make use of this concept in Sect. 1.2.3.2 (for further details see Pickstone 2000, 2007; Chang 2011d).

---

[20] This is not to deny the value of Truth as a regulative principle—or, less grandiosely, a rhetorically or motivationally effective purpose.

[21] This explication occurs in the 1969 Postscript included in the second edition of *The Structure of Scientific Revolutions* (Kuhn 1970, 180–191).

1.2.1.2 Problem-Fields

Having clarified my framework of analysis, let me come to the task at hand: to compare the relative merits of the phlogistonist[22] and oxygenist systems of chemistry as they existed in the late eighteenth century. In attempting to answer these questions, it is important to keep a clear head about *whose* judgment we are considering. There are various viewpoints to consider: what each of the competing sides in the dispute thought, and what we ourselves (various people) as historiographical and philosophical inquirers think. Which of these judgments we focus on depends on our purpose in making the inquiry. My present purpose is to assess whether there were good reasons for abandoning the phlogistonist system, not only whether past scientists themselves thought so. Therefore, I must give my own judgments in the end, though it would be silly to ignore the historical actors' judgments completely. The significance of that distinction will become clearer as I proceed with my concrete assessment.

I will try to be *comprehensive* in my appraisal, because what we have seen so far indicates that there never was a simple isolated "clincher" or a fatal flaw on either side that settled, or should have settled, the argument. That means there were probably a number of important factors relevant to the decision, all of which need to be considered carefully and weighed up against each other. So let us start by asking, very generally: what were the important scientific problems facing the phlogistonist and oxygenist systems, and how well did each system solve them? Answering these questions will require that we address two further questions: by which standards, or epistemic values, should the solutions to those problems be evaluated; and how exactly should each epistemic value be applied in the concrete situations under consideration? All of these questions were raised very clearly by Kuhn (1970, 1977) in his discussions of paradigm-choice in the face of incommensurability (see Sect. 1.3.3 for further thoughts on incommensurability in the Chemical Revolution).

In *The Structure of Scientific Revolutions* it was one of Kuhn's main points regarding incommensurability that different paradigms have different lists of problems they consider legitimate and important.[23] This most indisputable element of incommensurability often got swept aside in the excited discussions about its other aspects,[24] and it seems that Kuhn himself abandoned the concern with problem-fields in the reference-based discussions of incommensurability that he gave later in life. But I do not see how we can even start a reasoned debate about theory-choice or paradigm-choice without identifying the problem-fields recognized by the

---

[22] I should say "phlogistic" to be more faithful to contemporary usage, but I prefer "phlogistonist", which works well in parallel with "oxygenist" (rather than "oxygenic" or some such formulation to fit with "phlogistic"). My terms conform well enough to standard usage among historians, as can be seen in the discussion of the "phlogistonists" in Brock's standard-setting *Fontana/Norton History of Chemistry* (1992, 78).

[23] On the early Kuhn's emphasis of this aspect of incommensurability, see Hoyningen-Huene (1993), 208–209.

[24] Among the philosophical commentators on the Chemical Revolution, Andrew Pyle (2000, 104) is one of the exceptional people who stress the importance of this factor.

Table 1.1 The division of the problem-field in the Chemical Revolution

Problems considered important by both sides	Problems considered (very) important only by phlogistonists	Problems considered (very) important only by oxygenists
Understanding of combustion, calcination/reduction, and respiration	Explaining properties of compounds (including metals) in terms of properties of ingredients	Theory of heat and changes of state
Theory of acids	Mineralogy; geology	Chemistry of salts
Constitution of various substances	Meteorology	
	Nutrition; ecology	

competing sides. Theories are not simply "tested against evidence"; we must always choose *where* they ought to be tested against evidence—which is to say, where we most wish them to be empirically successful. The focus on problems is particularly congenial to a pragmatist view of knowledge, which will inform my discussion in Chaps. 4 and 5. For now, it is sufficient to recall John Dewey's (e.g., 1938) inclination to analyze knowledge in terms of processes of *inquiry*.

Table 1.1 shows the division of the problem-field in the Chemical Revolution. It is important to have this laid out in front of us, as that was literally the battlefield of the Chemical Revolution. Some problems were commonly acknowledged to be important by both sides, and others were considered important by only one side. Both the oxygenist system and the phlogistonist system assigned great importance to the understanding of three processes commonly recognized as closely related to each other: combustion, calcination, and respiration. The theory of acids was also something that both sides considered important and debated seriously.[25] Closely related to these issues was the *constitution* of various substances including water, metals and calxes, and various kinds of "airs", and also of various non-metallic substances that had been considered to be phlogiston-rich. Kirwan (1789, 6–7) thought that constitution was the key to the whole dispute: "The controversy is therefore at present confined to a few points, namely, whether the *inflammable principle* [phlogiston] be found in what are called phlogisticated acids, vegetable acids, fixed air, sulphur, phosphorus, sugar, charcoal, and metals." Similarly, in his latter-day defence of the phlogiston theory published from his exile in Pennsylvania, Priestley ([1796] 1969) focused almost exclusively on constitution—of metals and water most of all, but also of carbon, nitrogen and fixed air. In these works Kirwan and Priestley were responding to Lavoisier's contention that he had refuted the phlogistonist conceptions on the constitution of these substances.

In contrast to the above problems, some others were not universally considered important. The phlogistonists shared a widespread preoccupation of pre-Lavoisier

---

[25] See Brock (1992), 125. Richard Kirwan's classic exposition of the phlogiston theory was titled "An Essay on Phlogiston *and the Composition of Acids*". In that text, Kirwan (1789, 38) considered that Lavoisier's work on acidity was an important contribution: "With respect to the nature and internal composition of acids, it must be owned that the theory of chemistry has been much advanced by the deductions and reasonings of Mr. Lavoisier."

chemistry, which was "to account for the qualities of chemical substances and for the changes these qualities underwent during chemical reactions." (Kuhn 1970, 107)[26] More specifically, phlogistonists sought to explain the key properties of many substances in terms of the "principles", particularly phlogiston, that entered into their composition (see Sect. 1.2.3.2 for more on this). One salient case was the explanation of why metals (which were compounds for phlogistonists) had a set of common properties (Kuhn 1970, 148). Actually by the onset of the Chemical Revolution this was no longer a research problem in the phlogiston paradigm, as it was accepted almost as common sense that metals had their common metallic properties (including shininess, malleability, ductility, electrical conductivity) because of the phlogiston they contained.[27] The oxygenist side seems to have rejected not so much this answer as the question itself; chemistry reclaimed this stretch of territory only in the twentieth century. There were also some other problems neglected by the oxygenists that were considered important by some phlogistonists, although not all phlogistonists were equally concerned about them. These included various problems in mineralogy, geology, meteorology, and the circulation of nutrients in the environment.

On the other side, there were some problems considered much more important by oxygenists than by phlogistonists. Thermal phenomena were noted by all chemists and various phlogistonists tried to give some account of the nature of heat, but it was Lavoisier, building on Joseph Black's work on latent heat, who really brought heat (caloric) centrally into chemistry. Closely related to the theory of heat were questions regarding what we now call changes of state, about which Lavoisier had a very definite theory, to which he gave a very prominent place in his system of chemistry.[28] A similar case is the chemistry of salts. This was a common preoccupation of eighteenth-century chemistry, but the phlogiston theory had relatively little to offer here. In contrast, this area of research held out much promise for the new chemistry on the basis of Lavoisier's notion that "oxygen formed the glue or bond of dualistic union between acid and base to form salts".[29] Lavoisier displayed his enthusiasm for this line of research by devoting an entire third of his *Elements of Chemistry* to it.[30]

Having surveyed the problem-field of the Chemical Revolution, we can now tackle our main question: how do the performances of the two sides compare? First

---

[26] We shall see in Sect. 1.3.4 how Lavoisier himself was quite pre-Lavoisierian in this regard!

[27] See Kirwan (1789), 168.

[28] See the very first chapter of Lavoisier ([1789] 1965); Black's ideas were widely known but only published posthumously (see Black 1803). Some individual phlogistonists were also deeply concerned about heat and changes of state (e.g., Cavendish on boiling and evaporation), but these thoughts did not tend to affect their chemical theories. An apparent exception to the last statement is Jean-André De Luc, whose opposition to Lavoisier's chemistry seems to have been motivated by his ideas regarding the theory of rain; however, De Luc's chemistry was idiosyncratic and it is tricky whether to classify him as a phlogistonist, though he was clearly against Lavoisier.

[29] Brock (1992), 216.

[30] This is part 2 (of 3) of Lavoisier's text ([1789] 1965, 173–291), with the title "of the combinations of acids with salifiable bases, and of the formation of neutral salts."

of all, can we say which side was focusing on more important problems? My own judgment is that all of the problems listed in Table 1.1 above were important and worthwhile. Now, when we make such judgments, we ride roughshod over the historical actors' judgments of what was important. Historians will tend to be wary of such judgments, but it is not one we can avoid. Philosophers wishing to make an epistemic assessment certainly cannot avoid it, and even historians must judge what is worth writing about. Simply going along with the historical actors' judgments is not a solution, especially in cases like this where important scientists at the time disagreed with each other. The most pernicious option is to go simply with the historical winner's judgment.[31]

Accepting that all the problems considered important by either side were important (at least *prima facie*), we can ask how well each side did in solving them. Rather predictably, and just as Kuhn said, each side tended to provide good solutions to the problems that it considered important, and not such good solutions to other problems; that does not give us much of a basis for preferring one side to the other. It does seem to me that there were a larger number of important problems dealt with by the phlogistonist system but not by the oxygenist system, than conversely. Other things being equal, that circumstance would tend to recommend the phlogistonist system over the oxygenist system. Possibly more important is whether either side did clearly better in solving the common problems, which both sides recognized as important. In the Chemical Revolution there was no shortage of common problems, but each side thought it was doing quite well on all of them, and rather better than the other side. Proponents of the two systems offered substantively different solutions to common problems, and understood each other quite well, but disagreed on the evaluation of the relative qualities of those solutions. This brings us to the question of the criteria of judgment, or, the epistemic values that were in operation in the debates.

1.2.1.3 Divergent Epistemic Values

The phlogistonist and oxygenist systems gave very different solutions to most of the problems they shared, and they differed sharply in their judgments of who gave better solutions to these problems. Here we come into a different dimension of incommensurability, which has one important source in divergent epistemic values. Let us examine this situation more closely.

The most important factor was an opposition between *simplicity* and *completeness*. Oxygenists, particularly Lavoisier himself, prized simplicity greatly—especially the kind of simplicity that one could describe as elegance. Phlogistonists, particularly Priestley, saw more importance in completeness, wanting to account for all the observed phenomena in a given problem-area and for all the observed aspects of those phenomena. Lavoisierians liked to focus their attention on paradigmatic cases in which their theoretical conceptions worked out beautifully, setting the messier cases aside. In contrast, Priestley and some of his fellow phlogistonists endeavored

---

[31] For my reasons against aligning our judgments with the historical winners', see Chang (2009b).

to explain all the major phenomena they produced and observed, even if the explanations got cumbersome in the more difficult cases. It is not that either side failed to recognize the desirability of either simplicity or completeness, but there were clear differences in the degree of emphasis, or preoccupation, with those competing values.

A good example to illustrate this point is the calcination and reduction of metals. The red calx of mercury, which Priestley had initially used to produce dephlogisticated air, was taken up by Lavoisier and his colleagues as the paradigm case showing that calcination and reduction were processes of oxidization and de-oxidization. Mercury could be turned into this red calx by heating in ordinary air; the calx could then be turned back into metal simply by a higher degree of heat (produced by a large burning lens), yielding oxygen and producing or absorbing no other detectable substances. This wonderful exhibition of oxidization and reduction was cited over and over by the Lavoisierians. Priestley protested ([1796] 1969, 24): "But this is the case of only this particular calx of this metal". In his view, the Lavoisierians were distorting the whole picture by focusing on one exceptional case. Other metals behaved differently; Priestley (p. 31) pointed out that no calx of iron could be revived "unless it be heated in inflammable air, which it eagerly imbibes, or in contact with some other substance which has been supposed to contain phlogiston." Even for mercury, there was another type of calx[32] "which cannot be completely revived by any degree of heat, but may be revived in inflammable air, which it imbibes, or when mixed with charcoal, iron-filings, or other substances supposed to contain phlogiston." (p. 24)

This divergence between simplicity and completeness also played an important role in the debates regarding combustion. The difficulties of Lavoisier's theory of combustion, which I discussed briefly in Sect. 1.1, deserve some detailed attention here, especially as they tend to be neglected even in well-informed historical treatments. As background, recall that Lavoisier understood combustion as involving a *decomposition* of oxygen gas into "oxygen base" and caloric, the oxygen base combining with the combustible substance, and the caloric being released. The sensible heat generated in combustion came from the oxygen gas, and it was essential that the oxygen enabling combustion was in a gaseous state to begin with, since it was the abundance of combined caloric which put a substance into the gaseous state. The production of light in combustion was explained in a similar way, though more vaguely.

Let me return to Thomas Thomson's summary of objections to Lavoisier's theory of combustion (Thomson 1802, vol. 1, 354–358).[33] Thomson reckoned that, following Lavoisier's view, "one would naturally suppose, that when the product [of combustion] is a gas, all the caloric and light which existed in the oxygen gas would be necessary for maintaining the gaseous state of the product". But, for example, when charcoal is burned the product is a gas, yet the combustion still yields a great deal of heat and light. According to Thomson, Lavoisier was aware of this problem, but did not provide a convincing solution. Thomson also noted a converse problem: "One would

---

[32] Priestley explained that this was "that which remains after exposing turbith mineral to a red heat"; the modern editor of his text adds that "turbith mineral" is "basic mercuric sulphate".

[33] See also the discussion of Thomson's arguments by Partington and McKie (1937–1939), 340–342, and by Morris (1972).

naturally suppose that in every case of combustion the oxygen employed must be in the state of a gas. But this is very far from being the case". For example, "a very rapid combustion" takes place when nitric acid is poured on certain oils, but oxygen only enters that reaction in a liquid state, not a gaseous one. Or consider the explosion of gunpowder, which happens without the help of ambient oxygen gas, the oxygen being present in the solid state in the nitre (saltpeter) contained in the gunpowder itself. Thomson also pointed out that from Lavoisier's theory "one would naturally expect that caloric and light would be emitted during the condensation of other gases as well as oxygen: but this never happens unless oxygen be concerned." For example, when hydrogen and nitrogen gases combine, there is no heat or light emitted; ammonia gas and hydrochloric acid gas combine to make a "concrete salt", producing very little heat and no light. Thomson also noted that there was an emission of a good deal of caloric and light (i.e., combustion, to all appearances) in some reactions that did not involve any oxygen (nor any gases) as reactants: for example, when sulphur combines with certain metals, and when phosphorus and lime combine with each other.

Such problems did not move Lavoisier and his colleagues to modify their doctrines, not to mention abandon them. It is not that Lavoisier and his colleagues were *blind* to anomalies. For example, as Seymour Mauskopf (1988) relates in illuminating detail, Lavoisier was very interested in the chemistry of gunpowder, and made a few different attempts to explain its workings in terms of his theory. (Lavoisier's interest in gunpowder is no surprise after all, as he was a commissioner of the Royal Gunpowder Administration from 1775 and in that capacity set up his residence and laboratory at the Paris Arsenal.[34]) These attempts were not very successful, and no less than Claude-Louis Berthollet (1748–1822) used the gunpowder case against Lavoisier's theory of combustion, citing precisely the difficulty that Thomson later reported. Lavoisier was himself never quite satisfied with the clever yet clumsy defence he was able to give, although Berthollet was pacified on this point after his "conversion" to Lavoisier's system for other reasons.[35] After that stumble in the middle of Paris, the Lavoisierian theory of combustion marched on and out, unaltered. Thomson reported that Luigi Valentino Brugnatelli (1761–1818) had solved some of the difficulties with his concept of "thermoxygen", which is oxygen that combines with other substances while retaining its light and caloric. This troubled Thomson as a blatantly *ad hoc* move, but it was a way for Lavoisierian theory to keep its main doctrine while addressing an important anomaly; such ideas were entertained by Lavoisierians when necessary, but they were not admitted to the core of their system.[36]

---

[34] Guerlac (1975), 65–66.

[35] Mauskopf (1988, 110–111) highlights Berthollet's objection, and relates Lavoisier's ultimate discontent (p. 115). See Le Grand (1975) on Berthollet.

[36] As Lakatos said, fluctuations in the "protective belt" consisting of auxiliary hypotheses do not and should not affect the "hard core" of a research programme. It is interesting to note that Brugnatelli's solution here was probably only a rehash of the idea that Lavoisier himself had come up in order to fend off Berthollet. In fact, as Partington and McKie (1937–1939, 341–342) point out, Berthollet reminded Thomson that this issue had been addressed by Lavoisier. See Mauskopf (1988, 113–114) for an explanation of Lavoisier's work on this.

Thomson's conclusion on Lavoisier was clearly negative: "Upon the whole, it cannot be denied that Lavoisier's theory does not afford a sufficient explanation of combustion." Nor was Thomson alone in this sort of judgment. Numerous other chemists who accepted that oxygen combined with combustibles remained skeptical of Lavoisier's explanation of the heat and light in combustion; these "late phlogistonists" often maintained a system in which oxygen and phlogiston happily co-existed, the latter still being given the role of explaining what we would now identify as the energy relations in combustion (see Sect. 1.2.2 for further details).[37] But Lavoisier was more concerned about maintaining the simplicity of his theory than fitting all the known facts. Later celebratory historiography of the Chemical Revolution has nearly managed to obliterate this aspect of the story from collective memory.

In addition to simplicity and completeness, there were also broader types of epistemic values at play. A sort of epistemic conservatism was one of the values upheld by many phlogistonists, while oxygenists were taken with the idea of reform or novelty in itself. There is an interesting passage from Cavendish illustrating this point (1784, 152): "it will be very difficult to determine by experiment which of these opinions is the truest; but as the commonly received principle of phlogiston explains all phenomena, at least as well as Mr. Lavoisier's, I have adhered to that."[38] Cavendish's temperament shown here is surely an informative contrast to Lavoisier's youthful enthusiasm declaring to himself in 1773 that his investigations were "destined to bring about revolution in physics and chemistry", before he had published even his first attack on phlogiston.[39]

On the other hand, it cannot be that the leading phlogistonists were simply opposed to scientific change, as they surely delighted in making new discoveries and crafting some new theoretical ideas, too. Many arguments made on behalf of phlogiston were motivated not by conservatism but by *pluralism*, in reaction against Lavoisierian dogmatism. This is quite contrary to the common notion that the phlogistonists were blinded by dogma. In Sect. 1.2.2 I will say more about this, but for now Priestley will serve as an emblem of the scientific pluralism found on the phlogistonist side. His 1796 defence of phlogiston is a moving testimony in this regard. Declaring that "free discussion must always be favorable to the cause of truth", he reminded the reader of the non-dogmatic path he had walked in science:

> No person acquainted with my philosophical publications can say that I appear to have been particularly attached to any hypothesis, as I have frequently avowed a change of opinion, and have more than once expressed an inclination for the new [Lavoisierian] theory, especially that very important part of it *the decomposition of water*. (Priestley [1796] (1969), 21)

---

[37] See Partington and McKie (1937–1939), part 4; also Allchin (1992).

[38] Cavendish added that there was one other consideration in addition to this prudence or conservatism: "it is more reasonable to look for great variety in the more compound than in the more simple substance", in relation to plants and their composition.

[39] Lavoisier quoted in Donovan (1988), 219. Donovan cautions against reading Lavoisier's "revolution" to mean what we now commonly associate with the term; my current point is not affected by the precise meaning of "revolution".

He dedicated his book to "the surviving answerers of Mr. Kirwan" (the post-Lavoisier leaders of French chemistry, namely Berthollet, Laplace, Monge, Guyton de Morveau, Fourcroy and Hassenfratz), and requested an answer from them to his objections to the oxygenist system. Priestley drew an ominous parallel between the politics of science and the larger politics that had put a premature end to Lavoisier's life at the guillotine in 1794, for his part in the business of tax-collecting: "As you would not . . . have your reign to resemble that of *Robespierre*, few as we are who remain disaffected, we hope you had rather gain us by persuasion, than silence us by power."[40]

I do not think that all this was retrospective self-fashioning or a loser's spiteful plea for survival. Priestley had expressed similar epistemic views even at the height of his fame and success. For example, in the 1775 letter in which he announced the discovery of dephlogisticated air (oxygen), Priestley wrote (1775, 389):

> It is happy, when with a fertility of invention sufficient to raise *hypotheses*, a person is not apt to acquire too great attachment to them. By this means they lead to the discovery of new facts, and from a sufficient number of these the true theory of nature will easily result.

This passage directly followed his proposal that "nitrous acid is the basis of common air, and that nitre is formed by a decomposition of the atmosphere", to which he added: "But I may think otherwise to-morrow." One can almost hear an echo of Montaigne finishing his thoughts with "though I don't know".[41] In contrast, there was a clear absolutist impulse on the oxygenist side, perhaps most egregiously manifested in the ceremonial burning of Stahl's phlogistonist text. As described by Justus Liebig (1851, 25), this was "a festival in which Madame Lavoisier, robed as a priestess, committed to the flames on an altar, while a solemn requiem was chanted, the phlogistic system of chemistry."

What are *we* to conclude, having seen these arguments from both sides? My sense is that Lavoisier did clearly fail in giving a complete account of combustion. Does that mean his theory was simply wrong, and therefore its pleasing simplicity was immaterial? Or, to be a bit more subtle: shouldn't we say, following Bas van Fraassen's view of theory-assessment (1980, 87), that Lavoisier's theory of combustion was not empirically adequate, and no amount of simplicity, which is only a pragmatic virtue, could redeem it? The matter is not so straightforward, because empirical adequacy comes in bits and pieces (in various problem-areas treated by each theory), and moreover comes in degrees in each little bit and piece. The discomfort of a strict hierarchy of values between empirical adequacy and the pragmatic virtues becomes clear when we ask whether a tiny advantage in empirical adequacy would

---

[40] Priestley [1796] (1969), 17–18. He signed off his dedication with unflagging loyalty to the cause of the French Revolution: "I earnestly wish success to the arms of France, which has done me the honour to adopt me when I was persecuted and rejected in my native country. With great satisfaction, therefore, I subscribe myself Your fellow-citizen, Joseph Priestley."

[41] See Bakewell (2010), 43, and Chapter 7, for a nice exposition of this aspect of Montaigne's thought. Perhaps there is a fruitful comparison to be made between Montaigne's *Essays* and Priestley's *Experiments and Observations on Different Kinds of Air*. A Cartesian quest for certainty is clearly eschewed in both.

not be outweighed by a great deal of additional simplicity or other virtues.[42] The problem is worsened when we recognize that empirical adequacy itself is not a single-valued variable in any case; how do we weigh up an advantage in empirical adequacy in one area against a disadvantage in another area? All I can say, at this stage, is that we cannot say anything definitive about the oxygen–phlogiston choice, at least until we have taken in a comprehensive view of all the relevant problems and all the epistemic values relevant in evaluating the solutions.

Where different values jostle with each other, the epistemologist is in uncomfortable territory. How are we to say which of the values cherished by different historical scientists were more valuable? Do we have any right to make such judgments? In my view, this is not a question of rights, as we are not going to be *doing* anything to the past actors themselves by our judgment. Rather, I think the question is about the present, and I believe that we have a *duty* to ourselves to make such judgments. As philosophers of science, or as any responsible citizen who considers matters of knowledge, we need to make judgments on the basis of epistemic values in the present. Then it is impossible to avoid applying these judgments in some ways to the past, just like it is impossible to keep our present ethical values entirely out of our studies of history in general. (It will not do to say that we have no right to judge whether a past act, say a genocide, was right or wrong.) Conversely, it is important to remain alert to the judgments we do have about epistemic values operative in past science, because they will affect our judgments about present science. What we celebrate and condemn in past science, however implicitly or subtly, cannot remain safely separate from how we deal with present science. So it is not irrelevant to ask questions such as whether simplicity or completeness was the superior virtue in the phlogiston–oxygen choice. My honestly biased way of posing the question would be: which attitude was (and is) more rational or scientific, between adapting theories to new phenomena that we learn about, and giving dogmatic dominance to a favoured theory?

1.2.1.4   Divergent Instantiations of the Same Value

Having addressed different weights given to different epistemic values by the opposing sides in the Chemical Revolution, we must now examine a further complication raised by Kuhn (1977, 331), that even one and the same epistemic value may be interpreted and instantiated in divergent ways, leading to very different conclusions and even mutual accusations of betrayal of the value in question. There are a few significant cases of divergent value-instantiation in the Chemical Revolution.

Both sides valued unity, and each side cited the kind of unity it was able to achieve as persuasive evidence in its own favor. There was some convergence in this, since both systems unified combustion, calcination and respiration in similar ways. But beyond that there was significant divergence in what was unified, and

---

[42] In Chap. 5 I will come back more strongly to the question of whether empirical adequacy really ought to be considered a value that is more important than others.

how. Lavoisier's caloric theory linked together explanations of combustion and of changes of state. His ideas on oxygen connected combustion and acidity, since many combustion-products were acidic. On the phlogistonist side, there was a pleasing theoretical unity concerning the behavior of metals: their common properties, their calcination/reduction, and their reaction with acids. The phlogiston theory was also more conducive to a grand unity of all imponderable substances, as manifestations of "elementary fire": phlogiston, electricity, light, magnetism, and so on (for the connections between phlogiston and electricity, see Chap. 2, and Sect. 1.2.4.1).

Similarly but even more strikingly, the adherence of both sides to the value of *systematicity* is manifested in each side accusing the other of being arbitrary and haphazard. From the phlogistonist side, the accusation against the Lavoisierians was that they failed to adhere to the rule of assigning like causes to like effects. Both Priestley ([1796] 1969, 33) and Kirwan (1789, 281–282) used this argument in constitutional debates, to combat the oxygenist refusal to recognize the common presence of phlogiston in various substances. For his part, Lavoisier had a clear disdain for the continual complications and mutually conflicting changes that various phlogistonists introduced to their theories in their attempts to meet the challenges posed by various new phenomena:

> Why, therefore, need we have recourse to an hypothetical principle, the existence of which is ever supposed, and has never been proved; which in one case must be considered as heavy, and in another as void of weight, and to which, in some cases, it is necessary even to suppose a negative weight; a substance which in some instances passes through the vessels, and in others is retained by them; a being which its maintainers dare not rigorously define, because its merit and its convenience consist even in the uncertainty of the definitions which are given of it?[43]

Now, as I will discuss further in Sect. 1.3.1, Lavoisier's statement is misleading to the extent that it implies that each phlogistonist or each version of the phlogiston theory held these mutually contradictory beliefs about the nature of phlogiston; rather, there were different versions of the phlogiston theory, in some cases successive in time, that did not agree with each other in all their details. (My own comparative assessment here is between Lavoisier's system and the best versions of the phlogistonist system.) Still, Lavoisier was making a reasonable demand that the phlogiston system as a whole should be developing in a more systematic way. In contrast to Lavoisier, no leaders on the phlogistonist side had the will or the wherewithal to make everyone on their side sing from the same hymn sheet.

An examination of the arguments on both sides also reveals their common allegiance to what I will call "empiricism": a commitment to avoid invoking extraneous hypotheses, to stay close to observable facts and ideas derived from those facts. In the passage quoted above, Lavoisier and his colleagues denounced phlogiston as a hypothetical entity whose existence was "ever supposed, never proved" (unashamed of their own *lumière*, *calorique*, and *radical muriatique*). In their own theory, they

---

[43] This passage occurs in a report to the Paris Academy by Lavoisier, Berthollet and Fourcroy, which Lavoisier quotes in his comments on Kirwan's treatise on phlogiston (1789, 15). See also Lavoisier (1786).

claimed, "nothing is admitted but established truths"; theirs was "a doctrine which explains all the facts of chemistry without any supposition."[44] Most phlogistonists were no less adamant about their empiricism. I have already quoted Priestley as saying that he did not have strong attachments to hypotheses, and regarded them mostly as means for eliciting new facts.

Having reviewed those divergent value-instantiations, the question of our own judgment raises its head again: which do we think was the more valuable kind of unity, the more useful kind of systematicity, and the more genuine kind of empiricism? In each case, I cannot see any irrefutable way to argue that one was more important than the other. That is my honest judgment, although it may raise a worry about not being able to close the debate.

1.2.2 What Really Happened in the Chemical Revolution?

The evidential assessment made in the last section leaves us with a very uncertain verdict. It seems clear that each of the oxygenist and the phlogistonist systems had its own merits and difficulties, and that there were different standards according to which one or the other was better supported by empirical evidence. In a way, this is only an indication that evidential support is not a straightforward matter of logical or probabilistic connections between theory and observation, but a complex relationship mediated by epistemic values, which can be divergent and contextual. My own judgment is that both systems were partially successful in their attempts to attain worthwhile goals, and that there was no reason to clearly favour one over the other (I will say more about the implications of that judgment in Sect. 1.2.4). But if there was no clear justification for the choice of oxygen over phlogiston, then why *did* chemists make that choice? Why did the Chemical Revolution happen? In Sect. 1.1 I only touched briefly on this question. One reason I did not enter into a full discussion of it is that I think philosophers and historians have wasted a lot of time and energy by tackling an illusory question. There is great futility in the enterprise of explaining why the vast majority of chemists quickly went over to Lavoisier's side—because actually that is not quite what happened. I would like to set out a different and more accurate story of the Chemical Revolution, before I launch into further discussions of its causes, its rationality, and its consequences.

There is an extensive literature on the Chemical Revolution, which I will not be able to survey with any comprehensiveness here; instead I refer the reader to John McEvoy's (2010) up-to-date and thorough critical review of this literature. My own aim here is to advance a particular revisionist thesis: the Chemical Revolution did *not* consist in a swift and nearly universal conversion of the chemical community to Lavoisier's theory. I have made that argument more fully elsewhere (Chang 2010), so I will present a shorter summary. First of all we need to resist being taken in by triumphalist declarations emanating from Lavoisier, from his contemporary advo-

---

[44] This passage is from the preface to the French translation of Kirwan's treatise on phlogiston: Kirwan (1789), xiii.

cates, and from his posthumous glorifiers. But declarations of a clean victory can be found in some quite unexpected places, too. For example, take the opening sentence of Priestley's 1796 defence of the phlogiston theory: "There have been few, if any, revolutions in science so great, so sudden, and so general, as the prevalence of what is now usually termed *the new system of chemistry*, or that of the *Antiphlogistians*, over the doctrine of Stahl, which was at one time thought to have been the greatest discovery that had ever been made in the science." (Priestley [1796] 1969, 1) Perhaps that was just an exaggerated lament from the loser,[45] but strangely the same idea can also be found in the works of some very expert historians, such as Robert Siegfried (1989, 31): "Of all the well known revolutions in the history of science, the chemical is perhaps the most dramatic . . . . Only 20 years separate Lavoisier's first explorations of the chemistry of gases and the public capitulation of Richard Kirwan, the last significant European defender of the phlogistic views." The impression of unanimity is also voiced by Larry Holmes (2000, 751): "all but Priestley himself eventually came over to the side of the French chemists". McEvoy notes (2010, 18–19): "the suddenness, brevity and pace of the Chemical Revolution" as some of the key factors which "marked it in the minds of many commentators as arguably the best example of a classic revolution in the history of science." Similar ideas can be found in the most careful of the philosophical commentators, too: Andrew Pyle (2000, 105) judges that "by 1800, all its [phlogiston theory's] defenders had effectively been swept away", and Alan Musgrave (1976, 205) says that by 1796 "the Chemical Revolution was over".

A reasonably close look at the primary literature from about 1790 onward, however, should make it evident that there were numerous chemists who declined to jump on the Lavoisier bandwagon, whom I will call "anti-anti-phlogistonists". Although Lavoisier was sanguine enough to declare by 1790 (in a letter to Benjamin Franklin) that "a revolution has taken place in an important area of human knowledge", he knew full well that there were still many who were unconverted, especially in Germany and Britain; even the French chemists were still "divided", as he acknowledged in the same letter.[46] Many of the anti-anti-phlogistonists were respectable and respected chemists, not old men driven by sheer conservatism or dogmatism. There were at least three different types of dissenters in the period *after* the publication of Lavoisier's *Elements of Chemistry* in 1789, which is often seen as the point at which the Chemical Revolution was irreversibly consolidated (see Table 1.2).

First of all, there were indeed some die-hards. Priestley tops this list, but he is only a small part of the picture. Scheele did not survive long enough for his "die-hard" credentials to be tested truly, but up to his death in 1786 he showed no sign of relinquishing the phlogiston theory. One of the most striking figures is Jean-André De Luc, whose objection was based on his theory of rain postulating the transmutation of atmospheric air into water, as well as his general dislike of the rash and the revolutionary in science and politics alike (e.g., De Luc 1803). De Luc maintained

---

[45] Or Priestley may have felt disheartened by the rather sudden change of heart by some of the German phlogistonists in the mid-1790s (for this, see Hufbauer 1982).

[46] Quoted in Guerlac (1975), 112; Donovan (1993), 184.

Table 1.2 Varieties of anti-anti-phlogistonists, in the order of birth in each category

Die-hards ("elderly holdout", some not so elderly)	Fence-sitters	New anti-Lavoisierians
James Hutton (1726–1797)	Pierre-Joseph Macquer (1718–1784)	Benjamin Thompson, Count Rumford (1753–1814)
Jean-André De Luc (1727–1817)	Henry Cavendish (1731–1810)	George Smith Gibbes (1771–1851)
Antoine Baumé (1728–1804)	Georg-Christoph Lichtenberg (1742–1799)	Thomas Thomson (1773–1852)
Johann Christian Wiegleb (1732–1800)	Lorenz Crell (1745–1816)	Johann Wilhelm Ritter (1776–1810)
Joseph Priestley (1733–1804)	Claude-Louis Berthollet (1748–1822)	Humphry Davy (1778–1829)
Torbern Bergman (1735–1784)	Johan Gadolin (1760–1852)	
James Watt (1736–1819)	Friedrich Gren (1760–1798)	
Balthazar-Georges Sage (1740–1824)		
Carl Wilhelm Scheele (1742–1786)		
Jean-Claude Delamétherie (1743–1817)		
Jean-Baptiste Lamarck (1744–1829)		
Adair Crawford (1748–1795)		
Johann Friedrich Westrumb (1751–1819)		
Robert Harrington (1751–1837)		

close connections with various anti-Lavoisier figures in Germany, particularly Göttingen, and also with Priestley's associates in the Lunar Society of Birmingham, including James Watt.[47] In 1796 Priestley identified the Lunar group as the only remaining adherents to phlogiston that he knew of, in addition to Adair Crawford, who had just died (Priestley [1796] 1969, 20); he may have been unaware of Robert Harrington, who would go on to issue a "death warrant" of French chemistry (1804). On the German side, Karl Hufbauer (1982, 140–144) notes that most chemists there either converted to the Lavoisierian side or at least gave up any active resistance by 1796, but allows that there were some remaining phlogistonists, including Johann Christian Wiegleb and Johann Friedrich Westrumb, who were "virtually ostracized". And then there were people like Torbern Bergman in Sweden and James Hutton in Scotland, whose concerns were mineralogical and geological above all else. Hutton, for example, had a notion of the circulation of phlogiston in the environment, which smacks of modern ecology's understanding of the cycles of carbon and energy, according to Douglas Allchin (1994). Even right there in Paris there

---

[47] Middleton (1965, 115–131) gives a discussion of De Luc's theory of rain. See De Luc (1803), 1–306, for his detailed objections to the new chemistry, first in itself and then in relation to meteorology. For broader contexts of De Luc's work, see Heilbron (2005) and Tunbridge (1971).

remained significant anti-Lavoisierian figures, including Jean-Claude Delamétherie, the editor of the prestigious *Journal de physique* (called *Observations sur la physique* before 1794), who followed Priestley's ideas and cultivated a connection with De Luc; Arthur Donovan (1993, 174) considers him to have been "the most determined and effective of Lavoisier's French opponents".[48] There was also Jean-Baptiste Lamarck, whose idiosyncratic chemical ideas are understood by Leslie Burlingame (1981) as belonging to the natural-historical tradition of French science. To the list of French die-hards Perrin (1981, 62) also adds Antoine Baumé and Balthazar-Georges Sage.

The second category of dissenters sought compromise, or deliberate neutrality. Allchin, in his aptly titled paper "Phlogiston After Oxygen" (1992), makes a persuasive case that many chemists admitted the existence of oxygen for gravimetric considerations, while keeping phlogiston for what we would call energy considerations. J. R. Partington and Douglas McKie, in their classic series of papers on the phlogiston theory (1937–1939, 125–127, 143–148), already pointed to a large number of people in this category, many of them German or German-speaking, including Friedrich Gren, Lorenz Crell, Jeremias Richter and Johan Gadolin. Hufbauer's (1982) study of the German chemical community in the eighteenth century has elaborated further on that point. More generally, people often accepted Lavoisier's theory only partially, picking and choosing what made sense to them. The old phlogistonist Pierre-Joseph Macquer was taking this kind of approach when he died in 1784, and even Lavoisier's close colleague and ally Claude-Louis Berthollet remained skeptical about some of Lavoisier's ideas, especially his theory of acids.[49] There were many others who clearly saw some merit in Lavoisier's chemistry but did not consider it sufficient to reach a clear verdict in favor of it. As discussed above, Cavendish (1784, 150–153) gave a clear-headed view of how both theories could explain the phenomena he observed, while expressing a preference for staying with phlogiston. Alfred Nordmann (1986, 239–241) explains how Georg Christoph Lichtenberg made a strong case that there was not enough knowledge yet for a decisive verdict even in the 1790s, and how annoyed he was by the Lavoisier group's attempt to legislate the language of chemistry, by which act they forced other people to make a premature choice. Donovan (1993, 168) articulates the Lavoisierian gambit as follows: "His strategy [in "Réflexions sur le phlogistique"] was to force the reader to choose between two alternative theories he offered as the only possible choices"; thereby he "attempted to score a knockout, a *coup de science.*"

Even more interesting is the third category of dissidents, who fully acknowledged that Lavoisier's system had become established but also sensed that its time was passing quickly. Very suggestive in this connection is the following snippet of scientific conversation that I happened to stumble upon, from the year 1800. William Herschel (1738–1822) had just detected infrared radiation coming from the sun, which he saw as caloric rays separated from light rays by means of the prism. Joseph

---

[48] On Delamétherie's opposition to Lavoisier, see Guerlac (1975), 105–106.

[49] On Macquer, see Holmes (2000, 752); on Berthollet, see Le Grand (1975).

Banks (1743–1820) wrote to congratulate Herschel on this momentous discovery, but had one piece of advice: "I think all my friends are of the opinion that the French system of Chemistry, on which the names lately adopted by their Chemists are founded, already totters on its base and is likely soon to be subverted. I venture therefore to suggest to you whether it will not be better for you . . . to use the term Radiant Heat instead of Caloric; by the use of which latter word it should seem as if you had adopted a system of Chemistry which you have probably never examined." Herschel accepted Banks's advice happily: "I have the honour of your letter and shall be very ready to change the word caloric for radiant heat, which expresses my meaning extremely well."[50] Banks was a botanist and not a well-known chemist, but if the longtime President of the Royal Society and "all his friends" were predicting the imminent demise of the French chemistry in 1800, then there must be something that we have missed out on in our usual historiography. What did Banks have in mind, exactly? It's impossible to be sure, but we can make a fairly good guess. I have already discussed the difficulties of Lavoisier's theory of acidity, his theory of combustion, and his theory of heat. Some of Lavoisier's new-generation critics were located in Banks's London at the time, including Humphry Davy and Count Rumford, in addition to old dissidents like Cavendish and De Luc.

Perhaps the most interesting case of the new generation of anti-Lavoisier chemists was Humphry Davy (1778–1829). A mere lad of about 10 when Lavoisier's *Elements of Chemistry* was published, Davy was one of those who grew up with the Lavoisierian orthodoxy but came to reject it upon further consideration. He made his name not only with laughing gas and electrochemistry (more on the latter in Chap. 2), but by putting a nail in the coffin of Lavoisier's theory of acids with his argument that chlorine was an element and that muriatic acid (hydrochloric acid) did not contain oxygen, only hydrogen and chlorine.[51] After the acceptance of Davy's work, Lavoisier's oxygen theory of acidity fell clearly out of favour among chemists. Along with Rumford, Young and Cavendish, Davy was also one of those who mounted serious challenges to the Lavoisierian caloric theory of heat, whose dominance was never total.[52] As Siegfried (1964) reports in some detail, Davy actually entertained various systems of chemistry involving the revival of phlogiston. David Knight remarks (1978, 4): "there were widespread hopes and fears until at least 1810 that Davy would restore it [the phlogiston theory] and overthrow the French doctrines." Among those who expressed such hope in print was Sir George Smith Gibbes, doctor and chemical lecturer in Bath, later to be physician to Queen Charlotte; in 1809 Gibbes opined that Davy's discoveries had confirmed that Lavoisier was wrong after all.[53] Another who entertained such expectations was the

---

[50] Banks to Herschel, 24 March 1800, and Herschel to Banks, 26 March 1800, both quoted in Lubbock (1933), 266–267.

[51] See Gray et al. (2007).

[52] By the time the energy concept and early thermodynamics toppled the caloric theory altogether in the 1840s and the 1850s, Lavoisier's basic picture of the universe was really a thing of the past.

[53] See Golinski (1992, 213), who calls Gibbes "perverse" for this.

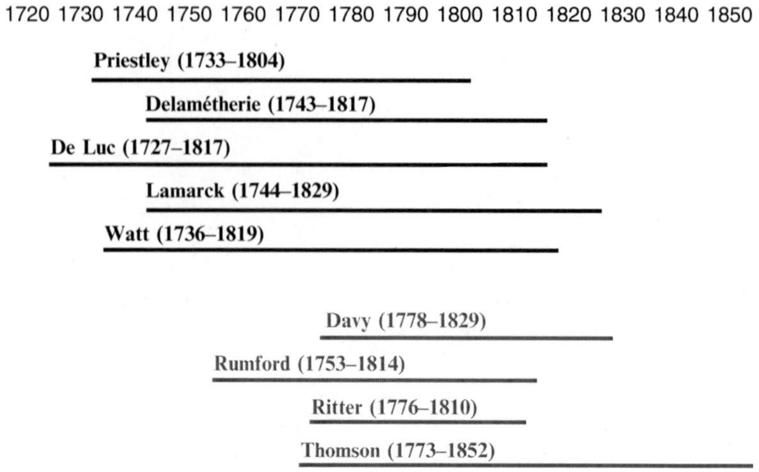

1720 1730 1740 1750 1760 1770 1780 1790 1800 1810 1820 1830 1840 1850

Priestley (1733–1804)

      Delamétherie (1743–1817)

De Luc (1727–1817)

      Lamarck (1744–1829)

Watt (1736–1819)

               Davy (1778–1829)

          Rumford (1753–1814)

            Ritter (1776–1810)

            Thomson (1773–1852)

Fig. 1.3 The overlap between old and new anti-anti-phlogistonists (salient figures from the first and third groups from Table 1.2)

young Michael Faraday (1791–1867), who wrote in 1812, citing Davy as an authority: "I would wish you not to be surprised if the old theory of Phlogiston should be again adopted as the true one".[54]

Taking all of that into consideration, what can we say the Chemical Revolution really consisted in? We still have to admit that a considerable number of chemists did become fully "converted" to Lavoisier's chemistry at least for a time,[55] and that it achieved a clear dominance in the textbooks. However, we also need to acknowledge that there were common cases of partial or half-hearted conversions, in many cases with a retention of phlogiston. Add to that not only the die-hard phlogistonists, but also the younger generation of dissidents who actually had their scientific education after Lavoisier's victory. A very interesting thing about those two generations is that they in fact overlapped significantly in time, the new generation coming up before all the die-hards had given up (see Fig. 1.3). Knight (1978, 29) actually understates the case when he says, in reference to a couple of later episodes: "As had happened with gothic architecture, this phlogiston survival was almost contemporaneous with the phlogiston revival".[56] Phlogiston was never truly and successfully revived, and there are indeed many senses in which Lavoisier and his colleagues brought about a "revolution" in chemistry, but it was not a sudden and clear-cut affair. It was a many-sided struggle that neither ended in unanimous agreement nor established any immutable orthodoxy.

---

[54] Michael Faraday to Benjamin Abbott, 11 August 1812, in James (1991), 17.

[55] See McCann (1978) for a thorough count.

[56] Knight refers to Stevenson (1849) and Odling (1871); the latter work will be discussed further in Sect. 1.2.4.

1.2.3 Weights, Composition, and Chemical Practice

Having modified the description of the manner in which the oxygenist system triumphed over the phlogistonist system, I now return to the task of explanation. Even in my revisionist history it is the case that a clear majority of chemists did eventually abandon phlogiston, and that fact remains to be explained. (If you are inclined to exclaim "It was all social!", see Sect. 1.3.2.) In short, my answer is that the victory of oxygen over phlogiston only came as part of a much larger development, namely the slow and steady ascendancy of what I will call "compositionism" in chemistry.

1.2.3.1 The Importance of Weight

There is something of crucial importance about the Chemical Revolution that we have not fully considered yet. Many well-informed historians and chemists have argued that Lavoisier's most important contribution to chemistry was his emphasis on weight: his recognition of its importance, and his use of precision measurements to trace it through chemical changes. According to this view, oxygen was not really the center of Lavoisier's chemical system, much as Lavoisier himself was enamored with it; rather, it was a consistent focus on weights that made the new system superior to the phlogistonist system. It does seem that what oxygenists considered the most decisive set of arguments against the phlogistonist system was based on the consideration of weights in chemical reactions. In contrast, phlogistonists were not so centrally concerned with weight, though some of them did recognize it as a relevant property and even knew how to measure it extremely well.[57] Where did this difference come from, and how did the difference on weight-related issues affect the evidential status of the competing systems? The focus on weight did not enter much into the discussion in Sect. 1.2.1, because it is not really an epistemic value. An attempt to puzzle out what sort of thing it is will bring us to a whole other dimension of evidential reasoning.

One can get a stark view about which side was right in the Chemical Revolution, if one shares Lavoisier's preoccupation with weight and his conception of weight as a conserved quantity. Take the decomposition and recomposition of water. Lavoisier's view is crystal-clear: we take 100g of water, and we make 15g of hydrogen and 85g of oxygen out of it (I am using Lavoisier's own figures here)[58]; and then, we can

---

[57] Kirwan devoted the first chapter of his book on phlogiston (1789) to the consideration of weights. Cavendish was unsurpassed in the precision measurements of all things, including weights. Bergman even made quantitative analyses of metals for their phlogiston contents (Brock 1992, 180).

[58] These numbers were given in Lavoisier's commentary on Kirwan's work (Kirwan 1789, 16). I should note that his pre-Revolutionary "g" was "grain", not the metric "gram" that he actually helped to put into place.

put those *precise* amounts of hydrogen and oxygen together, and make 100g of water again. What better proof than this could one have for the idea that water is a compound made up of hydrogen and oxygen? In contrast, consider the phlogistonist story: hydrogen is meant to be "phlogisticated water" and oxygen "dephlogisticated water". Cavendish and Priestley thought that water was the base of all gases, but gave no good story about why phlogistication should make water less dense than dephlogistication does, or a precise measure of how much phlogiston went into the part of water that became hydrogen (or came out of the part that became oxygen). One can easily see the force of Lavoisier's account.

We must not be over-hasty in our judgment, however. Did chemists during the Chemical Revolution have good reasons to accept Lavoisier's weight-based arguments about the constitution of certain key substances? This question is crucial, because weight-based arguments regarding constitution provided perhaps the only evidence available at the time that could have rationally compelled the phlogistonists to give up. Priestley and some others clearly rejected the weight-based arguments. Why did they do so, and were they wrong to do so? Lavoisier's reasoning given above rested on two very significant assumptions:

(a) Weight is a good and proper measure of the amount of all chemical substances.
(b) Weight is conserved.

If these assumptions were accepted, then there would be a compelling case for the use of weights as the main source of evidence for constitutional ideas. But was there good evidence to support assumptions (a) and (b) themselves?

In considering that question, we must start by losing the prejudice that weight is *of course* the important thing to focus on, and that weightless things do not belong in good chemistry. Assumption (a) did not hold universally even within Lavoisier's own chemistry, since the first two in his list of simple substances, light and caloric, did not have weight. Assumption (b) is also tricky. Contrary to common intuitions, there never was any deep metaphysical reason why weight must be conserved (and indeed, $E = mc^2$ proclaims that it is not). Of course, not all measurable quantities are conserved. Some turn out like temperature; temperature values cannot be added up to each other meaningfully, and they are not conserved. In any case, Priestley and some others were not ready to accept the precise weight-conservation claimed by Lavoisier in the decomposition and recomposition of water. This was particularly problematic given that the force of Lavoisier's argument was meant to derive from the rigorous precision of his experiments. William Nicholson (1753–1815), who did accept Lavoisier's system, gave some detailed reasons to doubt the claimed degrees of accuracy in Lavoisier's weight measurements. This was one aspect of Lavoisier's work that made him annoyed and doubtful:

> I must beg leave to observe, that these long rows of figures, which in some instances extend to a thousand times the nicety of experiment, serve only to exhibit a parade which true science has no need of: and, more than this, that when the real degree of accuracy in

experiments is hidden from our contemplation, we are somewhat disposed to doubt whether the *exactitude scrupuleuse* of the experiments be indeed such as to render the proofs *de l'ordre demonstratif.*[59]

For those with modern knowledge, it is nearly impossible not to sympathize with Nicholson here, in the face of Lavoisier's utter conviction that 85:15 was the correct oxygen–hydrogen ratio in water, rather than anything like 8:1.

1.2.3.2 Compositionism vs. Principlism

Let us recap the argument given so far: if one accepts the primacy of Lavoisier's accounting of weights, then there is indeed very strong empirical evidence for the oxygenist system over the phlogistonist system; but why should one accept the primacy of weight considerations? At the core of that question is whether and why one should accept assumptions (a) and (b) above. We have seen that they are not self-evident. Nor were they proven by direct empirical evidence. Rather, they were *presumed* by Lavoisier and his followers, and used as part of the fabric of the experimental practices by which they produced empirical evidence. The tracking of chemical changes by means of weight was a key epistemic activity within the oxygenist system, and it cohered well with many of Lavoisier's other practices. But it did not fit well into the phlogistonist system, although it was not entirely absent there. The arithmetic of chemical weights arose within a tradition of chemical knowledge that I call *compositionism*, which is to be contrasted to the *principlism*[60] underlying the phlogistonist doctrines. Each system incorporated a significant metaphysical doctrine about the fundamental ontology of chemical substances, which differed from each other sharply. I am going to argue that an increasing tendency toward compositionism created a climate that was mostly congenial to the oxygenist system and distinctly unfavorable to the phlogistonist system. I believe that this was the most important factor responsible for the oxygenist victory.

In identifying the compositionist and principlist traditions, I am drawing from the well-established work of many historians of chemistry, especially Siegfried, although the exact terminology is my own.[61] (I have given a previous exposition of these ideas (Chang 2011d), which I develop slightly here.) When I say "traditions" above I am using the term in a loose sense. To be more precise within my own

---

[59] This is from Nicholson's preface to Kirwan (1789), xi. Jan Golinski (1995) notes the same point and explores the contexts for such claims of precision.

[60] The term "principalist" has been used in some secondary sources, but I think it is more correct to spell it as "principlist", as we are referring to principles, not principals.

[61] See, for example, Siegfried and Dobbs (1968), Siegfried (1982), Klein (1994), and Siegfried (2002). Klein (1996) clearly demonstrates the interaction between the experimental and the theoretical dimensions.

analytical framework, I should say that these traditions are instances of what I characterized as *system-types* in Sect. 1.2.1.1. The phlogistonist system was a particular instantiation of principlism, and the oxygenist system was a particular instantiation of compositionism. (However, as will be explained further in Sect. 1.3.4, the situation is more complicated than that, since there were some significant principlist elements in Lavoisier's system and some compositionist elements in the phlogistonist system, too.)

A fundamental epistemic activity of the compositionist system-type was describing chemical substances as either elements, or compounds made up of those elements.[62] In addition, there were the more experimental activities of decomposing compounds into their elements, and recomposing them from those elements. When one could do both decomposition and recomposition,[63] that was regarded as the best proof of the presumed composition of a substance. These practices required the presumption that the components were stable units that are preserved through chemical reactions. That presumption also grounded the activity of explaining chemical reactions as the rearrangement of distinct and stable building-blocks which retain their identity throughout even when their properties are not manifest in a state of combination.

For those who have been educated in modern science, compositionism may seem like either common sense or a necessary condition of chemistry itself. So it helps to know what alternatives there were to compositionist chemistry, and the main eighteenth-century contender was principlism. This is a system-type formed around the concept of *principles*, namely fundamental substances that impart certain characteristic properties to other substances. The defining epistemic activities in principlism were: classifying substances according to observable properties; explaining the properties of substances by reference to principles; and effecting transformations of substances by the application (or withdrawal) of principles. Like compositionism, principlism was a system-type that had many instantiations, which all shared these three core activities. It is important to note that the principlist ontology presumed an asymmetry between principles and the other substances that are transformed by them, principles being active and the others passive. There are some old echoes in principlism, including the old metaphysics of the substratum of elements modified by the influence of principles, even of matter being given form. It may make sense to trace principlist chemistry back to the Aristotelian notion of the mixt,

---

[62] As explained in Chang (2011d), compositionism constitutes an analytical "way of knowing" in John Pickstone's terminology, because it rests on a commitment to "reduce complex . . . objects to configurations of elements" (Pickstone 2007, 494). Also implied in Pickstone's characterization of the analytical way of knowing is the activity of laboratory analysis, or better, decomposition: applying various methods to take a substance physically apart into its constituent parts.

[63] These are terms used by Lavoisier himself; see, for example, his remarks quoted in Kirwan (1789), 16. I avoid the terms "analysis" and "synthesis" here, because "analysis" could also mean the detection of substances without actual decomposition, and "synthesis" came to take on other significance later.

which Duhem attempted to revive philosophically at the start of the twentieth
century (see Bensaude-Vincent and Simon 2008, 125). Or we may indeed trace
principlism through the alchemical notion of the transmutation of substances;
Priestley once did write Benjamin Franklin: "You will smile when I tell you, that I do
not absolutely despair of the transmutation of metals."[64] However, perhaps the instance
of principlism most vivid to the modern historians of science is the phlogistonist
system of chemistry, originating from Johann Becher (1635–1682) and Georg
Ernst Stahl (1660–1734) and practiced up to the time of Lavoisier, most famously
by Priestley.

It is not that chemical operations in themselves could unequivocally be classified
as principlist transformation or decomposition–recomposition; that would be to
assume that the experimental practices were completely non-theoretical. Many
chemical operations were easily amenable to both interpretations; and this is not a
mere matter of armchair interpretation, since the conceptions did affect the actual
experiments, too. In this vein it is interesting to note an early debate regarding the
cogency of the practice of analyzing substances by the application of fire: some had
compositionist confidence that fire would only effect decompositions, while others
showed principlist worries that the fire would transform the substances being analyzed (Debus 1967; Holmes 1971).

How did compositionism become dominant? Various historians of chemistry
have written instructively about the origin of compositionism, tracing it back to
experimental practices of decomposition–recomposition as well as ideas about the
"elective affinities" between various chemical substances. As Siegfried (1982,
2002) and Klein (1994, 1996) have explained in detail, the origin of compositionist
chemistry goes back at least to the mechanical philosophers of the seventeenth century, becoming fully operative by the late eighteenth century. Robert Multhauf
(1962, 1996, chs. 14–16) has elucidated the connections between the origin of compositionist concepts and the development of certain processes in metallic and heavy-
chemicals industries. Larry Holmes (1971) and Allen Debus (1967) have stressed
the linkage with methods in analytical chemistry. The origin of compositionism was
multifarious, and its dominance was the result of the embedding of various practices
that occurred at glacial speed. In this ascendancy of compositionism, principlist
thinking gradually lost its appeal as it conflicted with the building-block ontology of
compositionism, in which all pieces of matter had equal ontological status. It is
probably fair to say that the firm establishment of compositionism in chemistry
occurred with the eighteenth-century doctrine of affinity, epitomized in Etienne-
François Geoffroy's affinity table published in 1718.[65] Geoffroy's system of chemistry

---

[64] Priestley to Franklin, 13 February 1776, quoted in Schofield (2004), 124. Schofield also quotes
Franklin's witty reply, dated 27 January 1777, saying that Priestley should lose the philosopher's
stone should he ever find it, as it will only aid mankind's wickedness in slaughtering one another—
conversation fit for Revolutionary times!

[65] On the history of affinity chemistry in connection with compositionism, see Klein (1994, 1996),
Klein and Lefèvre (2007), Kim (2003) and Taylor (2006).

was firmly based on the compositionist framework, representing compounds as combinations of indestructible parts, and chemical reactions as *rearrangement* of those parts. To that Geoffroy added the concept of affinity and its relative strength, which enabled the explanation of why certain chemical combinations took place in preference to others.

I believe it was the general adoption of compositionism that gave assumptions (a) and (b) above their appearance of self-evidence, and thereby led to the triumph of Lavoisier's weight-based chemistry. And it is not that weight was the only possible parameter to track chemical components, but it was the only one that worked out well enough for Lavoisier and his contemporaries.[66] Was compositionism itself supported by sufficient empirical evidence? That is the wrong question to ask, as noted already: assumptions like (a) and (b) *enable* the production of empirical evidence; they are not themselves tested by the evidence.[67] Principlist and compositionist ways of thinking were linked up with different experimental practices. Most importantly, principlist thinking was linked with the laboratory practices of *transformation*; compositionist thinking was linked with the laboratory practices of *decomposition and recomposition*. When I say "linked", what I mean is that the conceptual and the experimental aspects reinforced and shaped each other, rather than one causing the other in a unidirectional way; this expresses what I mean by the "coherence" of a system of practice. Priestley's pneumatic chemistry was a fundamentally transformative experimental practice, as an examination of his main treatise on the subject shows clearly. Before he started making the discoveries of all the various different airs that made him famous, he was reporting on things like "air *infected* with animal respiration, or putrefaction" and "air *infected* with the fumes of burning charcoal" (Priestley 1774, Contents). Priestley would not have had any particular ideas at that point of any substance *combining* with the air in the process of respiration or combustion. Rather, respiration was seen as a transformative operation, and that was that. When he was recounting the previous works that formed the background to his own research, Priestley extolled the virtue of Joseph Black, who had discovered that the input of fixed air *rendered* calcareous substances mild (Priestley 1774, 3).

This way of thinking also combined well, at least initially, with the Stahlian notion that phlogiston was the principle that imparted inflammability and metallicity to substances. As he proceeded with his own investigations, Priestley strongly adopted this principlist way of thinking about phlogiston, speaking routinely of how air was transformed by the addition of phlogiston in a "regular gradation from dephlogisticated air, through common air, and phlŏgisticated air, down to nitrous

---

[66] Whether something works out or not in this way is how nature exercises its guiding hand on science, as I will explain further in Chap. 4. Such guidance can only be obtained pragmatically in specific situations; Chap. 3 will show that in later and different settings, volume and even specific heat also successfully performed similar tracking functions as weight.

[67] Hence they qualify as "metaphysical principles" or "ontological principles" as I termed them in Chang (2008, 2009c). This is a neo-Kantian point of view, which has in recent times been championed by Michael Friedman (e.g. 2001).

air" (1775, 392). This series was not an idle piece of theorizing, but something rooted in his minute laboratory operations. In his numerous experiments leading to (and based on) these conceptions, Priestley took note of all sorts of properties that were given to substances as they were modified (such as color, smell, elasticity, and chemical reactivity with various other substances) but weight was not something he noted very frequently. Changes of weight seemed capricious in relation to phlogistication, which meant that weight was not taken as a reliable variable for the purpose of extracting stable patterns in nature's behavior.

What kind of experimental practices were linked to Lavoisier's thinking? He was also driven by pneumatic chemistry, but in a very different way from Priestley. Lavoisier's fascination, right from his early "crucial year", was about how air was absorbed and given out by solid substances in certain chemical reactions. His thinking was based on the presumption that chemical substances had *components*: stable units that are preserved through chemical reactions. This focus on the tracing of input and output, combined with Lavoisier's quantitative bent, resulted in his "balance-sheet" method (Poirier 2005),[68] or his "algebraic" inclination (Kim 2005). Having thus stepped into the compositionist way of thinking, Lavoisier sought and developed experiments conforming to the old compositionist practice of decomposition-and-recomposition, which was the most convincing method of ascertaining compositions.

In this context it is interesting to examine Lavoisier's experimental work on heat, in which he collaborated with Pierre-Simon Laplace (1749–1827).[69] Although Lavoisier and Laplace acknowledged that heat had no detectable weight (and in this paper they do not even commit strongly to the material reality of caloric), it is clear that they assumed heat to be a conserved quantity, and aimed to find a quantitative measure of it to make it amenable to compositionist thinking. They invented the ice calorimeter for this purpose, behind which was Lavoisier's chemical notion that liquid water was a compound of a definite amount of caloric (latent heat) per unit amount of ice. The ice calorimeter turned out to be difficult to use in practice, and compositionist practice concerning heat did not become very fruitful in Lavoisier's chemistry.

The balance was more successful as an instrument of quantification. It allowed a convenient and precise tracking of chemical components that had weight, and fostered a powerful system of compositionist practice. The balance, as it were, was the perfect instrument for managing the balance-sheet. This arithmetic of chemical weights was the key to Lavoisier's filling-out of the schema given by compositionism. In his chemical accounting of weights, chemical reactions were studied by keeping

---

[68] Thinking of the balance-sheet of weights in chemical reactions must have been pleasing to the commercial–bourgeois sensibilities of Lavoisier and his middle-class scientific community, though the oxygenists did not all share the same class background, and I have no means of supporting a real causal link here in any case.

[69] Lavoisier and Laplace [1783] (1920). See Guerlac (1976) for an informative discussion of this collaboration, and also Chang (2004), 134–136, and references therein.

track of the weights of the ingredients and products, and composition was determined with the help of the assumption that if a substance became heavier after a reaction then it must have become compounded with another substance. This weight-based compositionist system had its striking success in pneumatic chemistry where, before Lavoisier, the chemical role of various gases had been noted but their weights had not received sufficient attention. The success of Lavoisier's weight-based compositionist practice reinforced the conviction that weight was the most important variable in chemical reactions; however, those not operating in that system of practice would have found it difficult to share that conviction.

To summarize: the clear evidential advantage of the oxygenist system on the basis of weight considerations only holds if one accepts compositionism; phlogistonists disregarded weight-based arguments because they were principlists. The Chemical Revolution makes much more sense when we see it as a ripple riding on a large wave, which was the very gradual establishment of compositionism. It is important to see beyond the clash between phlogiston and oxygen. If we should want to conceive of the Chemical Revolution as the event that gave rise to "modern chemistry", we must follow Robert Siegfried and Betty Jo Dobbs (1968) in identifying it as a compositionist revolution, whose endpoint was not Lavoisier, but Dalton. More will be said about that in Chap. 3.

1.2.4   What Good Is Phlogiston?

Having refined the descriptive account of the Chemical Revolution and considered the explanations for it, I am now ready to address the normative question: was it right for late eighteenth-century and early nineteenth-century chemists to reject the phlogistonist system of chemistry? (see Chap. 5 if you are inclined to say: "It doesn't matter—there is enough evidence *now*!") In Sect. 1.2.1.1 I argued that the phlogistonist system had a slight advantage over the oxygenist system because it covered a broader problem-field. In Sect. 1.2.1.2 I examined the stand-off between divergent epistemic values, especially with simplicity on the oxygen side and completeness on the phlogiston side; my own sympathy there was with completeness, though that did not amount to a decisive verdict. In Sect. 1.2.1.3 I examined the divergent instantiations of the shared epistemic values of unity, empiricism and systematicity, and saw that I could not come down on either side. In Sect. 1.2.3 I noted that weight-based arguments regarding constitutions were clearly on the side of oxygen, but that there were no compelling immediate reasons of evidence for phlogistonists to accept the *premises* of those arguments. On balance, in all honesty, I cannot see that there were good enough reasons for a decisive rejection of the phlogistonist system. I am convinced that the death of phlogiston, however slow it might have been, was premature, as I have discussed in further detail in Chang (2011b).

Let us now follow through the implications of that conclusion. If I really stand by my verdict, then I have to live with its consequences. If I want to argue that the phlogiston should not have been killed, then I am obliged to say what good would have come of keeping it (see Sect. 1.3.5 on a more careful statement of this necessity for counterfactual thinking). Here I need to make a more serious attempt to answer the four questions raised toward the end of Sect. 1.1:

(1) Was there any knowledge that scientists lost when they rejected the phlogistonist system?
(2) Was there any knowledge that could have come from keeping the phlogistonist system, whose development was delayed or prevented because of its demise?[70]
(3) Was there a beneficial effect of having both phlogiston and oxygenist systems present, in terms of what was produced by interactions between them?
(4) Would there have been further beneficial interactions between the oxygenist system and the phlogistonist system, if the latter had been maintained?

Questions (1) and (3) concern historical actuality, and questions (2) and (4) concern potentiality. Questions (1) and (2) are about the merits of the phlogistonist system in itself, compared with the merits of the oxygenist system in itself; questions (3) and (4) concern the merits of an interactive–pluralistic way of doing science in comparison to a monistic way. Also, given what was said in Sect. 1.2.3, each of the questions needs to be dealt with at two different levels: first in relation to the competition and interaction between the phlogiston and oxygenist systems, and secondly in the wider context of the competition between principlism and compositionism. I will deal with questions (1) and (2) together in Sect. 1.2.4.1 below, and with (3) and (4) in Sect. 1.2.4.2.

1.2.4.1   Benefits of Phlogiston

There was one clear area of "Kuhn loss" in the Chemical Revolution, which I have mentioned already. Phlogistonists explained the common properties of metals by saying that all metals were rich in phlogiston[71]; this explanation was lost through the Chemical Revolution, as it does not work if we make the familiar substitution of phlogiston with the absence of oxygen (or, as Lavoisier had it, a strong affinity for oxygen).[72] As Paul Hoyningen-Huene puts it (2008, 110): "Only after more than a 100 years could the explanatory potential of the phlogiston theory be regained in modern chemistry. One had to wait until the advent of the electron theory of metals".[73]

---

[70] One also ought to ask whether an actual revival of an unjustly dismissed system of knowledge could produce some knowledge now. I think that is indeed possible; however, I do not highlight this question here because in the case of the phlogistonist system I think its full potential has actually been realized in the end, as I explain in Sect. 1.3.5. But I may be wrong in this judgment.

[71] Kuhn (1970), 157.

[72] Or I should say "oxygen base", to be precise. For Lavoisier's view on this, see his commentary in Kirwan (1789), 15–16. For a detailed discussion of whether a translation between the phlogiston and the oxygen theories can be worked out, see Chang (2012b).

[73] At least, it is interesting that this came to be regarded as a legitimate scientific problem again, over a century later!

Not only that, but the phlogistonist account actually has a close resonance with the modern notion that all metals share metallic properties because they all have a "sea" of free electrons. If we were to be truly whiggish, we would recognize phlogiston as the precursor of free electrons.[74] The phlogiston–electricity connection is actually not at all a retrospective fabrication by whiggish historians or philosophers. Allchin (1992, 112), following William M. Sudduth (1978), identifies no fewer than 23 people who postulated a close relationship between phlogiston and electricity in the eighteenth century. There were some good motivations for this identification (even aside from the common desire, expressed for example by James Hutton (1794), to find a grand unity among all the imponderable fluids): for example, it was found that electricity could be used to reduce calxes to metals, which was a role performed by phlogiston. For such reasons, the English chemist John Elliott (1780, 92) even proposed that phlogiston should be re-named "electron".[75] (Later on, when the electrolysis of water in 1800 ended in a puzzle about why the oxygen and hydrogen gases were produced at separate places, Johann Wilhelm Ritter's answer was that hydrogen gas was a compound of water and negative electricity, and oxygen a compound of water and positive electricity; this lined up exactly with Cavendish's earlier notion that hydrogen was phlogisticated water, on making the identification of phlogiston with negative electricity; this episode will be explored in more depth and detail in Chap. 2).

To show that it is not only mad philosophers of science or completely bygone scientists who have had these wild thoughts about phlogiston, I refer to the great American chemist Gilbert Newton Lewis (of the "octet rule" and the still-current definition of acidity). In his Silliman Lectures at Yale University, Lewis (1926, 167–168) declared that the phlogistonists made "the next great step in chemical classification" after the work of Boyle, "through a study of the phenomenon which we know as reduction and oxidation, but which was first called phlogistication and dephlogistication". Lewis thought that the demise of phlogiston in the Chemical Revolution constituted a great lost opportunity for chemistry:

> If they [the phlogistonists] had only thought to say "The substance burning gives up its phlogiston to, and then combines with, the oxygen of the air," the phlogiston theory would never have fallen into disrepute. Indeed, it is curious now to note that not only their new classification but even their mechanism was essentially correct. It is only in the last few years that we have realized that every process that we call reduction or oxidation is the gain or loss of an almost imponderable substance, which we do not call phlogiston but electrons.[76]

Lewis's statement here is a good reminder that the modern meaning of "oxidation" has nothing inherently to do with oxygen; in this area of chemistry, it might have been more sensible to stay with the basic conceptual structure behind the terminology of phlogistication/dephlogistication (gain/loss of electrons), instead of the rather confusing terminology of reduction/oxidation. In other words, if the idea of phlogiston

---

[74] For a full consideration of such a whiggish view, which lines up its judgments with the best modern science but not any science from the past triumphant in its own time, see Chang (2009b).

[75] This quirky fact I owe to Partington and McKie (1937–1939, 350).

[76] I thank Patrick Coffey for alerting me to this passage.

had been maintained, it would have been easier for chemists and physicists to come to grips with that mysterious substance playing crucial roles in chemical transformations, which is abundant in metals among other places. This potential is clearly suggested in Allchin's (1997) report of success in using the concept of phlogiston in teaching about redox reactions to modern-day students.

Let me keep up the whiggish fantasy for a little bit, to see where else it might lead. If my thoughts about the missed theoretical opportunities seem too speculative, we can at least admit that the preservation of the phlogiston concept would have stimulated some experimental enquiries that were not taken up under the Lavoisierian umbrella. If phlogiston had survived, and its association with electricity maintained, I am confident that nineteenth-century scientists would have made attempts to isolate the electric fluid from phlogiston-rich substances such as metals, using any plausible means at their disposal. Would it not have occurred to someone to hit the surface of a metal with powerful ultraviolet rays (already discovered in 1802) in an attempt to disengage phlogiston? As soon as there were sensitive enough electrometers, the photoelectric effect would have been detected. What about trying to run an electric current between two electrodes across a near-vacuum, a very familiar sort of thing from the traditional practice of drawing sparks from static electricity? In fact Davy, and also Jöns Jakob Berzelius, had the experimental knowledge that electricity could be passed through a vacuum. Davy even assessed that the passage of electricity was easier in vacuum than in air, and for Berzelius all this was evidence for the materiality of electricity (see Russell 1963, 145). But electrical discharge in rarefied gases, despite its striking visual manifestations, did not receive serious and widespread scientific attention till the 1870s, when Eugen Goldstein identified "cathode rays" as such (see Darrigol 2000, 274ff). Is it too irresponsible to speculate that cathode rays would have been discovered and investigated very early on if the phlogiston–electricity line of investigation had been encouraged further? Elliott would have been pleased to congratulate my imaginary investigators for the experimental isolation of the "electron".

We might also say that the phlogiston theory gave a good explanation of the production of flame in combustion. As Knight (1978, 33) points out, Davy was unhappy about the neglect of light in Lavoisier's theory, writing in his youthful piece published in 1799 that there were two defects of Lavoisier's theory, namely the assumption of material caloric and "the total neglect of light". Allchin (1992, 111–112) tells us the same about Hutton and others. But did phlogiston do any better here? In modern theory, flame is a plasma, which is mostly a mixture of positive ions and electrons. As I have just mentioned, in a whiggish understanding of the phlogistonist theory of metals, there is a clear reason to identify phlogiston with free electrons. That fits in nicely here, if we take the release of flame as a result of the dissociation of phlogiston (electrons) from the combustible substance. In the early nineteenth century William Brande (1814) at the Royal Institution carried out experiments showing that flame was subject to electrostatic attraction, but this work failed to open up a new line of research; progress would have been facilitated much better in a phlogistonist–electronist framework. The phlogistonist account, of course, needs modification here, as it tended to treat flame as just phlogiston rather than phlogiston mixed in with dephlogisticated stuff and also did not specifically explain why the plasma should glow, but still it was less hopeless than the Lavoisierian story that flame was some mixture of two imponderable chemical substances, *calorique* and *lumière*, both disengaged from oxygen gas.

If phlogiston had been retained, it would also have served as a reminder that there was more to chemical reactions than the grouping and re-grouping of gravimetric building-blocks. Whiggishly speaking, phlogiston served as an expression of chemical potential energy, which the weight-based compositionism of the oxygenist system completely lost sight of. The Lavoisierian tradition was actually quite unstable on this count. For example, Lavoisier sowed the seed of the destruction of his own theory of combustion, by putting so much emphasis on weight and then assigning no weight to caloric. Lavoisier's theory of combustion in fact never got very far in explaining the release of heat and light in combustion, without the concept of energy available. To start thinking about energy, chemists should not have needed to wait for help from the likes of Mayer, Joule and Helmholtz and things like the link between heat and mechanical work. If phlogiston had lived, it would have given chemists a productive open end to start thinking about something like energy. I have already noted in Sect. 1.2.2 that a number of phlogistonists tried to preserve phlogiston for something like energy considerations, while admitting oxygen to their systems for weight-related considerations. This move, whether somehow remembered or freshly re-invented, was not lost on some of the Victorian chemists. In the 9th edition of the *Encyclopaedia Britannica* (1876), F. H. Butler identified phlogiston as another name for potential energy: "The supposed subtraction of phlogiston in the calcination of metals . . . was yet a loss of potential energy, by virtue of the combination of the metal with the [oxygen] gas; and the gain of phlogiston was an increase of potential energy, attendant on the removal of oxygen." Butler recognized this notion as a distinct advancement made in the phlogistonist period: "It was only in the latter part of the eighteenth century that such airy nothingness became commonly regarded as an intimate and necessary constituent of various solid and fluid bodies."[77]

William Odling made the same point in a most interesting paper from 1871. Although not a household name today, Odling was one of the leading theoretical chemists of Victorian Britain, and at that time the Fullerian Professor of Chemistry at the Royal Institution. According to Odling (1871, 319), the major insight from the phlogistonists was that "combustible bodies possess in common a power or energy capable of being elicited and used", and that "the energy pertaining to combustible bodies is the same in all of them, and capable of being transferred from the combustible body which has it to an incombustible body which has it not". Lavoisier had got this wrong by locating the energy in the oxygen gas in the form of caloric, without a convincing account of why caloric contained in other gases would not have the ability to cause combustion. Odling (p. 322) thought that

---

[77] Vol. 5 (1876), p. 461. Butler's piece was the "Historical Introduction" to the entry on "Chemistry", jointly authored overall by "Prof. [Henry] Armstrong, R. Meldola and F. H. Butler".

"the Stahlians, though ignorant of much that has since become known, were never-theless cognisant of much that became afterwards forgotten." He also cited Alexander Crum-Brown (1866), another leading Victorian chemist, as having the same view that "there can be no doubt" that potential energy was what the earlier chemists "meant when they spoke of phlogiston" (p. 322). In fact Odling and Crum-Brown had been anticipated by Liebig two decades earlier (1851, 49–50), before the widespread use of the energy concept in chemistry: "Many chemists, even at the present day, find it impossible to do without certain collective names, analogous to the word phlogiston, for processes which they regard as belonging to the same class, or determined by the same cause. But . . . they employ, since the time of Berthollet, terms which designate what are called 'forces.'"

Admitting that phlogistonists tended to conceive of phlogiston as a material substance, Odling questioned whether this was meant in such a standard way (pp. 323–324): "though defining phlogiston as the principle or matter of fire, . . . they [Stahlians] thought and spoke of it as many philosophers nowadays think and speak of the electric fluid and luminiferous ether", the still-surviving imponderable fluids of the mid-nineteenth century. In any case, Odling thought this substance-talk could be pardoned (p. 323):

> That Stahl and his followers regarded phlogiston as a material substance, if they did so regard it, should interfere no more with our recognition of the merit due to their doctrine, than the circumstance of Black and Lavoisier regarding caloric as a material substance, if they did so regard it, should interfere with our recognition of the merit due to the doctrine of latent heat.

Although phlogiston was clearly not exactly chemical potential energy as under-stood in 1871, Odling (p. 325) argued that "the phlogistians had, in their time, pos-session of a real truth in nature which, altogether lost sight of in the intermediate period, has since crystallized out in a definite form." He ended his discourse by quoting Becher: "I trust that I have got hold of my pitcher by the right handle." And that pitcher (or *Becher*, cup?), the doctrine of energy, was of course "the grandest generalization in science that has ever yet been established."

All in all, I think it is quite clear that killing phlogiston off had two adverse effects: one was to discard certain valuable scientific problems and solutions; the other was to close off certain theoretical and experimental avenues for future scien-tific work. Perhaps it's all fine from where we sit, since I think the frustrated poten-tial of the phlogistonist system was quite fully realized eventually, by some very circuitous routes. But it seems to me quite clear that the premature death of phlogiston retarded scientific progress in quite tangible ways. If it had been left to develop, I think the concept of phlogiston would have split into two. On the one hand, by the early nineteenth century someone might well have hit upon energy conservation, puzzling over this imponderable entity which seemed to have an elusive sort of reality which could be passed from one ponderable substance to another.[78]

----

[78] As Fox (1971) documents in great detail, the caloric theory was having a good deal of difficulty by the early nineteenth century, which would have fostered willingness to consider alternatives.

In that parallel universe, we would be talking about the conservation of phlogiston, and how phlogiston turned out to have all sorts of different forms, but all interconvertible with each other. This would be no more awkward than what we have in our actual universe, in which we still talk about the role of "oxygen" (acid-generator, *Sauerstoff*) in supporting combustion, and the "oxidation" number of ions. On the other hand, the phlogiston concept could have led to a study of electrons without passing through such a categorical and over-simplified atomic theory as Dalton's. Chemists might have skipped right over from phlogiston to elementary particles, or at least found an alternative path of development that did not pass through the false simplicity of the atom–molecule–bulk matter hierarchy. Keeping the phlogiston theory would have led chemists to pay more attention to the "fourth state of matter", starting with flames, and served as a reminder that the durability of compositionist chemical building-blocks may only be an appearance. Keeping phlogiston alive could have challenged the easy Daltonian assumption that chemical atoms were physically unbreakable units.[79] The survival of phlogiston into the nineteenth century would have sustained a vigorous alternative tradition in chemistry and physics, which would have allowed scientists to recognize with more ease the wonderful fluidity of matter, and to come to grips sooner with the nature of ions, solutions, metals, plasmas, cathode rays, and perhaps even radioactivity.

1.2.4.2 Benefits of Phlogiston–Oxygen Interaction

The actual and potential merits of the phlogistonist system in itself are not the only things we should consider. As I will discuss further in Chap. 5, there may be important benefits arising from plurality itself. More specifically, here I want to consider the actual and potential benefits of having the phlogistonist system in interaction with the oxygenist system (or having principlism in interaction with compositionism). Might there not be something to be gained from cross-fertilization or from competitive jostling? We easily grant such benefits in the realm of cross-cultural interactions; why not in science?

There are some concerns to be dispelled, before I consider the possible benefits of maintaining multiple systems. (These points will be discussed in their full detail and generality in Chap. 5, but some preliminary remarks are in order here.) First of all, wouldn't the co-existence of different systems cause confusion and prevent effective research? There is little evidence of such harmful confusion caused by the co-existence of phlogiston and oxygen, and there isn't any convincing reason to think that the situation would have got much worse later if phlogiston had been kept. Second, doesn't the maintenance of too many competing scientific systems dissipate valuable resources? That is a valid general concern, but at least on the

---

[79] This "Daltonian" assumption is not truly Dalton's, since Dalton thought that atoms had an internal structure: a small hard core of ordinary matter, surrounded by an "atmosphere of caloric". But since Dalton thought that the material core of each atom was indestructible, the "Daltonian" designation is fair enough.

phlogistonist side it would not have been a serious worry, since the advocates of phlogiston were mostly amateur scientists working individually, often using inexpensive equipment procured with their own resources (unlike Lavoisier and his colleagues relying effectively on state support). Third, won't arguments about fundamentals divert scientists' energy and attention, preventing them from launching into specialist research? Again, in the case of the phlogiston–oxygen debate there is no indication that the scientists who engaged in the debate on fundamentals were any less productive in producing specialist research than those who did not debate fundamentals.

Having neutralized those concerns for the moment, let me turn to the question of productive interaction. There is a simple-minded point that contains the germ of something much more sophisticated: how would Lavoisier have done what he did, if Priestley hadn't made oxygen and showed him how to do it, and if Cavendish hadn't made water from hydrogen and oxygen and let Blagden tell Lavoisier about it?[80] Maurice Crosland (1983, 238) is quite categorical on this point: "Certainly without the benefit of the work done on gases by British men of science, and notably by Cavendish and Priestley, Lavoisier would not have been able to build chemistry anew." Of course, we cannot rule out the possibility that Lavoisier and his friends might have stumbled on those particular experiments on their own eventually, but the point is more general. Priestley, Scheele, Cavendish and other phlogistonists of the late eighteenth century far outstripped any other group (including Lavoisier and his colleagues) in their ability to make new experimental discoveries in pneumatic and metallic chemistry, the very discoveries that Lavoisier built on so effectively. The point is more about principlism, rather than phlogiston *per se*. Aside from the genius of the individuals involved, I believe there was something about the principlist practice of transformations that was very conducive to these experimental innovations and discoveries, as it encouraged chemists to apply various powerful agents (thought to be, or to contain, various principles) to various substances, to see what happens. On the other hand, it was the compositionist re-interpretation of these results, rather than their assimilation into principlist theories, that brought about the undoubtedly great achievements of Lavoisier and his colleagues. I submit that such in-tandem progress would not have been possible without the co-existence of the principlist and the compositionist traditions, separate but interacting. Was the fruitful interaction a mere accident? At least I like to think of it as an affordance of serendipity by plurality.

It may be less exciting to think about how the oxygenist system challenged and enriched the phlogistonist system, but it is important to raise the topic at least, for the sake of symmetry and completeness. There is no doubt that the phlogistonist system was challenged and stimulated by the opposition from Lavoisier's theory.

---

[80] There is an open question about a similar point regarding Scheele and oxygen, since we cannot be sure how much Lavoisier knew about what Scheele had done. As Anthony Butler (1984) argues, there is evidence that Scheele sent Lavoisier a letter about oxygen at about the same time as Priestley's visit to Paris.

The hybrid systems discussed in Sects. 1.2.2 and 1.2.4.1 were healthy developments arising out of this opposition. Even for those who kept to the phlogistonist system without admitting oxygenist ideas explicitly, their phlogistonism came to be developed in more compositionist ways with more attention to weights. Such hybrid systems were not compatible with the post-Revolutionary insistence that phlogiston be expunged altogether from any system of chemistry. The consolation, on the other side, is that the benefits already reaped by the oxygenist system from its interaction with the phlogistonist system were there to stay, and not eliminated by the demise of phlogiston itself.

What about further benefits that could have come from maintaining the phlogiston–oxygen interaction for longer? Actual history provides a guide and a starting point for our thinking on that question. I have already mentioned the extinction of the hybrid systems, which accepted oxygen for weight considerations and retained phlogiston for energy considerations. These hybrids would have marked just the right path in Odling's view (though Odling does not seem to have been aware that there was actually a hybrid tradition at the time of the Chemical Revolution): "Chemists nowadays are both Stahlian and Lavoisierian in their notions; or have regard both to energy and matter." (1871, 323) That kind of complete view of chemistry had already been in place, before it was rooted out by the Lavoisierian orthodoxy. If the interaction between phlogiston and oxygen had been kept up, it seems to me that chemists would have been able to maintain a very dynamic hybrid tradition throughout the nineteenth century.

A productive tension between principlism and compositionism would have helped maintain a healthy pluralism in chemistry. As compositionism developed into a purer form and became increasingly dominant in its march from Lavoisier to Dalton and beyond, there was growing temptation to think of the foundations of chemistry as a matter of stark choice: subscribe to a simple-minded atomism, or renounce any ontological discourse about chemical substances. This is what some commentators have picked up as the nineteenth-century opposition between atomism and positivism. If chemists had resisted such dichotomies and remained more aware of phlogistonist–principlist successes while holding on to basic compositionism, they would have been able to develop a more flexible view of "elements" and a more nuanced take on "imponderables", leading to an easier incorporation of electricity and thermodynamics into chemistry. Conversely, there is also something I call the "lacuna effect". There are many things that the phlogistonist system failed to explain satisfactorily, but the oxygenist system did not even try to explain. In such cases one benefit of keeping the phlogistonist system would have been to serve as a reminder of the unsolved problems.

I would like to close with a plea for imagination, regarding what might constitute new scientific knowledge and where it might come from. I have serious intentions in my discussion of the phlogistonist system of chemistry, about its forgotten merits and the benefits it could have brought into the nineteenth century. However, even if every single one of my specific ideas about the lost future of phlogiston turned out to be futile, I would still be hopeful for a broader effect of following my musings: the freeing of the reader's scientific imagination.

1.3   Choice, Rationality, and Alternatives

In the third section of each chapter, I anticipate various objections and try to position my views more carefully with respect to existing literature in the history and philosophy of science. The sub-sections here may not have systematic relations to each other.

1.3.1   Rationality

In my earlier discussion in Sect. 1.1, I quickly dismissed philosophical attempts to portray the Chemical Revolution (as it is commonly conceived) as a rational event. If such arguments were successful, they would show that it would have been a rational thing for all chemists to make a quick conversion to Lavoisier's system; *a fortiori*, we would have to conclude that the actual rejection of the phlogiston theory, which was made less swiftly than these philosophers imagine, was certainly not too hasty. This would invalidate my judgment that phlogiston was killed prematurely, so I must consider this possibility more carefully. I consider these arguments fully in another place (Chang 2010), so I will just present their highlights here.

Before we get into the arguments, a brief word is necessary about what is meant by "rational". There is no general agreement about the meaning of rationality, nor would I dream of manufacturing one by writing several lines about it here. But I think I can make a few helpful points that should elicit general assent. Firstly, rationality is *not* a matter of truth; rather, rationality is about good ways of making judgments and decisions, given what one knows or believes at the time. Our most rational judgments may well deviate widely from the ultimate truth (if there be such a thing) because of the limitations of what we have to go on. Secondly, rational thinking or discourse follows some rules or methods that are agreed within the relevant community, to the extent that there is conscious deliberation at all. Thirdly, the minimal condition of rationality is instrumental: at least, a rational action must either achieve some stated aim of the agent, or at least be intended by the agent as contributing toward a certain aim.

I will examine the three best arguments for the rationality of the Chemical Revolution that I have found in the literature. The first is due to Philip Kitcher (1993, 272), who sets out to demolish the view that "there was no cognitively superior reasoning available to the participants, which would have decided the issue in favor of Lavoisier". He wishes to "argue that this fashionable picture is a myth", less adequate than the old view that "the phlogiston theory crumbled under the cumulative force of Lavoisier's evidence". An improved version of this old view is what Kitcher tries to provide, more successfully in my view than anyone else. He is clearly aware of the various merits of the phlogiston theory, and grants that there was initially no clear difference between the empirical adequacy of the phlogiston theory and Lavoisier's theory (p. 273). However, Kitcher argues, the phlogiston theories were unable to deal with the new empirical evidence that emerged in the 1780s.

As in many other arguments (starting with Lavoisier's own) designed to show the factual inadequacy of the phlogiston theory, Kitcher focuses on weight relations. Kitcher avoids the common mistakes of assuming that the phlogistonists simply ignored the evidence, or that they fled into the idea of the negative weight of phlogiston (which only a very small number of people entertained). Rather, he correctly notes (p. 277): "they do something that is far more reasonable: to wit, accept Lavoisier's claim that something from the air is absorbed and try to combine this concession with the traditional idea that phlogiston is emitted." Kitcher argues that this defensive strategy ran into dead-ends eventually, and notes correctly that Kirwan's theory ended up in tangles, even inconsistencies.[81] But he gives no detailed account of Priestley's latter-day defence of phlogiston ([1796] 1969, 1803), which did not have the same inconsistencies as Kirwan's. Nor does he consider the later version of the phlogiston theory advanced by Henry Cavendish (1784), which was free of any contradictions or inordinate complexities as far as I can see.

And there is a far more important point to note: the relevant question of empirical adequacy is a comparative one, not an absolute one. The question is not whether the phlogiston theory was absolutely flawless (to which the answer is "of course not"), but whether it was better or worse than its competitors. We really need to lose the habit of treating "phlogiston theory got X wrong" as the end of the story; we also need to ask whether Lavoisier's theory got X right, and whether it didn't get Y and Z wrong. There has been a great tendency, among philosophers and historians alike, to ignore and minimize the things that Lavoisier's theory could not explain (or got wrong by modern standards). I have made my best attempt to counter this tendency in Sect. 1.1, and in Sect. 1.2.1 I have given a balanced assessment of the competing merits of the two systems. Kitcher is much more careful on this point than most detractors of phlogiston; still, his concession that Lavoisier's analysis was "not free of problems" is quickly mumbled through in half a paragraph, followed by a longer apologetic footnote (p. 278, and footnote 70 there). After that he resumes a lengthier discussion highlighting the problems with which the phlogiston theories had the most difficulty.

The second argument that I wish to discuss is given in Musgrave's 1976 paper "Why did oxygen supplant phlogiston?", which I think is still, on the whole, the best philosophical paper there is on the subject. According to Musgrave, the Chemical Revolution was a perfectly rational affair, and also a case vindicating Lakatos's philosophy of science.[82] He tells us that the phlogiston program was highly progressive up to the confirmation of Priestley's prediction in 1783 that a calx would be reduced by heating in inflammable air, which he at the time considered to be pure phlogiston (see Fig. 1.4). Priestley and his friends were very pleased to see the calx

---

[81] Kirwan (1789) and Kitcher (1993), 283–288. For a more detailed account of the fate of Kirwan's theory, see Mauskopf (2002).

[82] Alan Musgrave (private communication) says that he actually started his work on the Chemical Revolution with the intention of showing that the Lakatosian methodology did not work. As a good Popperian, Musgrave was duly impressed by this failed attempt at a refutation!

Fig. 1.4 The reduction of a
calx in inflammable air

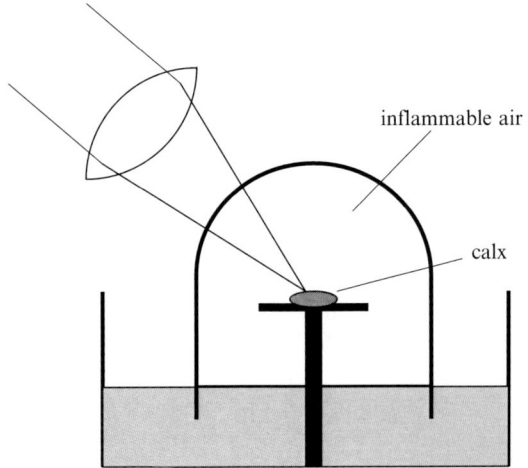

inflammable air

calx

literally imbibing phlogiston and thereby turning into metal. And then Musgrave
(1976, 201) describes how Lavoisier dramatically turned this apparent phlogistonist
triumph upside down by exploiting Cavendish's new work on the production of
water by the combustion of inflammable air. Characteristically, Lavoisier began his
counter-offensive by noting that the calx in Priestley's experiment would have lost
some weight in turning back into metal, as in other cases of reduction. Then he
deduced that the lost weight would have gone into the water that must have been
produced in the experiment, by the combination of the oxygen from the calx, and
the inflammable air (which he then re-named "hydrogen"). Ironically it was Priestley
himself who confirmed Lavoisier's prediction (or retrodiction) that water must be
(must have been) produced, by performing the experiment over mercury, instead of
water as in the original setup.

But as Lakatos might have predicted from the general nature of research programs, the phlogiston program was not conclusively defeated at this point, thanks to
Cavendish and Priestley's ingenious *post hoc* modification of their theory to the
effect that inflammable air was not phlogiston but phlogisticated water, while oxygen
(or, dephlogisticated air) was dephlogisticated water. However, Musgrave argues
(pp. 203–206), from this point on the phlogiston theory was forever on its back foot,
twisting itself to accommodate inconvenient new findings but not managing to make
any successful novel predictions. Although "a degenerating programme can soldier
on, and phlogistonism did just that", at that point the Lakatosian verdict kicks in: it
is irrational to hold on to a degenerating research program. So Musgrave concludes
that it was rational for chemists to abandon the phlogiston program after 1783 or so,
and most chemists were indeed rational, leaving behind the "elderly hold-outs" like
Priestley and Cavendish.

Musgrave's argument is certainly attractive, but as I argued briefly in Sect. 1.1,
there is a fundamental difficulty: where are the successful novel predictions made

by the oxygen program after the phlogiston program stopped making them? Musgrave (p. 201) counts Lavoisier's deduction that water must have formed in Priestley's 1783 experiment as a novel prediction, but this is a difficult claim to sustain. Lavoisier's analysis was only made in retrospect, though it can be said that in the logical sense his theory "predicted" the production of water, which Priestley had failed to observe in the original experiment. But, as Musgrave acknowledges, the same "prediction" was also made by Cavendish's theory. And Lavoisier had not predicted the production of water in the experiment of exploding hydrogen and oxygen together. His hypothesis about the composition of water was itself a *post hoc* adjustment made in order to explain this unexpected result. Lavoisier had expected the product of this reaction to be an acid since it contained oxygen, and he had actually tried to produce an acid without success, and without detecting the water produced in it, either. Musgrave (p. 199) himself tells us all of that, with perfect clarity. Lavoisier's account of the composition of water started its life not as a novel prediction, but as a classic *ad hoc* hypothesis. It lacked use-novelty as well as temporal novelty, as the result of this experiment was used directly in Lavoisier's construction of his hypothesis.

Were there any successful novel predictions made by Lavoisier? Musgrave (p. 203) gives us one: "water . . . should, since it contains oxygen, support slow combustion and yield hydrogen. Iron filings immersed in water did indeed rust and hydrogen was collected." But, again, this was just as deducible from Cavendish's 1784 version of the phlogiston theory: if iron gives its phlogiston to water, that should produce phlogisticated water, which is hydrogen. Similarly with Lavoisier's famous decomposition of water vapor by hot metal: the transfer of phlogiston from metal to water would turn the metal into calx, and turn the water into inflammable air (phlogisticated water), just as it happens when a metal is dissolved in an aqueous solution of an acid. So, even though these were successful novel predictions, they were shared by the phlogiston and oxygen programs alike, and did not provide a reason to choose one over the other. Meanwhile, there were some distinctly un-progressive aspects of the oxygen research program in the 1780s and beyond. As mentioned earlier, Lavoisier confidently predicted in vain that muriatic acid (hydrochloric acid, HCl, in modern terms) would be decomposed into oxygen and the "muriatic radical"; two other non-existent radicals, fluoric and boracic, can be seen in Lavoisier's table of simple substances (Fig. 1.2). Lavoisierian responses to similar anomalies of prussic acid (HCN) and sulphuretted hydrogen (H_2S) not containing any oxygen also had no progressive outcomes. And in neutralizing Berthollet's challenge about the combustion of gunpowder, Lavoisier again only managed *ad hoc* hypotheses unaccompanied by successful novel predictions. Lavoisierians also made pretty un-progressive responses to the discovery that not only oxygen but also chlorine gas supported combustion, but no other known gases did. So, if we stick to Lakatos' criterion of progressiveness, I think the verdict between phlogiston and oxygen is actually quite unclear. In the end, Musgrave does not give us a convincing rational reconstruction of the post-1783 phase of the Revolution.

The third case I will consider is Andrew Pyle (2000), who provides perhaps the most sophisticated of simplicity-based arguments. The sophistication is already evident in his recognition that weight-gain in calcination was not "a knock-down refutation" of the phlogiston theory (p. 109). Pyle is with Musgrave in emphasizing that up to about 1783 Lavoisier's theory had little overall advantage. So it makes sense that few people "converted" up to that point, and that Lavoisier himself did not launch an aggressive campaign until he had arrived at his new hypothesis about the composition of water. But Pyle's explanation of why most chemists did convert to the oxygen theory quickly after 1783 is not satisfactory. One problem is that he only selects out the parts of the story that look rational to him, such as the conversion of Kirwan, setting aside the diehard phlogistonists and some of the fence-sitters. But even if we allow Pyle's selection of events for the moment, his argument about their rationality is quite precarious, resting on a rather subtle point about simplicity.

Pyle notes (2000, 108–109), quite rightly, that the phlogiston theorists had to concede that while the metals lost phlogiston in the process of calcination, something else (such as water or fixed air) became combined with the metal to give it extra weight. But why invent and hold on to such complicated stories, when there was a simpler story that did the job? Pyle also clearly notes that "mainstream" phlogiston theory after 1783 was of a hybrid nature, that is, acknowledging a clear chemical role for oxygen (by whatever name), while maintaining the existence of phlogiston. And then, in the midst of this highly nuanced discussion, Pyle (p. 113) suddenly descends into a simple-minded point about simplicity:

> By 1800, the old phlogiston theory was dead, and the outstanding dispute was between Lavoisier's theory and a spectrum of compromise-theories. How might such a debate be settled? Here the factor of simplicity comes into play on the side of Lavoisier. His theory of combustion is objectively simpler than compromise theories in that it represents combustion in terms of 3 factors rather than 4.

I take it that the three factors that Pyle identifies in Lavoisier's theory are: the combustible, oxygen base, and caloric. On the phlogiston side, the factors involved must be all of those, plus phlogiston. I am not sure why Pyle thinks that phlogistonists necessarily needed caloric rather than using phlogiston to account for heat (there were diverging opinions on this point among phlogistonists), and why he is letting Lavoisier off the hook by ignoring the fact that he also postulated the existence of *lumière*, the substance of light, which was the very first item in his table of simple substances (Fig. 1.2). Depending on how one counts, the substance-count could easily be 4–3 in favor of the phlogistonists. In any case, it does not seem right to choose the fundamental theory of chemistry on the basis of whether it postulates X or $X + 1$ substances. We would first need a good story about why that kind of simplicity is so important.

Pyle (p. 114) also reinforces a slightly different simplicity-based argument in favor of Lavoisier, which is more about the constancy and uniformity of opinion rather than simplicity as such. As mentioned in Sect. 1.2.1.4, this is related to an argument that Lavoisier himself made with much rhetorical effect, that phlogiston was a "veritable Proteus", which changed its form just as needed. And no two phlogistonists could agree about what it really was. At first glance it does seem terrible that phlogistonists could not even agree amongst themselves while Lavoisier's school had a unified stance, but on more careful consideration this is not an argument that carries much weight. It has no force when we are trying to consider the rationality of *each* phlogistonist's position. (Should Scientology rationally convince Christians to give up Christianity because there are so many mutually conflicting variants of the latter?) In any case, any judgment based on the lack of constancy and uniformity on the phlogistonist side needs to be moderated by a recognition that the anti-phlogistic camp was not as tidy as Lavoisier might have wished. We have seen how Lavoisier himself made major modifications in some of his views as he went along. I have also mentioned the reluctance of Berthollet and others to go completely with Lavoisier's theory of acidity. Important sympathizers and converts including Black and Lichtenberg objected to the new nomenclature. There were also disagreements on the nature of caloric among the oxygenists, whether it wasn't really the same substance as light, or whether it was made up of particles or not. On the constitution of alkalis there was as much uncertainty and flip-flopping on the oxygenist side as on the phlogistonist side. On all sides there was considerable ontological discomfort and indecision in general about the imponderables, and there were many who used neither caloric nor phlogiston, preferring their own ideas about the nature of "elementary fire" and such. When we recognize the general fluidity and diversity of the field, the fluidity and diversity of the phlogiston theory do not seem so egregious.

My own sense about the rationality of the Chemical Revolution is as follows. The Chemical Revolution, *as it actually happened*, was a fairly rational affair, in the sense that there was reasoned debate about the choice between the competing systems for the most part. The evidential situation was not clear-cut, and the response was accordingly diverse, which is also quite rational. The main irrationality I see in the picture is not in the refusal of some chemists to go along with Lavoisier, but in the readiness of too many others to do so, which will be discussed in Sect. 1.3.2 below. It was perhaps irrational to retain terms like "oxygen" after the rationale for their naming had disappeared. Irrationality increased in the later retrospective glorifications of Lavoisier, though in one sense those were rational, too: they served the (political) purposes of those who made them![83]

1.3.2 Social Explanations of the Chemical Revolution

On the opposite side of the spectrum from the rationality-obsessed philosophers are those who argue that the Chemical Revolution can really only be explained by social factors. I think interest-based explanations for following Lavoisier only work easily for those caught up in the narrow politics of Parisian science, and even that picture gets seriously murky after the Revolution breaks out. Musgrave (1976, 206–207)

---

[83] See Bensaude-Vincent (1983, 1996) and Kim (2005) for a discussion of these purposes.

gives an effective rebuttal of three social factors that are often invoked as explanations of scientists' various decisions in the Chemical Revolution: nationality, age, and nomenclature. As his arguments are too brief, I will add some further observations drawing from other sources.

As far as I am aware, the most careful and comprehensive sociological–bibliometric study of the Chemical Revolution is the one by H. Gilmann McCann (1978, 117), which concludes that age and nationality *were* "significant causes of paradigm choice, more important than any other variables except passage of time." Although I think McCann was successful in his own aim, which was to advance and test a detailed (quasi-Kuhnian) theory of the structure of scientific revolutions, his type of analysis cannot generate the kind of explanation that I am looking for here. It is certainly valid to observe that younger chemists were more likely to go with Lavoisier than older ones, but that is more or less subsumed by the uninteresting general observation that the young are, on the whole, more prone than the old to adopting new stuff. As for nationality, to say that the French were more likely than others to adopt the "French chemistry" is too easy, and it does nothing to explain why the new chemistry won the intra-French struggle in the first place. The Revolution having started in France, there are certainly interesting things to say about each of the other national contexts, especially in the case of the German lands, and Karl Hufbauer (1982) gives a detailed and nuanced story. It should be duly noted that the founders of the phlogiston theory were Becher and Stahl, both Germans, and that accordingly there was a nationalistic favouring of the phlogiston theory in Germany. Stahl's work in general had even provided "the rallying platform of the German chemical community" in an earlier generation (pp. 8–11). Yet not all German chemists were German nationalists. Advocating Lavoisier served a useful political purpose for those who were "cosmopolitan" or "Francophile", and for those not identifying so strongly with the German chemical community, the "Lavoisierian bandwagon" could be quite attractive (p. 97). And the phlogistonist bias was eventually turned on its head, as the adoption of the new chemistry in the end became a requirement for those who belonged to the new generation of the German chemical community (pp. 140–144). But when it comes to the question of why there was this shift toward the new chemistry in Germany, Hufbauer's explanation seems to be quite internalistic: it was the debate around the production of oxygen from mercury oxide that constituted the turning of the tide (ch. 8, esp. p. 139), and there is nothing very sociological in Hufbauer's account of that event.

For giving a social explanation of why there was a fad for the new chemistry, I think our best bet is to refer to the very conscious and well-organized campaign for the new chemistry that Lavoisier and his colleagues ran, utilizing all kinds of institutional and rhetorical means at their disposal. This is a matter of academic politics rather than larger socio-political forces, and factors like the influence of the new chemical nomenclature belong here. Lavoisier's was certainly an impressive and effective campaign. I think this is the most important explanatory factor on the social–institutional side of the story, and it has been documented quite thoroughly in the secondary literature. Henry Guerlac (1975, ch. 11), Arthur Donovan (1993, ch. 7), and Maurice Crosland (1995) give good overviews of this "campaign" or even "propaganda" (Crosland 1995, 116) for the "new chemistry" or "French chemistry". Key elements in the campaign included a new nomenclature, a new definitive textbook (*Traité élémentaire de chimie*), a new journal (*Annales de chimie*),[84] public demonstrations of experiments (Guerlac 1975, 101–102; Duveen 1954), and a concerted demolition of the opposition such as Kirwan's text.

It is, however, important not to get carried away too much with the observation about the importance of Lavoisierian campaign. If the demise of phlogiston had been simply due to the Lavoisier-fad, then phlogiston might have returned after the Revolutionary execution of Lavoisier in 1794, or at least after the subsequent dissolution of the well-disciplined band of French scientists around him. Lavoisier's ruthless campaign also made enemies as well as converts, as Crosland (1995) documents in some detail. This type of resentment may not have been irrelevant politically to Lavoisier's own death, especially through the grudge that Jean-Paul Marat bore for the high-handed dismissal of his attempts at scientific work by Lavoisier.[85] But phlogiston never did return in great force, and most of the new anti-anti-phlogistonists discussed above were not phlogistonists (hence my playful but literal terminology). It may have been Lavoisier and his friends who killed phlogiston, but there was a greater force at work which *kept* it dead: post-Lavoisierian chemistry was resolutely compositionist (see Sect. 1.2.3), and that is what prevented the return of phlogiston—until it eventually turned into entities more in keeping with compositionism, such as energy (conserved) and electrons (ponderable, though only with a very slight weight).

1.3.3   Incommensurability

I have made various comments about incommensurability in Sect. 1.2. My ideas about incommensurability have been developed more systematically in another place (Chang 2012b), so I will only present a brief summary here. The Chemical Revolution was one of the stock examples to which Kuhn referred repeatedly in *The Structure of Scientific Revolutions*, although he did not make a separate in-depth study of it. At almost every major juncture in *Structure* the Chemical Revolution appears as an illustration, including here: "after discovering oxygen Lavoisier worked in a different world." (Kuhn 1970, 118) Accordingly, the Chemical Revolution has been understood by those sympathetic to the idea of Kuhnian scientific revolutions as a prime exhibition of incommensurability. Paul Hoyningen-Huene (2008, 101, 114) shows in detail how well the Chemical Revolution fits the Kuhnian model of scientific revolutions, and goes as far as to suggest that this was because

---

[84] This was set up by Lavoisier and colleagues against the phlogistonist *Journal de physique* of Delamétherie. Crosland (1994) is the most extensive source; see also Court (1972).

[85] Donovan (1993) and Poirier (1996) gives good accounts of Marat's role.

the Chemical Revolution was actually constitutive of Kuhn's thinking about revolutions. All the same, in many other accounts of the Chemical Revolution incommensurability does not feature as a key element, and some authors have made an explicit denial of incommensurability in the Chemical Revolution. In the course of my earlier discussion, I hope I have provided a well-rounded picture of a very real case of incommensurability.

My view of incommensurability rests on a distinction articulated by Hoyningen-Huene and Howard Sankey (2001, ix–xv) between *semantic* and *methodological* incommensurability, which they characterize as follows:

> The thesis of semantic incommensurability derives from the claim of Kuhn and Feyerabend that the meaning of the terms employed by theories varies with theoretical context.
>
> According to the thesis of methodological incommensurability, there are no shared, objective methodological standards of scientific theory appraisal. Standards of theory appraisal vary from one paradigm to another. There are no external or neutral standards which may be employed in the comparative evaluation of competing theories.[86]

I think there was a significant degree of methodological incommensurability in the Chemical Revolution, although only minimal semantic incommensurability.

On the side of semantic incommensurability: at the operational and phenomenal levels, in most instances there were sufficiently theory-neutral terms which both sides could resort to when they needed to communicate clearly to each other in describing experiments and observations. There are some exceptions: for example, Priestley seems to have spoken of "inflammable air" not only in referring to hydrogen but sometimes when he must have had carbon monoxide, whose identity was not confirmed until about 1800. But in that case there was no difficulty in distinguishing, when necessary, "light inflammable air" from "heavy inflammable air". At the theoretical level there was a degree of semantic incommensurability, since simple-minded direct translations between the phlogiston and oxygen theories do not succeed (for one, "phlogiston" cannot be translated into one oxygenist term). However, significant common similarity-relations and similarity-groupings can be identified, based on the correspondence between phlogiston content and affinity for oxygen, although with some exceptions and interpretive ambiguities. On the whole, there were no significant communication-breakdowns due to semantic incommensurability that would have prevented meaningful and rational debate.

In my view, methodological dimensions of incommensurability are what made the Chemical Revolution a truly underdetermined case of paradigm-choice. I have discussed these dimensions, which included problem-field (Sect. 1.2.2.1), epistemic values (Sects. 1.2.2.2 and 1.2.2.3), and what might be called practice-based metaphysics (Sect. 1.2.3.2). The last dimension was not discussed explicitly by Kuhn, but I think there is an affinity between my discussion and Jed Buchwald's (1992)

---

[86] Hoyningen-Huene and Sankey note that Feyerabend agreed with this methodological thesis as well, but did not put it under the rubric of incommensurability, reserving the latter term only for the semantic thesis. They also note a "perceptual" dimension to Kuhn's concept of incommensurability (p. ix), but do not enter into a detailed discussion of it.

explication of the later Kuhn's notion of incommensurability, in which instruments are seen as generators of taxonomic schemes. I suspect that the Chemical Revolution illustrates a common trend in scientific development: methodological incommensurability is widespread, and its presence in the course of scientific development raises many interesting philosophical issues.

1.3.4 Between Principlism and Compositionism

In Sect. 1.2.3.2 I presented a view of the phlogistonist and oxygenist systems respectively rooted in principlism and compositionism. But that picture is too simple, and here I would like to present a more nuanced picture. Lavoisier's thinking was not entirely compositionist, and the phlogistonist system was not entirely principlist, either.

Even as he heralded a compositionist revolution, Lavoisier himself was steeped in principlism when it came to the theory of acids (see Sect. 1.2.1.2), as he regarded oxygen as the principle of acidity—strongly enough to base his neologism "oxygen" on that idea. This has been pointed out by many historians. Brock (1992, 112–113), for example, explains the irony that Lavoisier's oxygen theory of acidity was a direct descendant of none other than Stahl's idea that vitriolic acid was the "universal acid", or the principle of acidity. Bergman and others had later proposed fixed air (carbon dioxide) as the universal acid and, in a way, Lavoisier only went one level deeper by adding that it was the oxygen in fixed air that made it acidic. Berthollet objected to the principlism inherent in this way of thinking. One could also argue that Lavoisier's caloric was another principle, which imparted the property of fluidity and elasticity to matter. Carlton Perrin (1973, 97–101) takes this point further, and argues quite persuasively that the first five simple substances in Lavoisier's table were all principles (*lumière*, *calorique*, *oxygène*, *azote* (nitrogen) and *hydrogène*; Fig. 1.2 above): "The distinguishing feature of members of the first sub-group appears to be that each of them is a key principle in Lavoisier's chemistry. That is, each of them is a substance which conveys important generic properties."[87] The "*-gène*" suffix is a giveaway for this, and the Lavoisierians did consider the name "*alcaligène*" for *azote*. Lavoisier refers to this first sub-group as "elements" and "principles" both, perhaps reflecting his mixed-up position situated uncomfortably between principlism and compositionism. As time went by, the principlist aspects of the Lavoisierian system were gradually removed.

On the other side, the phlogistonist system had an interestingly ambiguous relationship to compositionism. Although the very notion of phlogiston as the giver of inflammability and metallicity was rooted in principlism, some of the most convincing demonstrations of the reality of phlogiston took place in the form of the classic

---

[87] Siegfried (1982, 37) called this group of substances Lavoisier's "taxonomic garbage"!

compositionist activity of decomposition-and-recomposition. This was already the case with Stahl's experiments that showed sulphur to be composed of sulphuric acid and phlogiston, and continued with the impressive Cavendish–Priestley experiments that demonstrated the composition of metals as calx combined with phlogiston. It makes sense that these experiments would have been regarded as the most convincing pieces of evidence that the phlogistonist side could offer, in the general climate of intensifying compositionism. The same compositionist corruption of phlogiston can be seen in the development of affinity chemistry; it was not unusual for phlogiston to be included in affinity tables, signaling its use in compositionist activities. The more principles became like ordinarily material building-blocks, the more difficult it became for them to function properly as principles.

One may wonder why the phlogistonists were themselves not tempted to go over to compositionism altogether. One step in that direction would have been to quantify phlogiston in the same way Lavoisier and Laplace quantified caloric by means of the calorimeter. In fact Priestley did something very much like this, when he used the "nitrous air test" to measure the degree of phlogistication of gases. One may further wonder: if a principle is a material substance, when it combines with another substance why shouldn't that combination be a simple joining of equal-status substances as the compositionists imagined? Those principlists who were no longer rooted in the ancient metaphysics of form and substance easily went over to this quasi-compositionist way of thinking, and used the term "principle" almost synonymously with "element". Lavoisier himself might have been just such a lapsed principlist. This incongruity, in my view, was at the heart of the demise of the phlogiston system. The Chemical Revolution may be seen as an internal collapse of the phlogistonist system resulting from compositionist corruption.

What tended to hold principlists back from turning entirely compositionist were those experiments in which a certain chemical substance did seem to impart characteristic properties to all the compounds that it formed, in that regard dominating the other substances. So there was an observed asymmetry between principles and other substances at the level of properties, although this no longer implied that principles were somehow essentially different kinds of substances from other substances. Especially if a substance could be used effectively as a tool in the laboratory (as was the phlogiston contained in charcoal in the reduction of metals), then it was apt to be regarded as a principle. In that vein, phlogiston continued to be regarded as a principle that imparted combustibility (conceived in a broad sense). This consideration of experimental practice also makes sense of Lavoisier's lingering principlism. Regarding oxygen, he was strongly impressed by the experience of turning things acidic by burning them. Regarding caloric, the same operational confidence came from turning solids into liquids, and liquids into gases, by the application of heat.

The blurring of the line between principlism and compositionism can also be seen in the development of affinity chemistry, which began long before the Chemical Revolution and continued well beyond it (see Kim 2003 and Taylor 2006 for definitive treatments of affinity theory in France and Britain, respectively). Affinity could be made compatible with either tradition, and was taken up by both. It seems that the origin of the affinity concept, with Etienne-François Geoffroy (1672–1731), had much to do with the birth of compositionism, as Ursula Klein (1994) shows. But, as noted above, it was not unusual to include phlogiston in affinity tables. Weight was not considered very important to affinity chemistry originally, though there was no problem with assigning weight to most affinity-governed substances. Lavoisier and Kirwan could happily debate affinities with each other, and Berthollet grafted a full affinity theory onto the Lavoisierian program, though the post-Lavoisier mainstream of chemistry moved to an affinity-less compositionism which Mi Gyung Kim (2005, 173) pinpoints as a "stoichiometric atomism" that created a strain in French chemistry.

1.3.5   Counterfactual History

There was much engagement in counterfactual history earlier in the chapter, especially in Sect. 1.2.4. Many historians are understandably wary of counterfactual thinking, concerned that it is invalid because it is not based on actual evidence, and also pointless because it does not serve a clearly defined purpose. They say counterfactuals may be interesting territory for novelists, but not something historians should take seriously. In the face of this skepticism, Greg Radick and his co-authors have done a valuable service by pulling together various worthwhile considerations in a recent Focus section in *Isis* on "Counterfactuals and the Historian of Science". As Radick (2008, 547) says: "Whenever [historians of science] attempt to do more than chronicle the scientific past—when they seek to explain its shape, judge its significance, clarify its challenges to understanding—they make claims about what might or might not have happened." For myself, there are three specific reasons to engage in counterfactual history, which were implicitly given in the course of the discussion in Sect. 1.2.4. Here I would like to articulate them more explicitly and systematically.

The first reason is to support causal claims. I follow Geoffrey Hawthorn's (1991) contention that counterfactual reasoning does aid the causal understanding of history. Asserting "X caused Y" commits one to a counterfactual statement in the spirit of "if X hadn't been there, Y wouldn't have happened" (though typically something weaker than that categorical statement). This is a commonplace among philosophers of causation, except for the most robust of Humeans. Depending on the particular situation and the particular theory of causation one subscribes to, the relevant counterfactual statement may be a bit different. For example, what we believe may be "if I could have turned X into X_c, Y would have become Y_c", or "if X hadn't been there, Y would still have happened because the situation was overdetermined and Z would have caused Y". But the point is that believing in a casual story that goes beyond mere correlation doesn't really make sense unless one is also willing to commit to some counterfactual assertions as well.

In the case at hand, I wished to argue that the premature death of phlogiston retarded scientific developments. How does one support such causal statements? The most satisfying method would be to intervene directly in the situation in order to see what happens when we change the parameters that are alleged to be causally efficacious, but it is by definition impossible to intervene in the past. The next best thing would be to simulate an intervention, by comparing observations gathered in contrasting situations where the key variables have naturally varied. Such comparative study can be done credibly when there are many independent domains engaged in similar activities (such as various nations at various times trying to make economic development), but in the history of science that tends to be more difficult. There was just one Chemical Revolutions, not multiple ones in some of which phlogiston made a quick disappearance and in others it lasted longer. In such situations, one can either give up on making any causal statements, or try to rely on *counterfactual* comparative situations. So we make an imaginary intervention in the situation, and see if we can reason out what would happen in such a situation. What if Priestley had been a wily and charismatic character who remained in Britain and fostered a continuing tradition of phlogistonist chemistry? How would chemistry have developed then? And so on.

Some would object that we would have no basis on which to continue the reasoning, and this is actually the same kind of problem faced by physicists who indulge in thought-experiments, whether they be something as innocuous as imagining a frictionless plane on which to slide something down, or as outrageous as the young Einstein thinking about observing a light wave while traveling at the speed of light. The least that this kind of exercise can do is articulate our tacit assumptions. And if our counterfactual reasoning can be regarded as reasonable extrapolations from what we know empirically (of the kind that we have to make anyway even in everyday life), then I think it can be used as legitimate support for causal claims. I hope that my thinking falls into this category when I say, for example, that if nineteenth-century scientists thought that metals were full of phlogiston they would have attempted to extract it by various means at their disposal (such as heat and ultraviolet rays). This I take as a reasonable generalization from the way curious scientists have behaved over the centuries. If so, this piece of counterfactual reasoning can be used as support for the causal claim that the demise of phlogiston retarded scientific progress.

The second purpose of counterfactual reasoning is the theme on which I ended Sect. 1.2: to free up our imagination. Our thinking tends to get limited by what we know, and by what we think are the limits of possibility. A healthy dose of counterfactual reasoning can open up our minds in various beneficial ways. For the philosopher of science, being able to imagine a happy counterfactual state of affairs will be a useful antidote to Panglossian tendencies to glorify the actual past and present of science, which is like the confidence, from someone who has never traveled, that his own country is the best possible place in the world. Counterfactual imagination can break our normative complacency, and force us to engage in a more thorough and searching evaluation of actual scientific developments; I hope that this sort of benefit shows in my discussion in Sect. 1.2.1. Even for the historian, counterfactual thinking can be useful in aiding revisionist work. Historical observation and data-gathering are theory-laden activities as much as their scientific counterparts are. There will be things we do not notice in the past record of science because our view is restricted by what we think we know, scientifically and historically, thinking that there was no other possible way of doing good science except how it was actually done. If we go in with a bit more counterfactual imagination, we may more easily see valuable and interesting aspects of past science that we did not catch before; I hope that Sect. 1.2.2 has demonstrated such a benefit.

Finally, there is one other purpose of counterfactual history that is important to my thinking generally but has not been highlighted in the discussion so far. Ultimately I am interested in *taking*, not just imagining, paths that were not taken by actual history. Especially in that sense Radick (2008) is correct in noting the similarity between the spirit of my work and Steve Fuller's (2008a) activist time-travel in counterfactual history.[88] Now, there is nothing we can actually do about the opportunities that were lost in the actual past (Fuller's time-travelling exercise in persuading Oresme must remain in the imagination), but we can do something about it in our actual present.

As I will explain further in Chap. 5, Sect. 5.3.4, my main aim in this direction is actually to attain more and better scientific knowledge, and that is the most ambitious part of my program of "complementary science" (Chang 2004, esp. chapter 6). But in order to have some sense of whether and how to make live a possible path of development that was actually not taken in the past, we need to first imagine what it *would have been* like if that path had been taken. In that context, counterfactual history functions as a preliminary step, a feasibility study, or even a plan of action, for a more active step. Counterfactual reasoning can serve as reconnaissance down the imagined paths-not-taken, in order to determine which of those paths are so promising as to deserve the significant scientific labor of clearing them. This book fails to enter into the active final stage of complementary science, for various reasons, and for the most part I have had to content myself with armchair-and-keyboard philosophy. My next major project in complementary science, regarding the history and understanding of batteries, will attempt the active stage fully; the initial steps that I have taken are described elsewhere (Chang 2011c), and some intimation is also given in Chap. 2 of this book. In another direction, regarding the anomalous variations in the boiling point of water, the passive work done in my philosophical history of thermometers (Chang 2004, chapter 1) led directly to the active work in the laboratory (Chang 2007b, also summarized in Chang 2011c).

Coming back to phlogiston: my passive historical judgment is that there was no convincing reason to kill off phlogiston. The active next step beckons: should we bring it back, then? But we have to keep in mind that there are two different issues here. When I say that phlogiston should not have been killed 200 years ago, it is first of all a judgment about actual history: I think the phlogistonist system had achieved a great deal and was still doing valuable service when it was terminated. That, by

---

[88] I should note that even Fuller's discussion in his *Isis* piece contains one traditional element: assuming that there is such a thing as "the" viewpoint of the present and "the" viewpoint of the past. But I imagine that he would not object to a more pluralistic framing of issues, which would also not alter the points he is making there.

itself, does not say whether and in what ways the phlogiston system would have continued to generate useful contributions. And unless I had a firm positive sense on that counterfactual matter, it would make little sense to launch into the Quixotic business of bringing phlogiston back. Therefore it was important that I engage in counterfactual history, exploring the imagined benefits of having phlogiston present at various points in the two centuries of science after its effective demise.

The results of my counterfactual investigations were reported in Sect. 1.2.4. The initial counterfactual returns were encouraging, and I became fairly convinced that a robust survival of phlogiston would have accelerated developments in chemistry and physics. However, as my historical research progressed, I was also pleasantly surprised to learn that the work of reviving phlogiston had actually been done already by a number of other scholars. Not only have there been relatively maverick attempts to employ phlogiston again for various scientific purposes, reaching from Davy in the early nineteenth century to Allchin in the late twentieth. Even more important is the recognition that some important aspects of the phlogiston concept were actually brought back, under different names, in order to help remedy the shortcomings of Lavoisierian–compositionist chemistry as it weathered the nineteenth century. When Odling and others saw phlogiston as the predecessor of chemical potential energy, and when Lewis saw phlogiston as the predecessor of electrons, what they were doing may have been whiggish but it was certainly not pointless. And their insights provide a sufficient answer as to why I am not going to try to bring phlogiston back to modern chemistry—it is already here!

References

Allchin, Douglas. 1992. Phlogiston after oxygen. *Ambix* 39: 110–116.

Allchin, Douglas. 1994. James Hutton and phlogiston. *Annals of Science* 51: 615–635.

Allchin, Douglas. 1997. Rekindling phlogiston: From classroom case study to interdisciplinary relationships. *Science and Education* 6: 473–509.

Ashbee, Ruth. 2007. The discovery of chlorine: A window on the Chemical Revolution. In *An element of controversy: The life of chlorine in science, medicine, technology and war*, ed. Hasok Chang and Catherine Jackson, 15–40. London: British Society for the History of Science.

Bakewell, Sarah. 2010. *How to live: A life of Montaigne in one question and twenty attempts at an answer*. London: Chatto & Windus.

Bensaude-Vincent, Bernadette. 1983. A founder myth in the history of science? The Lavoisier case. In *Functions and uses of disciplinary histories*, ed. Loren Graham, Wolf Lepenies, and Peter Weingart, 53–78. Dordrecht: Reidel.

Bensaude-Vincent, Bernadette. 1996. Between history and memory: Centennial and bicentennial images of Lavoisier. *Isis* 87: 481–499.

Bensaude-Vincent, Bernadette, and Jonathan Simon. 2008. *Chemistry: The impure science*. London: Imperial College Press.

Black, Joseph. 1803. *Lectures on the elements of chemistry*. Edinburgh: W. Creech.

Brande, William Thomas. 1814. The Bakerian Lecture: On some new electro-chemical phenomena. *Philosophical Transactions of the Royal Society* 104: 51–61.

Brock, William H. 1992. *The Fontana history of chemistry*. London: Fontana Press.

Brooke, John Hedley. 1980. Davy's chemical outlook: The acid test. In *Science and the sons of genius: Studies on Humphry Davy*, ed. Sophie Forgan, 121–175. London: Science Reviews Ltd.

Bridgman, Percy Williams. 1959. *The way things are*. Cambridge, MA: Harvard University Press.

Buchwald, Jed. 1992. Kinds and the wave theory of light. *Studies in History and Philosophy of Science* 23: 39–74.

Burlingame, Leslie. 1981. Lamarck's chemistry: The Chemical Revolution rejected. In *The analytic spirit: Essays in the history of science in honor of Henry Guerlac*, ed. Harry Woolf, 64–81. Ithaca/London: Cornell University Press.

Butler, Anthony R. 1984. Lavoisier: A letter from Sweden. *Chemistry in Britain* 20: 617–619.

Cavendish, Henry. 1766. Three papers, containing experiments on factitious air. *Philosophical Transactions of the Royal Society* 56: 141–184.

Cavendish, Henry. 1784. Experiments on air, Part 2. *Philosophical Transactions of the Royal Society* 74: 119–153.

Chang, Hasok. 2004. *Inventing temperature: Measurement and scientific progress*. New York: Oxford University Press.

Chang, Hasok. 2007b. *The myth of the boiling point*. http://www.cam.ac.uk/hps/chang/boiling, first posted on 18 October 2007.

Chang, Hasok. 2008. Contingent transcendental arguments for metaphysical principles. In *Kant and the philosophy of science today*, ed. Michela Massimi, 113–133. Cambridge: Cambridge University Press.

Chang, Hasok. 2009b. We have never been whiggish (about phlogiston). *Centaurus* 51: 239–264.

Chang, Hasok. 2009c. Ontological principles and the intelligibility of epistemic activities. In *Scientific understanding: Philosophical perspectives*, ed. Henk De Regt, Sabina Leonelli, and Kai Eigner, 64–82. Pittsburgh: University of Pittsburgh Press.

Chang, Hasok. 2010. The hidden history of phlogiston: How philosophical failure can generate historiographical refinement. *HYLE* 16(2): 47–79.

Chang, Hasok. 2011a. The philosophical grammar of scientific practice. *International Studies in the Philosophy of Science* 25: 205–221.

Chang, Hasok. 2011b. The persistence of epistemic objects through scientific change. *Erkenntnis* 75: 413–429.

Chang, Hasok. 2011c. How historical experiments can improve scientific knowledge and science education: The cases of boiling water and electrochemistry. *Science and Education* 20: 317–341.

Chang, Hasok. 2011d. Compositionism as a dominant way of knowing in modern chemistry. *History of Science* 49: 247–268.

Chang, Hasok. 2012a. Joseph Priestley (1733–1804). In *Philosophy of chemistry*, ed. Andrea I. Woody, Robin Findlay Hendry, and Paul Needham, 55–62. San Diego: North Holland/ Elsevier.

Chang, Hasok. 2012b. Incommensurability: Revisiting the Chemical Revolution. In *T. S. Kuhn's The Structure of Scientific Revolutions: Impact, relevance and open issues*, ed. Vasso Kindi and Theodore Arabatzis. London: Routledge.

Chang, Hasok, and Catherine Jackson, eds. 2007. *An element of controversy: The life of chlorine in science, medicine, technology and war*. London: British Society for the History of Science.

Court, S. 1972. The *Annales de chimie* 1789–1815. *Ambix* 19: 113–128.

Crosland, Maurice, ed. 1971. *The science of matter*. Harmondsworth: Penguin Books Ltd.

Crosland, Maurice. 1980. Davy and Gay-Lussac: Competition and contrast. In *Science and the sons of genius: Studies on Humphry Davy*, ed. Sophie Forgan, 95–120. London: Science Reviews Ltd.

Crosland, Maurice. 1983. A practical perspective on Joseph Priestley as a pneumatic chemist. *British Journal for the History of Science* 16: 223–238.

Crosland, Maurice. 1994. *In the shadow of Lavoisier: The* Annales de chimie *and the establishment of a new science*. Oxford: British Society for the History of Science.

Crosland, Maurice. 1995. Lavoisier, the two French revolutions and "the imperial despotism of oxygen". *Ambix* 42: 101–118.

Crowther, J.G. 1962. *Scientists of the Industrial Revolution*. London: The Cresset Press.

Crum-Brown, Alexander. 1866. Note on phlogistic theory. *Proceedings of the Royal Society of Edinburgh* 5: 328–330.

Darrigol, Olivier. 2000. *Electrodynamics from Ampère to Einstein*. Oxford: Oxford University Press.

Davy, Humphry. 1812. *Elements of chemical philosophy*. London: J. Johnson and Co.

De Luc, Jean-André. 1803. *Introduction a la physique terrestre par les fluides expansibles; précédée de deux memoires sur la nouvelle théorie chymique, considérée sous différens points de vue. Pour servir de suite et de développement aux* Recherches sur les modifications de l'atmosphère, 2 vols. Paris/Milan: La Veuve Nyon; J.Luc Nyon.

Debus, Allen G. 1967. Fire analysis and the elements in the sixteenth and the seventeenth centuries. *Annals of Science* 23: 127–147.

Dewey, John. 1938. *Logic: The theory of inquiry*. New York: Holt, Reinhardt & Winston.

Djerassi, Carl, and Roald Hoffmann. 2001. *Oxygen*. Weinheim: Wiley-VCH.

Donovan, Arthur. 1988. Lavoisier and the origins of modern chemistry. *Osiris (2nd Series)* 4: 214–231.

Donovan, Arthur. 1993. *Antoine Lavoisier: Science, administration and revolution*. Oxford: Blackwell.

Duveen, Denis I., and Herbert S. Klickstein. 1954. A letter from Berthollet to Blagden relating to the experiments for a large scale synthesis of water carried out by Lavoisier and Meusnier in 1875. *Annals of Science* 10: 58–62.

Elliott, John. 1780. *Philosophical observations on the senses of vision and hearing; to which are added, a treatise on harmonic sounds, and an essay on combustion and animal heat*. London: J. Murray.

Fourcroy, Antoine. 1801. *Système des connaissances chimiques et de leurs applications aux phénomènes de la nature et de l'art*. Paris: Baudouin.

Fox, Robert. 1971. *The caloric theory of gases from Lavoisier to Reignault*. Oxford: Clarendon Press.

Friedman, Michael. 2001. *Dynamics of reason*. Stanford: CSLI Publications.

Fuller, Steve. 2008a. The normative turn: Counterfactuals and a philosophical historiography of science. *Isis* 99: 576–584.

Gago, Ramón. 1988. The new chemistry in Spain. *Osiris (2nd Series)* 4: 169–195.

Golinski, Jan. 1992. *Science as public culture: Chemistry and enlightenment in Britain 1760–1820*. Cambridge: Cambridge University Press.

Golinski, Jan. 1995. The nicety of experiment. In *The values of precision*, ed. M. Norton Wise, 72–91. Princeton: Princeton University Press.

Gray, Tamsin, Rosemary Coates, and Mårten Åkesson. 2007. The elementary nature of chlorine. In *An element of controversy: The life of chlorine in science, medicine, technology and war*, ed. Hasok Chang and Catherine Jackson, 41–72. London: British Society for the History of Science.

Guerlac, Henry. 1975. *Antoine-Laurent Lavoisier: Chemist and revolutionary*. New York: Charles Scribner's Sons.

Guerlac, Henry. 1976. Chemistry as a branch of physics: Laplace's collaboration with Lavoisier. *Historical Studies in the Physical Sciences* 7: 193–276.

Hacking, Ian. 1992. The self-vindication of the laboratory sciences. In *Science as practice and culture*, ed. Andrew Pickering, 29–64. Chicago: University of Chicago Press.

Hacking, Ian. 2000. How inevitable are the results of successful science? *Philosophy of Science* 67: 58–71.

Harrington, Robert. 1804. *The death warrant of the French theory of chemistry*. London: Longman.

Hartley, Harold. 1971. *Studies in the history of chemistry*. Oxford: Clarendon Press.

Hawthorn, Geoffrey. 1991. *Plausible worlds: Possibility and understanding in history and the social sciences*. Cambridge: Cambridge University Press.

Heilbron, John. 2005. Jean-André De Luc: Citoyen de Genève and philosopher to the Queen of England. *Archives des Sciences* 58: 75–92.

Holmes, Frederic L. 1971. Analysis by fire and solvent extractions: The metamorphosis of a tradition. *Isis* 62: 128–148.

Holmes, Frederic L. 2000. The "revolution in chemistry and physics": Overthrow of a reigning paradigm or competition between existing research programs? *Isis* 91: 735–753.

Hoyningen-Huene, Paul. 1993. *Reconstructing scientific revolutions: Thomas S. Kuhn's philosophy of science*. Chicago: University of Chicago Press.

Hoyningen-Huene, Paul. 2008. Thomas Kuhn and the Chemical Revolution. *Foundations of Chemistry* 10: 101–115.

Hoyningen-Huene, Paul, and Howard Sankey, eds. 2001. *Incommensurability and related matters*. Dordrecht: Kluwer.

Hufbauer, Karl. 1982. *The formation of the German chemical community (1720–1795)*. Berkeley/Los Angeles: University of California Press.

Hutton, James. 1794. *A dissertation upon the philosophy of light, heat and fire*. Edinburgh: Cadell, Junior, Davies.

James, Frank A.J.L., ed. 1991. *The correspondence of Michael Faraday*, vol. 1, 1811–1831. London: The Royal Institution of Great Britain.

Jungnickel, Christa, and Russell McCormmach. 1999. *Cavendish: The experimental life*. Lewisburgh: Bucknell University Press.

Kant, Immanuel. [1787] 1998. *Critique of pure reason* (trans: Paul Guyer and Allen W. Wood). Cambridge: Cambridge University Press.

Kim, Mi Gyung. 2003. *Affinity, that elusive dream: A genealogy of the Chemical Revolution*. Cambridge, MA: The MIT Press.

Kim, Mi Gyung. 2005. Lavoisier: The father of modern chemistry? In *Lavoisier in perspective*, ed. Marco Beretta, 167–191. Munich: Deutsches Museum.

Kirwan, Richard. 1789. *An essay on phlogiston and the composition of acids*. London: J. Johnson.

Kitcher, Philip. 1993. *The advancement of science: Science without legend, objectivity without illusions*. New York/Oxford: Oxford University Press.

Klein, Ursula. 1994. Origin of the concept of chemical compound. *Science in Context* 7(2): 163–204.

Klein, Ursula. 1996. The chemical workshop tradition and the experimental practice: Discontinuities within continuities. *Science in Context* 9(3): 251–287.

Klein, Ursula, and Wolfgang Lefèvre. 2007. *Materials in eighteenth-century science: A historical ontology*. Cambridge, MA: The MIT Press.

Knight, David. 1978. *The transcendental part of chemistry*. Folkestone: Dawson.

Kuhn, Thomas S. 1970. *The structure of scientific revolutions*, 2nd ed. Chicago: University of Chicago Press.

Kuhn, Thomas S. 1977. Objectivity, value judgment, and theory choice. In *The essential tension: Selected studies in scientific tradition and theory change*, 320–339. Chicago: University of Chicago Press.

Lavoisier, Antoine-Laurent. 1786. Refléxions sur le phlogistique, pour servir de développement à la théorie de la combustion et de la calcination, publiée en 1777. *Mémoires de l'Académie des Sciences* 1783(1786): 505–538.

Lavoisier, Antoine-Laurent. [1789] 1965. *Elements of chemistry*. New York: Dover.

Lavoisier, Antoine-Laurent, and Pierre-Simon Laplace. [1783] 1920. *Mémoire sur la chaleur*. Paris: Gauthier-Villars.

Le Grand, H.E. 1975. The "conversion" of C. L. Berthollet to Lavoisier's chemistry. *Ambix* 22: 58–70.

Lewis, Gilbert Newton. 1926. *The anatomy of science*. New Haven: Yale University Press.

Liebig, Justus. 1851. *Familiar letters on chemistry, in its relations to physiology, dietetics, agriculture, commerce, and political economy*, 3rd ed. London: Taylor, Walton, & Maberly.

Lubbock, Constance A. 1933. *The Herschel chronicle: The life-story of William Herschel and his sister Caroline Herschel*. Cambridge: Cambridge University Press.

Mauskopf, Seymour H. 1988. Gunpowder and the Chemical Revolution. *Osiris (2nd Series)* 4: 93–120.

Mauskopf, Seymour H. 2002. Richard Kirwan's phlogistic theory: Its success and fate. *Ambix* 49: 185–205.

McCann, H. Gilmann. 1978. *Chemistry transformed: The paradigmatic shift from phlogiston to oxygen*. Norwood: Ablex.

McEvoy, John G. 1997. Positivism, whiggism and the Chemical Revolution: A study in the historiography of chemistry. *History of Science* 35: 1–33.

McEvoy, John G. 2010. *The historiography of the Chemical Revolution: Patterns of interpretation in the history of science*. London: Pickering & Chatto.

Middleton, W.E. Knowles. 1965. *A history of the theories of rain and other forms of precipitation*. London: Oldbourne.

Morris, Robert J. 1972. Lavoisier and the caloric theory. *British Journal for the History of Science* 6: 1–38.

Multhauf, Robert P. 1962. On the use of the balance in chemistry. *Proceedings of the American Philosophical Society* 106: 210–218.

Multhauf, Robert P. 1996. Operational practice and the emergence of modern chemical concepts. *Science in Context* 9(3): 241–249.

Musgrave, Alan. 1976. Why did oxygen supplant phlogiston? Research programmes in the Chemical Revolution. In *Method and appraisal in the physical sciences*, ed. Colin Howson, 181–209. Cambridge: Cambridge University Press.

Nordmann, Alfred. 1986. Comparing incommensurable theories. *Studies in History and Philosophy of Science* 17: 231–246.

Odling, William. 1871. On the revived theory of phlogiston. *Proceedings of the Royal Institution of Great Britain* 6: 315–325.

Partington, J.R., and Douglas McKie. 1937–1939. Historical studies on the phlogiston theory (in 4 parts). *Annals of Science* 2: 361–404; 3: 1–58; 3: 337–371; 4: 113–149.

Perrin, Carlton E. 1973. Lavoisier's table of the elements: A reappraisal. *Ambix* 20: 95–105.

Perrin, Carlton E. 1981. The triumph of the antiphlogistians. In *The analytic spirit: Essays in the history of science in honor of Henry Guerlac*, ed. Harry Woolf, 40–63. Ithaca/London: Cornell University Press.

Pickstone, John V. 2000. *Ways of knowing: A new history of science, technology and medicine*. Manchester: Manchester University Press.

Pickstone, John V. 2007. Working knowledges before and after circa 1800: Practices and disciplines in the history of science, technology, and medicine. *Isis* 98: 489–516.

Poirier, Jean-Pierre. 1996. *Lavoisier: Chemist, biologist, economist*. Philadelphia: University of Pennsylvania Press.

Poirier, Jean-Pierre. 2005. Lavoisier's balance sheet method: Sources, early signs and late developments. In *Lavoisier in perspective*, ed. Marco Beretta, 69–77. Munich: Deutsches Museum.

Priestley, Joseph. 1774. *Experiments and observations on different kinds of air*. London: J. Johnson.

Priestley, Joseph. 1775. An account of further discoveries in air, in letters to Sir John Pringle, Bart. P.R.S. and the Rev. Dr. Price, F.R.S. *Philosophical Transactions of the Royal Society* 65: 384–394.

Priestley, Joseph. 1788. Experiments and observations relating to the principle of acidity, the composition of water, and phlogiston. *Philosophical Transactions of the Royal Society* 78: 147–157.

Priestley, Joseph. 1790. *Experiments and observations on different kinds of air, and other branches of natural philosophy, connected with the subject*, 2nd ed., 3 vols. Birmingham: Thomas Pearson.

Priestley, Joseph. [1796] 1969. *Considerations on the doctrine of phlogiston, and the decomposition of water (and two lectures on combustion, etc. by John MacLean)*. New York: Kraus Reprint Co.

Priestley, Joseph. 1803. *The doctrine of phlogiston established and that of the composition of water refuted*. Philadelphia: P. Byrne.

Priestley, Joseph. 1970. *Autobiography of Joseph Priestley, with introduction by Jack Lindsay*. Bath: Adam & Dart.

Pyle, Andrew. 2000. The rationality of the Chemical Revolution. In *After Popper, Kuhn and Feyerabend*, ed. Robert Nola and Howard Sankey, 99–124. Dordrecht: Kluwer.

Radick, Gregory. 2008. Why what if? *Isis* 99: 547–551.

Russell, Colin A. 1963. The electrochemical theory of Berzelius (in 2 parts). *Annals of Science* 19: 117–145.

Schofield, Robert E. 1963. *The Lunar Society of Birmingham: A social history of provincial science and industry in eighteenth-century England*. Oxford: Clarendon Press.

Schofield, Robert E. 1997. *The enlightenment of Joseph Priestley: A study of his life and work from 1733 to 1773*. University Park: Pennsylvania State University Press.

Schofield, Robert E. 2004. *The enlightened Joseph Priestley: A study of his life and work from 1773 to 1804*. University Park: Pennsylvania State University Press.

Siegfried, Robert. 1964. The Phlogistic conjectures of Humphry Davy. *Chymia* 9: 117–124.

Siegfried, Robert. 1982. Lavoisier's table of simple substances: Its origin and interpretation. *Ambix* 29: 29–48.

Siegfried, Robert. 1988. The Chemical Revolution in the history of chemistry. *Osiris (2nd Series)* 4: 35–52.

Siegfried, Robert. 1989. Lavoisier and the phlogistic connection. *Ambix* 36: 31–40.

Siegfried, Robert. 2002. *From elements to atoms: A history of chemical composition*. Philadelphia: American Philosophical Society.

Siegfried, Robert, and Betty Jo Dobbs. 1968. Composition, a neglected aspect of the Chemical Revolution. *Annals of Science* 24: 275–293.

Soler, Léna. 2008. Revealing the analytical structure and some intrinsic major difficulties of the contingentist/inevitabilist issue. *Studies in History and Philosophy of Science* 39: 230–241.

Stevenson, W.F. 1849. *The composition of hydrogen, and the non-decomposition of water incontrovertibly established*. London: James Ridgway.

Sudduth, William M. 1978. Eighteenth century identifications of electricity with phlogiston. *Ambix* 25: 131–147.

Taylor, Georgette. 2006. *Variations on a theme: Patterns of congruence and divergence among 18th century chemical affinity theories*. Ph.D. dissertation. London: University College London.

Thomson, Thomas. 1802. *A system of chemistry*, 4 vols. Edinburgh: Bell & Bradfute and E. Balfour.

Tunbridge, Paul A. 1971. Jean André De Luc, F.R.S. *Notes and Records of the Royal Society of London* 26: 15–33.

Uglow, Jenny. 2002. *The lunar men: Five friends whose curiosity changed the world*. London: Faber & Faber.

van Fraassen, Bas. 1980. *The scientific image*. Oxford: Clarendon Press.

Chapter 2
Electrolysis: Piles of Confusion and Poles of Attraction

Abstract However one might assess the arguments about the nature of water in the Chemical Revolution (Chap. 1), it may seem that the electrolysis of water (first performed in 1800) must have produced decisive evidence that it was a compound substance. But electrolysis came with a serious puzzle: if the action of electricity was breaking up each particle of water into a particle of oxygen and a particle of hydrogen, how did the oxygen and hydrogen gases emerge at electrodes that were separated from each other by macroscopic distances? The distance problem turned the electrolysis of water into a serious anomaly, rather than positive evidence, for Lavoisierian chemistry. Ritter and his followers argued that electrolysis was in fact a pair of syntheses: water was an element after all, and its combination with positive and negative electricity formed oxygen and hydrogen. This view was dismissed by the majority of post-Lavoisierian chemists, but never conclusively refuted at the time. Those who opposed Ritter proposed a plethora of different solutions to the distance problem, none of them completely convincing. The modern ionic theory only emerged in the last years of the nineteenth century, so there was nearly a whole century of electrochemistry taking place without a consensus on some very basic questions. Nonetheless, electrochemistry made significant progress. Its experimental practices were stabilized and standardized without recourse to agreed-upon fundamental theory. In the theoretical realm there was pluralistic progress, with several competing systems each making its distinctive contributions, in productive interaction with each other.

2.1 Electrolysis and Its Discontents

The Chemical Revolution did not deliver an unequivocal proof that water was a compound, at least according to my account given in Chap. 1. But in the year 1800, 6 years after Lavoisier's death, a marvelous new device arrived with a promise to do better by making a cleaner decomposition of water. This was the famous "pile"

H. Chang, *Is Water H₂O?: Evidence, Realism and Pluralism*, Boston Studies
in the Philosophy of Science 293, DOI 10.1007/978-94-007-3932-1_2,
© Springer Science+Business Media B.V. 2012

Fig. 2.1 The Voltaic pile

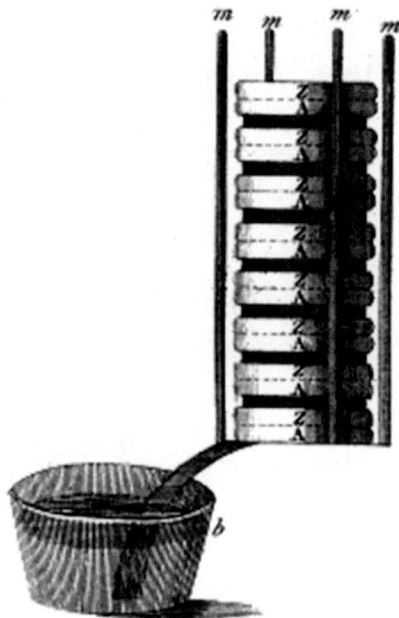

(or, the electrical battery[1]) invented by Alessandro Volta (1745–1827), Professor of Physics at the University of Pavia, who was already a renowned researcher on electricity with the invention of several important instruments to his credit. As shown in Fig. 2.1, Volta's new instrument was literally a pile of pairs of metallic plates—in Volta's original setup these were zinc, marked "Z" in the figure, and silver, marked "A", but almost any pair of metals could be used. These metallic couples were stacked up, separated by layers of wettable material (cardboard, leather, etc.) soaked in a solution—even plain water worked, but salt water was much better (Volta 1800, 404 and 406).

On connecting the top and the bottom of the pile with a conductor (such as a human body), various effects were produced. In the arrangement shown in the figure here, one could receive a shock by dipping one hand in the basin of water connected to the bottom, and touching the top of the pile with the other hand. With a pile consisting of 20 metallic pairs, Volta received "shocks that take the whole finger and make it quite painful" (pp. 407–408). With any conductor connecting the two ends ("poles") of the pile, a current of electricity flowed through the conductor. On placing wires from the battery on his skin, Volta reported, he felt a pain and quivering that lasted and intensified as long as the circuit stayed closed: "What proof more evident of the continuation of the electric current?" He thought he had called forth an "endless

---

[1] "Battery" in later usage, that is. Originally the term was used to refer to a set (battery) of storage-jars for holding static electricity. After Volta's invention, the term "galvanic battery" was used to describe a collection of cells producing electricity; over time, the term came to refer to single cells as well. In his original paper Volta (1800, 420) very sensibly proposed to call his instrument the electro-motive apparatus (*appareil électro-moteur*), but this name did not catch on.

circulation of the electric fluid", a "perpetual motion" (pp. 420–421). Investigating the effects of the pile further, the only detector of electric current Volta had was his own body, so he carried on with various self-experiments. Applying the current to various sense-organs, he experienced strange tastes, visual flashes, and crackling noises; only the sense of smell could not be excited (pp. 420–428).

Almost as soon as the news of Volta's pile reached England, William Nicholson (1753–1815) and Anthony Carlisle (1769–1840) in London used it to make a decomposition of water into hydrogen and oxygen gases (Nicholson 1800). Volta had sent his paper describing the pile in the form of a long letter in French to Joseph Banks (1743–1820), the President of the Royal Society, for publication in its *Philosophical Transactions*. Banks showed Volta's letter before its publication to his friend Carlisle, surgeon to the Prince of Wales. Carlisle was so excited about Volta's work that he made a pile for himself, and showed it to his friend Nicholson, an enterprising gentleman scientist. After stints working for the East India Company and as a commercial agent for Josiah Wedgwood's pottery firm, Nicholson was engaged in a pioneering and uncertain attempt to make a living out of scientific publishing (Lilley 1948–1950, 82). It is perhaps fortunate for science that he had launched a scientific periodical of his own just 3 years earlier. *A Journal of Natural Philosophy, Chemistry and the Arts* (affectionately known to this day as "Nicholson's Journal") invited contributions on all manner of scientific subjects from anyone who had something interesting to offer. Nicholson selected and edited the articles himself; he answered to no professional bodies, and the journal was only funded by subscriptions.

Nicholson used his own journal as the outlet for reporting on his joint work with Carlisle — with no fuss, no delays, and no need to get it past any referees. The news of Volta's pile created an instant sensation and excitement in all corners of scientific Europe, and results contained in Nicholson's paper of July 1800, particularly the electrolysis of water, were at the forefront of this excitement. As historian Samuel Lilley explains (1948–1950, 83–86), Nicholson's Journal became for a time the premier venue for publishing new ideas and experimental results on Voltaic electricity. Especially thanks to Volta's invention, electrochemistry became the perfect subject for what Lilley has called "popular research", which Nicholson wished to promote through his journal (pp. 93ff). Anyone not destitute could afford a few coins and other bits and pieces to rig up a Voltaic pile; the effects were fascinating and significant, ranging from entertaining bodily shocks to previously unknown chemical decompositions. Those sending excited electrochemical reports to Nicholson included old veterans like Joseph Priestley (1802), ambitious unknowns like the young Humphry Davy (1800a, b), and a number of now-forgotten others.

2.1.1 The Distance Problem

The decomposition of water using the Voltaic pile had serious and unexpected implications. Now, you might wonder what was really new or problematic in this, since Lavoisier many years earlier had famously made a demonstration of the decomposition of water, in which he passed steam through a hot gun-barrel and showed that hydrogen gas (or, inflammable air) was produced while the metal in the gun-barrel was oxidized. But Lavoisier's experiment was theoretically ambiguous, because it required the use of a substance whose own constitution was at the very heart of the debate—namely a metal, which the phlogistonists deemed full of phlogiston. As explained in Chap. 1, Cavendish and Priestley could easily say: the metal gave phlogiston to the water, turning it into inflammable air (hydrogen); having given up its phlogiston, the metal turned into a calx; this was all perfectly in line with the phlogistonist theory.[2] Electrolysis did not have the same interpretive ambiguity— or so it seemed initially.

Actually an electrical decomposition of water had been made before as early as 1789, but it is easy to see why Nicholson and Carlisle's work was better. The earlier decomposition was the work of two Dutchmen in Amsterdam, Adriaan Paets van Troostwijk (1752–1837) and Jan R. Deiman (1743–1808), who achieved the feat by passing sparks of static electricity through water repeatedly. Although this experiment is said to have contributed significantly to the acceptance of Lavoisier's new chemistry in the Netherlands, it was messy because the hydrogen and the oxygen were produced in a mixture. They did re-spark the mixture back into water, but they could not easily separate out the small volume of the gas produced to confirm the presence of hydrogen and oxygen by other means.[3] In the Nicholson–Carlisle setup, the gases came out neatly separated, each of them easily bottled up and tested. The modern setup shown in Fig. 2.2 (from Pauling and Pauling 1975, 357) is a direct descendant of one of Nicholson's setups, using two inverted cups to collect the gases (Nicholson 1800, 185). As George John Singer put it (1814, 339): "The decompositions produced by the Voltaic apparatus are effected with remarkable precision. The component parts of the bodies subjected to its action are separated at some distance from each other, and no observable change occurs in the intermediate space."

But the very cleanness of Nicholson–Carlisle electrolysis also revealed a deep problem. If the action of electricity was to break down each molecule of water into a particle of oxygen and a particle of hydrogen, why did the two gases not issue from the same place, but in different locations separated by a macroscopic distance, easily a few inches? And why did oxygen always come from the wire connected with the positive pole of the battery, and hydrogen from the negative? As Nicholson himself put it (1800, 183):

> it was with no little surprize [sic] that we found the hydrogen extricated at the contact with one wire, while the oxigen [sic] fixed itself in combination with the other wire at the distance of almost two inches. This new fact still remains to be explained, and seems to point at some general law of the agency of electricity in chemical operations.

---

[2] This is how Priestley put it (1788, 154; emphases original): "That water is decomposed when inflammable air is procured from iron by steam, is not probable; since the inflammable principle [phlogiston] may very well be supposed to come from the iron, and the addition of weight acquired by the iron may be ascribed to the *water* which has displaced it. Also when the *scale of iron*, or *finery cinder*, is heated in inflammable air, it gives out what it had gained, *viz.* the water." See also Priestley [1796] (1969), 30–33. To say that the metal absorbs water as it gives up phlogiston is quite like saying that the metal absorbs dephlogisticated water (which is what oxygen was, for Cavendish and Priestley).

[3] See Snelders (1979) and Snelders (1988), 135–137, for further details on this experiment and its impact.

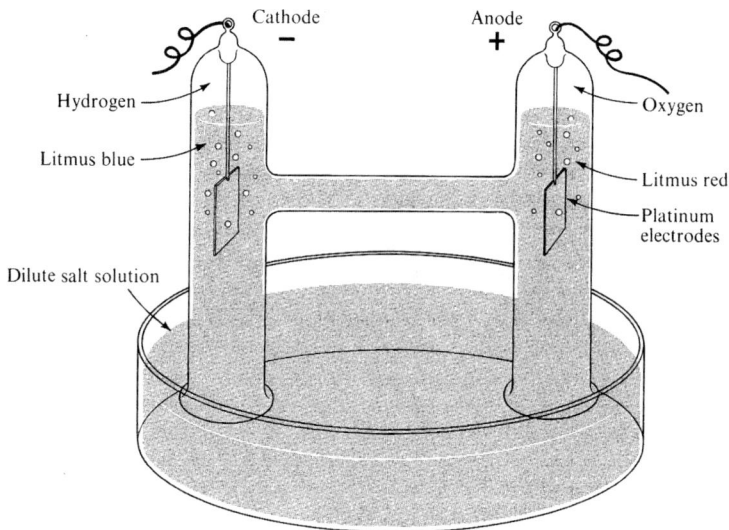

Fig. 2.2 The electrolysis of water, a modern view; as in most typical modern setups, the liquid being electrolyzed here is not pure water, but water with a little bit of ionic solute in it; note that the volume-ratio of the gases shown in this picture is inaccurate

Later, when Nicholson used gold and platinum electrodes, oxygen was obtained in the gaseous form, rather than oxidizing the electrode metal. As easy it was for others to reproduce the experiment, it was obvious to most commentators that there was a serious problem here. I will call this the "distance problem". Humphry Davy (1778–1829), then working at the Pneumatic Institution in Bristol, was one of those deeply puzzled. Within 2 months of the appearance of Nicholson's paper Davy began to publish on electrochemistry, firing an article off to Nicholson every month for the rest of the year 1800. By December he was not sure whether water had really been decomposed:

> Many new observations must be collected, probably before we shall be able to ascertain whether water is decomposed in galvanic processes. Supposing its decomposition, we must assume, that at least one of its elements is capable of rapidly passing in an invisible form through metallic substances, or through water and many connected organic bodies; and such an assumption is incommensurable with all known facts. (Davy 1800b, 400)[4]

The distance problem was not about to go away. Later experiments increased the distance to three feet (Singer 1814, 341). Davy himself came up with a double-cup arrangement that accentuated the distance problem, as hydrogen and oxygen were produced in clearly separate pots of water in this setup. In the first of his famous Bakerian Lectures to the Royal Society, Davy exhibited an experiment (shown in

---

[4] Note his use of the term "incommensurable", 160 years before Kuhn and Feyerabend! See also Davy [1801] (1839), 206: "The facts relating to the separate production of oxygen and hydrogen acid and alkali in water, are totally incommensurable with the usually received theory of chemistry".

Fig. 2.3 Davy's electrolysis
in separate cups

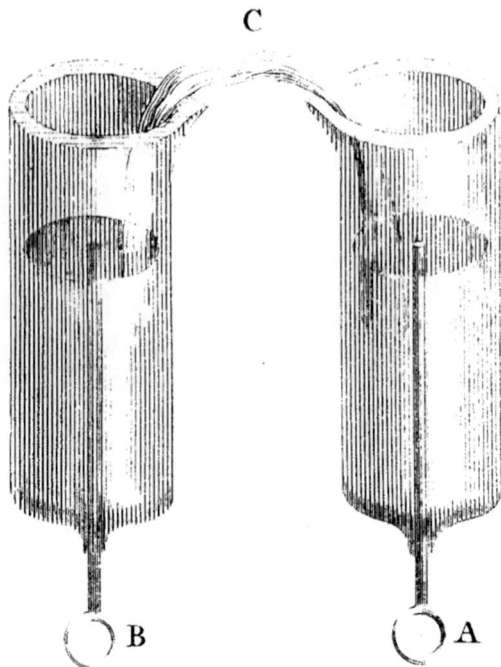

Fig. 2.3) with water in two cones of gold connected by strands of amianthus (fine asbestos); the idea of the asbestos bridge he attributed to William Hyde Wollaston (Davy 1807, 6, 3; Figure 2 from Plate 1).

Statements of the distance problem are ubiquitous in the literature of the time, though I will only cite several instances here. L. Pearce Williams, who among today's historians of science has given the clearest statement of the importance of the distance problem that I have seen, says that "such a strange and inexplicable effect immediately attracted universal attention." (Williams 1965, 227) Davy's disquiet was not atypical of the mood in Britain, and we will see the excited German responses shortly. In France, too, the problem was taken seriously, and tackled officially (in true French fashion) at the *Institut National*, which took the place of the *Académie des Sciences* during the Revolutionary and Napoleonic periods. The great naturalist Georges Cuvier (1769–1832), recently appointed in 1799 as Professor of Natural History at the *Collège de France*, was charged with the task of reporting on Voltaic phenomena to the *Institut*. Cuvier stated the distance problem unequivocally: "Why, then, do the oxygen and hydrogen, proceeding from the same particle of water, appear at two distant points? And why does each of them appear invariably on the wire which belongs to one of the two extremities of the pile, and never on the other?" (quoted in Wilkinson 1804, 148) The mineralogist and abbot René-Just Haüy (1743–1822) expressed the same worry in his official textbook of physics for the *Ecole Polytechnique*: if one molecule of water is decomposed, why do the gases appear in different places? If two molecules are decomposed, why does one give only hydrogen and the other oxygen? (Haüy 1806, vol. 2, 50–52).

It is interesting to note some retrospective assessments that were given after the initial period of electrochemical excitement, and I will quote two of them. The first is by George John Singer (1768–1817), who made a living from giving private lectures in his mother's house in London, and counted Michael Faraday (1791–1867) among his students.[5] Although Singer is long forgotten now, he was highly regarded in his own time, and his *Elements of Electricity and Electro-Chemistry* (1814) had the honor of being translated into French, German and Italian. Singer was clear that the distance problem had not been solved, despite 15 years of intense debate:

> The most difficult feature of all the Voltaic decompositions, is the invisible form, in which the separated elements of various compounds appear to traverse the fluid, and arrange themselves at the opposite wires . . . . without any apparent alteration of the interposed fluid. On the hypothesis of electric energy, the hydrogen is said to be attracted by the negative wire, because it is naturally positive; and the oxygen by the positive wire, because it is naturally negative; this does not explain how the same particle of water can have its elements liberated at so great a distance from each other. . . . (Singer 1814, 378–379)

We must keep in mind that Singer and his contemporaries had no concept of free ions floating around undetected in water.

The other retrospective I want to cite comes from Michael Donovan (1790–?) of Dublin, whose unjust neglect by historians of science is even more serious than Singer's. Donovan won a prize competition of the Royal Irish Academy in 1815 with an essay on the history and current state of galvanism (which was still a popular term, in honor of Luigi Galvani, covering both Voltaic electricity and animal electricity). Even though Donovan does not seem to have been an influential or even a familiar figure especially outside Ireland and Britain, his book is full of sound knowledge and keen insight. Donovan (1816, 45) noted that "the phenomenon was a subject of astonishment to philosophers, and no rational explanation could be given." The separate production of electrolytic products "had been long considered inexplicable" (p. 340). Although he did offer his own theory that was meant to resolve all major problems (which I will discuss briefly in Sect. 2.2.1.2), Donovan was clear that no consensus had been reached on his own or any other theory yet, and he was in fact quite scathing about the state of electrochemical theories up to his time:

> we are compelled to regret that the ardour manifested in the inquiry has been detrimental to itself. The invention of hypotheses, at too early a period, occupied the attention of those whose talents would have been so much more advantageously exerted in experiment: and in this we have to regret, not only mispent [sic] time and labour, but also the early bias acquired by the mind, which is ever unfriendly to the reception of true impressions. (Donovan 1816, 149–150)

Around the time of Singer's and Donovan's assessments of the situation, the distance problem had fascinated the young Faraday in his home-made experiments (Williams 1965, 22–23, 267). The problem still had not received a clear solution by the time Faraday turned his attention to it more powerfully 20 years later, starting in 1832. Williams reckons that Faraday's solution to this problem "was to provide him

---

[5] The biographical information on Singer is taken from the *Dictionary of National Biography* (1897), vol. 52, 211–312.

with the basic concept which led to all his famous discoveries", yet the majority of chemists did not agree with, or perhaps even really understand, Faraday's solution (see Sect. 2.2.3.2 for more on this). In fact there was no agreed solution to the distance problem at least until the establishment of the modern ionic theory of solutions first proposed by Svante Arrhenius (1859–1927) in the 1880s, according to which some molecules of water are already dissociated into ions before any external electricity is applied. The uncertainty surrounding the distance problem pervaded the whole first century of electrochemistry, as we can see through the eyes of the renowned physical chemist Wilhelm Ostwald (1853–1932) in his formidable history of electrochemistry published in 1896:

> Besides the fundamental fact of decomposition itself there was the very remarkable phenomenon that the products of decomposition appeared simultaneously at different points. . . . Science has ceaselessly pursued the answer to this question ever since and it has taken a long time and much work to solve the problem satisfactorily. (Ostwald [1896] 1980, 128–129/131[6])

2.1.2   Electrolysis as Synthesis

The distance problem threatened to blur completely the implications of electrolysis on the constitution of the decomposed substances. If electrolysis had initially been seen as a decisive confirmation of Lavoisier's theory (in the Dutch setup), it now produced one of its most difficult problems. So it is not surprising that some anti-Lavoisierians seized upon electrolysis. As Williams puts it, perhaps too dramatically (1965, 228): "A small group of German chemists pounced upon this strange behaviour of water undergoing electrochemical decomposition and used it as their main offensive piece in an attempt to blow up Lavoisier's new system of chemistry." In this campaign, "the foremost aggressor" was Johann Wilhelm Ritter (1776–1810), a prolific, imaginative and controversial scientist–philosopher. Born in Silesia and educated at the University of Jena, Ritter began publishing on galvanism at the age of 20, and considered galvanism a fundamental force that underlay the workings of both animate and inanimate parts of nature. On learning about Volta's invention, he plunged into a prolific series of experiments trying out all kinds of possible combinations of materials that worked as batteries, and making a thorough examination of their effects. The Prussian physicist and meteorologist Heinrich Wilhelm Dove (1803–1879) noted his appreciation of Ritter's work: "This gigantic task was undertaken by Ritter, . . . who almost sacrificed his senses to the investigation"; "he died early, . . . exhausted by restless labour, sorrow and disordered living." (quoted in Mottelay 1922, 380–381) Ritter was guided by not only a keen power of observation and instrument-design, but also by some large themes such as the union of polar opposites, a deep periodicity in nature, and an odd Thales-like belief that "everything is modified water". As Walter Wetzels (1990) explains, Ritter was fast becoming the

---

[6] In my references to Ostwald, the latter number cited (131 in this case) is the page number in the original German edition.

German Romanticists' favorite physicist (see Sect. 2.3.5 for further details). The Romantic poet Novalis (1772–1801) declared: "Ritter is indeed searching for the real soul of the world in nature. He wants to decipher the visible and tangible letters and to explain the positing of the higher spiritual forces" (quoted in Wetzels 1990, 210).

According to Ritter, what happened when electricity was passed through water was *synthesis*, not decomposition: at the positive pole of the battery, positive electricity combined with water and created oxygen; at the negative pole, negative electricity combined with water and created hydrogen.[7] Then the two gases naturally came out at separate places, which were the locations for the supply of the two types of electricity. So water was seen again as an element, and oxygen and hydrogen as compounds. Ritter wrote:

> The two types of gases evolved have up to now usually been regarded as the different con-stituent parts of the same water. Thus I was really able to prove by this experiment[8] that these gases were by no means due to a decomposition of water as one would very well believe according to the new [Lavoisier's] chemical theory. They were due to two processes completely different from each other. (quoted in Ostwald [1896] 1980, 156/161)

What I will call Ritter's "synthesis view" of electrolysis is now only remembered by a handful of historians. Even those who know about it do not tend to realize how natural the synthesis view might have seemed at the time. Earlier historians and scientists were not blinded to this fact. Arrhenius, in his textbook of electrochemistry, took it that Ritter's was the initially predominant view ([1897] 1902, 21):

> It became necessary to explain why the ions were only separated at the poles by the electric current. It was at first believed (Ritter) that hydrogen was formed by the union of water with negative electricity, and that oxygen resulted from the combination of water with positive electricity.

That is perhaps an overstatement of the initial prevalence of Ritter's view, but it is significant that Arrhenius, the originator of modern electrochemistry, regarded the synthesis view as sensible enough at least for its own time. Ostwald ([1896] 1980, 24/24) gives us a better sense of how the synthesis view would have been in line with the prevalent ontological conceptions at the time: "Generally electricity was regarded as matter and the products obtained from water as compounds of electricity either with water or with its constituents." The notion of electricity as a material substance was not seriously challenged at the time, even by the Lavoisierians. In electrolysis one pumps electricity into a liquid; isn't it reasonable, after all, to think that the electricity might combine with the substance being electrolyzed, thereby effecting chemical change? (Is the hydrogen side of Ritter's account really so different from our modern schoolbook story that electrons come in from the cathode, combine with hydrogen ions (H^+) in the water, and turn it into neutral hydrogen gas?)

Now, it is easy to see the attraction of Ritter's synthesis view of electrolysis for those who opposed Lavoisier's system. Compare the competing views of Lavoisier and Cavendish about the composition of water, and both of them with Ritter's view,

---

[7] Ritter may have been anticipated in the synthesis view by one of Nicholson's anonymous authors, as noted by Ostwald [1896] (1980), 148–149/152–153.

[8] This is the experiment with the V-tube arrangement that I will describe in Sect. 2.2.1.2.

Table 2.1 The alignment of Cavendish's and Ritter's views on the constitution of water

We observe:	Inflammable air	and Vital air	Combine, to make	Water
Lavoisier says:	Hydrogen	Oxygen	→	H–O [compound]
Cavendish says:	Phlogisticated water	Dephlogisticated water	→	Water [element]
Ritter says:	Negatively electrified water	Positively electrified water	→	Water [element]

all summarized in Table 2.1. Cavendish's and Ritter's schemes match up almost exactly if we identify phlogiston with negative electricity. So the phlogiston theory gave a way of making perfect sense of the electrolysis of water, avoiding the distance problem altogether. George Smith Gibbes, who was mentioned in Chap. 1, Sect. 1.2.2 as one of the new generation of anti-Lavoisierians, came up with just such a phlogistonist rendition of Ritter's view (see the exposition in Wilkinson 1804). Nicholson published in his journal several papers by authors agreeing with Gibbes's view.[9] Davy also considered this possibility seriously enough, in connection with his other neo-phlogistonist speculations.[10] Hans Christian Ørsted (1777–1851), who would become a scientific celebrity for his discovery of electromagnetism in 1820, shared Ritter's dislike of Lavoisierian theory for metaphysical reasons, which I will discuss further in Sect. 2.3.5. Ørsted played a key role in communicating Ritter's ideas to the Parisian community, through a series of letters published in Delamétherie's *Journal de physique* (Williams 1965, 229; Christensen 1995). Even half a century later, W. F. Stevenson (1849) was defending this sort of view. The connection between phlogiston and negative electricity is even more compelling to us, considering the appealing retrospective identification between phlogiston and electrons noted in Chap. 1, Sect. 1.2.4.

It is easy to imagine that Priestley would have found this topic irresistible, especially if we recall that electricity was the first subject for his scientific research, which got him hooked on science and gave him a measure of initial recognition before his work in pneumatic chemistry. From his exile in America Priestley followed the developments by reading Nicholson's Journal, and had enough results by September 1801 for a paper of his own. In this paper, printed in the March 1802 issue of Nicholson's Journal, Priestley (1802, 198) argued that the electrolysis of water did not proceed as the Lavoisierians would have it.[11] For one thing, although the application of Voltaic electricity did often produce hydrogen and oxygen, the

---

[9] See Golinski (1992), 213, for a brief discussion of Gibbes's continuing opposition to Lavoisierian theory.

[10] See Siegfried (1964), Brooke (1980), 150, and also Knight (1978), 52. Of Davy's own statements, note especially Davy (1808a), 33, occurring in the middle of the celebrated Bakerian Lecture in which he announced the discovery of potassium and sodium.

[11] This paper is passed over by most historians who discuss Priestley's work; one exception is a brief discussion given by Schofield (2004), 366.

proportions of the two were by no means always in the expected ratio of 2:1 by volume. He claimed that the production of oxygen[12] in electrolysis was only the release of the gas that had been held in solution in the water, not the result of a decomposition of water, which he still considered as a "wholly chimerical" hypothesis, though it was "almost universally received at present". In support of this view Priestley described experiments in which the production of oxygen stopped after a while if the experiment was conducted in a vacuum, or if the contact with external air was cut off by a layer of oil over the water. And it did not arise at all if the water was de-gassed beforehand (by boiling, I presume).[13] As for the other side of the reaction, it would have been easy enough to imagine that the negative electricity combined with the water and made hydrogen. After this counter-Lavoisierian blast on the first page of the article, Priestley then continued, true to his form, with reports of a plethora of strange and delightful experimental results, especially involving the dissolution of various metallic wires (even gold) serving as the positive electrode. These results are unexpected from a modern point of view, but they are not entirely implausible (see Sect. 2.3.2 for further discussion).

From all of that Priestley concluded that there was a link between negative electricity and phlogiston, and between positive electricity and oxygen.[14] He reminded the reader that he had already noted, years ago, "the similarity of the electric matter and phlogiston". In fact he was the discoverer of the fact that charcoal was a good conductor of electricity, about which he was theoretically pleased since charcoal was an indisputably phlogiston-rich body. It was tempting to identify negative electricity with phlogiston completely, but there was a discomfort as nothing corresponded exactly to positive electricity. So instead he said:

> These experiments favour the hypothesis of two electric fluids, the positive containing the principle of oxigen [sic], and the negative that of phlogiston. These united to water seem to constitute the two opposite kinds of air, viz. dephlogisticated and inflammable. (p. 202)

This view is very similar to Ritter's, though Priestley did not name Ritter and it is not clear whether he would have known about Ritter's work by this point. Despite some intricacies exhibited in relevant experiments, Priestley thought the basic connections were clear enough. He noted that if the facts of electrolysis were as the

---

[12] Priestley did use the terms "oxygen" and "hydrogen" (or, "oxigen" and "hidrogen") interchangeably with "dephlogisticated air" and "inflammable air" in this paper. To be precise: the term "hidrogen" only occurs in the marginal summaries, so Nicholson may have been responsible for that; however, "oxigen" occurs several times in Priestley's main text, freely mixed in with "dephlogisticated air".

[13] Priestley's description is ambiguous in this passage, as to whether these measures prevented the production of oxygen gas only, or both gases. On p. 201 he reports an experiment in which an oil-covering on the water also stopped the production of hydrogen (inflammable air). But the key point for the moment, which is clear throughout the paper, is that Priestley thought that the production of the two gases happened independently from each other. See Sect. 2.3.2 for further details.

[14] In one place in the article (p. 202, middle) he has the polarity switched, but I think that is a simple error.

mainstream opinion maintained, then they would have "amounted to a full proof of the new theory [Lavoisier's]". But they were not, and instead they gave "a sufficient proof of the doctrine of phlogiston." With his characteristic combination of defiance and politesse, he signed off: "Whether in this you will agree with me or not, I am, Dear Sir, Yours sincerely, J. Priestley." (p. 203)[15]

All of this came as a very unpleasant surprise to the Lavoisierians, who might have thought that at the start of the new century they were finally at the point of eliminating the opposition altogether. But the electrolysis of water, which had promised to be the last nail in the coffin of phlogiston theory, threatened instead to become a thorn in the side of Lavoisierian theory, even a shot in the arm for the dying phlogiston theory. It is not clear how much attention Priestley's paper received at the time (especially from those who did not read Nicholson's Journal), but Ritter's works certainly did attract attention. Johann Bartholomäus Trommsdorff (1770–1837), Professor of Physics and Chemistry at Erfurth, gave a vivid account of the feeling of the Lavoisierians at the time:

> We are at this time fully occupied in Germany with galvanic experiments. Mr. Ritter, a young man of very conspicuous talents, devotes himself entirely to this branch of natural philosophy, relatively to which he has made several very ingenious experiments. He flatters himself that he can prove, in a clear and satisfactory manner, that water is a simple body; and his friend, Professor Pfaff,[16] asserts that he has transformed this liquid into a corresponding quantity, either of oxygen gas, or of hydrogen gas. . . . Notwithstanding I cannot account, in a satisfactory manner, for the effects of the Voltaic pile, I am still far from concluding, with Ritter and Pfaff, that water is an indecomposed [sic] body, and that the destruction of the splendid edifice of modern chemistry is the inevitable consequence of their experiments. (quoted in Wilkinson 1804, 135–136)

Cuvier gave a similar view, more succinctly and more drastically:

> It [Ritter's view] appears, however, so contradictory to the totality of all the other chemical phenomena, that it would have been almost impossible to have admitted it, provided even no other satisfactory explanation could have been given of the experiment in question. (quoted in Wilkinson 1804, 151)

Cuvier was probably being honest in indicating that he and his fellow Lavoisierians could not possibly accept Ritter's view even if they themselves couldn't provide a good account of electrolysis.

2.1.3 Lavoisierian Rescue-Hypotheses

There was no shortage of hypotheses about the mechanism of electrolysis, all aimed at resolving the distance problem and thereby rescuing Lavoisierian chemistry.

---

[15] See also the discussion in Wilkinson (1804), 74–80.

[16] Christoph Heinrich Pfaff (1773–1852) taught from 1798 at Kiel University, where he would remain until his death (Hufbauer 1982, 223); on his electrochemical work, see Kragh (2003).

There were three options available to those who wished to defend the compound view of water. The first two were already articulated by Cuvier in his report of 1801 mentioned above, and the last emerged a few years later.

(a) *Imbalance.* Cuvier attributed this hypothesis to the mathematician Gaspard Monge (1746–1818), one of Lavoisier's colleagues that had participated in the demolition of Kirwan. Monge's view was that electrolysis resulted in an imbalance of substances around each electrode: "the galvanic action tends to abstract, in each of the waters, one of its constituent parts, leaving in it an excess of the other constituent part." (quoted in Wilkinson 1804, 150) The main problem with this view, although very logical, was that the presumed imbalance was not readily detectable, and it was difficult to see how an excess of hydrogen or oxygen in water could have no detectable effects. The imbalance postulated by Monge is actually not an outlandish idea,[17] but at the time it was not taken seriously by many.

(b) *Invisible transport.* According to this story, the electricity entering into the water grabs hold of one part of a water molecule, freeing up the other to be released there; then the electricity, along with its captive, rushes over to the other electrode, and releases the captive there; the electricity itself goes on back into the battery, completing the circuit. William Cruickshank (?–1810/11), military surgeon and chemist at Woolwich (now in London),[18] was probably the first person to introduce this idea. His version goes as follows:

> the galvanic influence (whatever it may be) is capable of existing in two states, that is, in an oxygenated and deoxygenated state. . . . Now when water is the fluid interposed, and the influence enters it from the silver side [the negative pole of a zinc–silver Voltaic pile][19] deoxygenated . . . it seizes the oxygen of the water, and disengages the hydrogen, which accordingly appears in the form of gas; but when the influence enters the zinc[-side] wire, it parts with the oxygen, with which it had formerly united, and this either escapes in the form of gas, [or] unites with the metal to form an oxyde. . . . (Cruickshank 1800b, 257–258)

Plausible as it may sound, the invisible-transport idea was challenged in very concrete ways. Cuvier's report noted the difficulty raised by the fact that electrolysis could be effected in separate bodies of water that were connected by sulphuric acid, or even a human hand (Wilkinson 1804, 150). Many chemists would have had trouble with the notion that oxygen or hydrogen could pass through sulphuric acid with impunity, and the possibility of it passing right through the human body would have seemed downright bizarre.

(c) *Molecular chains.* Considering the difficulties of the above two hypotheses, it makes sense that another alternative gained a wide following. The most popular

---

[17] See, for example, Pauling and Pauling (1975), 358; note that Fig. 2.2, taken from that text, indicates acidity and alkalinity around the anode and the cathode, respectively. See Sect. 2.2.2 for further details.

[18] See Coutts (1959) for some informative details on Cruickshank's life and work.

[19] Coutts (1959, 125) explains the convention in the designation of the parts of the battery used by Cruickshank.

GALVANISM.

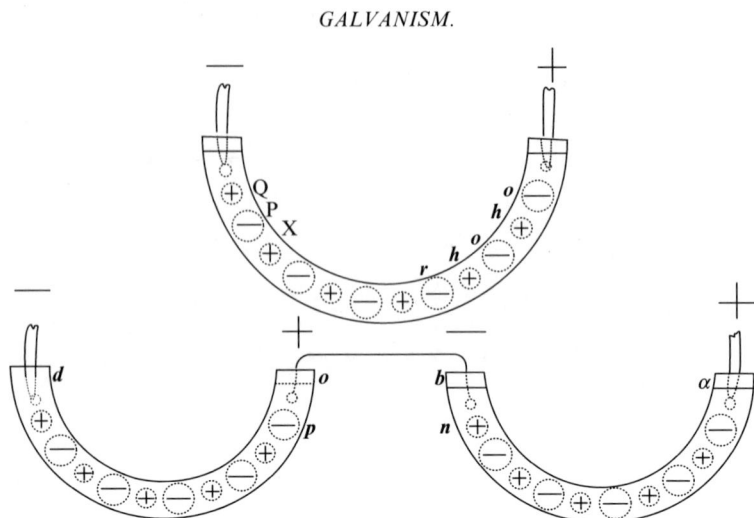

Fig. 2.4 The Grotthuss chain

version of this idea came from Christian Johann Dietrich (Theodor) von Grotthuss (1785–1822) from Courland (Latvia), who was only 20 years old when he published the idea that makes his name still recognizable to physical chemists. This hypothesis did not involve an invisible transport of lone particles of hydrogen and oxygen through a body of water, but an invisible chain of molecules within the body of water connecting the two poles. In this picture, each water molecule is electrically polarized, with hydrogen positive and oxygen negative. Grotthuss (1806, 335) called the Voltaic pile "an electrical magnet", and imagined that the molecules would connect up in a line, like a set of little bar magnets between the poles of a larger magnet, or like iron filings tracing the lines of magnetic force connecting the poles of a bar magnet. This molecular chain, in Grotthuss's own illustration, is shown in Fig. 2.4 (Grotthuss 1806, Plate IX).

When the battery is switched on, the decomposing action begins. The negative electrode grabs the hydrogen particle (electro-positive) right next to it, neutralizes it, and releases it. Having been deprived of its partner, the oxygen particle in that water molecule then goes and grabs the hydrogen particle next to it, forming a new water molecule. This partner-swapping is propagated throughout the chain, and it is matched perfectly by the action originating from the positive electrode. And then each of the newly-formed water molecules flips around, due to the electrical repulsion/attraction from the electrodes, so the initial sort of configuration is restored. Grotthuss's idea became very popular, to the extent that he complained that others (e.g. Biot) adopted it without reference to him (Partington 1964, 27). Davy (1807, 29–30) also proposed essentially the same view without naming Grotthuss.

Grotthuss's idea was certainly adequate for accounting for the facts well-known by then. However, there was a question raised about why the water molecules would line up into a chain in the first place. Grotthuss himself had a clear physical explanation/motivation, namely the bar-magnet analogy. But it is unlikely that everyone else shared that view. Even those who thought some sort of molecular chain to be necessary did not agree on the exact configuration of the chain, as I will explain further in Sect. 2.2.2. There were numerous other hypotheses as well, and many variations on the views introduced so far. It would be nearly impossible to catalogue all of these ideas, though Ostwald [1896] (1980), Mottelay (1922), and Partington (1964) have made heroic attempts. The main point here is that there were a variety of views, none commanding a consensus.

2.1.4 "No Winner" Is Not "No Win"

When we look back on the state of electrochemistry in the early nineteenth century, it is clear why there would have been no clear consensus on the microscopic mechanism of electrolysis. In fact, the insurmountable difficulty of microscopic theory in this field was clear enough to the practitioners at the time—no need for hindsight there. Electrolysis was an experimental technique far too much ahead of its time, if we are inclined to demand a firm theoretical basis for laboratory methods.

In order to make a credible hypothesis about the mechanism of electrolysis, some definite ideas were required on how the atomic particles presumably constituting water interacted with each other. It was necessary to have some sense of what bound the atoms together, in order to theorize about how electricity might unbind them. Electrochemical facts themselves suggested a crude electrostatic view, but that had clear limitations as I will explain further in Sect. 2.2.3.

Another difficulty was not knowing how electricity really operated, or indeed what it was. The dominant view was that it was an imponderable (or "subtle") fluid, much like phlogiston, caloric, or magnetism. But there was an unresolved disagreement between those who believed that there was only one fluid of electricity, whose relative excess and deficit manifested as positive and negative charge, and those who believed that there were two separate fluids, positive and negative. And for the two-fluid theorists, what did the designations of "negative" and "positive" really mean? It is easy to see how the choice between one-fluid and two-fluid theories would have changed the shapes of hypotheses about electrochemical mechanisms significantly. And then when it came to theorizing about how electricity acted on ordinary matter (and on each other, if there were two electric fluids), there was precious little to go on.

More specifically, there was a great deal of uncertainty about the nature of electricity issuing from the Voltaic pile. It was in fact a subject of excited debate at this time, whether Voltaic electricity (or "galvanism" as it was commonly called for some time) was the same thing as the traditional "common electricity" generated by friction. Although there was an emerging consensus on the identity of the two, it is easy to imagine how the lingering doubt on this issue would have added to the uncertainty surrounding the workings of electricity in general. The uncertainty about the nature of Voltaic electricity was compounded by the uncertainty about the workings of the Voltaic pile, as explained in Sect. 2.3.3. It is no wonder that the theory of the battery was so uncertain, with no firm knowledge of the substances constituting its main components. There was a big blank about the constitution of metals, the phlogistonist conception having been largely rejected and the question itself dismissed in Lavoisierian chemistry. There was continuing uncertainty about the nature of acids (and salts), with Lavoisier's theory of acidity on its way out. Water itself was of course an enigma, and the microscopic structure of aqueous solutions unfathomable.

When we now consider the choice among the early nineteenth-century theories of the mechanism of electrolysis, there is no great temptation to be biased in any particular direction. All these hypotheses are fundamentally wrong from the modern point of view. And yet, each has its own grain of truth. It is completely unreasonable to expect that the scientists at the time should have been able to come up with anything like the modern ionic theory. How should they have imagined the idea of a *spontaneous* dissociation of some, but not all, water molecules into ions? Also, why should they have imagined the dissociation to be not into what we might denote as H^+ and O^- ions, but H^+ and OH^- ions? The latter account was only firmed up by the end of the nineteenth century, after many twists and unlikely turns. The rationality of the old chemists has to be judged in terms of what they did with the available alternatives, or with alternatives that would have been unreasonable not to explore (see Sect. 2.3.6 for a further discussion of this point).

Given this situation, it seems to me that the nineteenth-century scientists were wise when they decided not to decide—or rather, not to declare a clear winner amongst a group of imperfect contenders. Leaving the ultimate truth undecided, electrochemists got on with their work, experimental and theoretical, as we will see in some detail in Sect. 2.2.2. That seems to me like the right and mature thing to have done, rather than giving in to the temptation of a clear choice—as in the case of the Lavoisierian bandwagon that made the Chemical Revolution. The legacy of Lavoisierian dogmatism is the only significant blemish on the pleasing pluralism of nineteenth-century electrochemistry: just one among the several serious theoretical alternatives was suppressed, namely Ritter's synthesis view, for its attempted anti-Lavoisierian resurrection of elementary water. The dogmatic nature of this suppression also meant that the electrolysis could not function as independent evidence for the compound nature of water until much later. (More positively, this stroke of dogmatism did create a useful common ground for all those who believed in compound water; see Chap. 5 on how such benefits can still be preserved in a more pluralistic regime of science.)

Even with all the uncertainty surrounding electrolysis, it would be wrong to underplay its importance and all the advancements that it stimulated. As I will discuss further in Sect. 2.2.3, nineteenth-century chemists and physicists got on with developing and using electrolysis, undeterred by the deep uncertainty and disagreements lying at the theoretical heart of electrochemistry. They did not have the luxury of first having a good fundamental theory with which they could make infallible interpretations of new experiments. Uncertainty was an ineliminable fact of life in that process, so several different systems of knowledge developed and flourished simultaneously, each making its own contributions and enriching and stimulating each other. I will discuss the philosophical merits of this way of proceeding in Sect. 2.2.3, and then again in Chap. 5. But before getting on to any philosophizing, it has to be admitted that currently most of us do not even have a well-rounded historical awareness of this period of electrochemistry, despite the labors of a handful of scholars. I blame this state of affairs on what I call the "closure obsession", which has been prevalent in the twentieth century not only among scientists, but among philosophers and historians of science as well. The closure obsession crops up in some unlikely places, too. For example, Thomas Kuhn, even as he famously disputed the common notion that scientific debates were matters of strict right and wrong, insisted on the necessity of a monopolistic paradigm in each field in order to enable normal scientific research. Much of nineteenth-century electrochemistry, like many other lesser-known phases of the history of science, did not have a unified theoretical basis; I suspect this is why it has been largely neglected by scientists, philosophers and even historians.

2.2   Electrochemistry Undeterred

In this part of the chapter I will take a deeper and broader look at the debate on the electrolysis of water. First, I will consider how well the various competing theories accounted for the phenomena. Did Ritter's synthesis view deserve to be eliminated? And how good were the various Lavoisierian rescue-hypotheses? After that consideration, I will use this debate on electrolysis as a lens through which we can gain a fresh and informative view of the development of electrochemistry in the earlier parts of the nineteenth century. Except for a few recognizably brilliant moments, such as Davy's discovery of the alkali metals and Faraday's elucidation of electrochemical equivalents, electrochemistry for most of the nineteenth century may seem to have been mired in disputes between competing theories, without a productive agreement until after Arrhenius. I would like to correct that impression, and bring out fruitful patterns of development hidden in the messy state of plurality.

2.2.1   How the Synthesis View Was Eliminated

The electrolysis of water created an unexpected threat to the notion of compound water. The threat came in the form of what I have dubbed the *synthesis view*, that the hydrogen and oxygen arising from the application of the Voltaic pile to water were actually compounds made up from elementary water and each of the two fluids of electricity. In the first section of this chapter (Sect. 2.1) we have seen the rise of this threat to Lavoisierian chemistry, and the vigorous reactions that it generated. I have also indicated that the dismissal of the synthesis view was not made on the basis of any agreed solution to the *distance problem*. Now I would like to take an in-depth look at this situation, in order to make a reasoned assessment of whether and how electrolysis served as positive evidence for the claim that water was a compound made up of hydrogen and oxygen. In this section I will examine the justification for the elimination of the synthesis view; in Sect. 2.2.2 I will evaluate the merits of the Lavoisierian accommodations of the distance problem.

It will be helpful to make use of some very basic philosophy of science in the initial framing of our discussion. As I go on I will want to propose some significant changes in the way philosophers think about evidence, but to get the discussion going the standard philosophical common sense will be proper and sufficient. So let us start by considering the testing of two competing hypotheses regarding the constitution of water according to the hypothetico-deductive (H–D) model. Call these hypotheses C (for "compound water") and E (for "elementary water"):

> C = "Water is composed of oxygen and hydrogen."
> E = "Water is an element."

For H–D testing we need to deduce predictions from the hypothesis in question, and compare them with observations. So we want to see what hypotheses C and E predict about the outcome of the application of Voltaic electricity to water. Promptly we find that from C or E alone we can make no useful predictions at all. And in the early nineteenth century there were no sufficiently developed theories of electricity to enable scientists to deduce what should happen if they passed a current of electricity through a body of water.

To broaden the philosophical framework just a bit now: if an observation is to serve as evidence for a hypothesis, some clear logical relationship needs to be established between the two; it would be ideal if the hypothesis could be inferred from the observation, but the converse will give us at least something, according to the H–D model. If we can't have that either, at least we will want some positive probabilistic relationship established between the observation and the hypothesis. For the time being, I am not interested in advocating any particular philosophical theory of confirmation. What is certain, whichever theory we go with, is that the establishment of the requisite inferential relationship between observation and hypothesis requires some additional assumptions in the present case, and in most scientific cases.

What we have here is not quite the familiar problem of auxiliary hypotheses blunting the force of refutations. Rather, it is a problematic *lack* of any credible auxiliary hypotheses. It is worth our while to pause here for a moment to highlight the positive enabling role of auxiliary hypotheses. For philosophers who have been introduced to the concept of auxiliary hypotheses through dealing with Duhemian holism or the difficulties of Popperian falsificationism, it is perhaps natural to regard them as annoying interfering factors muddying the logic of theory-testing. But as Duhem stressed, there can be no H–D testing of theories at all, except in the most trivial situations, without the help of appropriate auxiliary hypotheses that make the "D" in "H–D" possible. The need for auxiliaries is enhanced by the familiar gap between what is somewhat directly observable (e.g., that hydrogen and oxygen gases issue from the negative and positive electrodes) and the theoretical hypothesis that such observations are meant to confirm, which are often squarely in the realm of the unobservable (e.g., that a water molecule is composed of hydrogen and oxygen atoms). The situation is going to be similar for any other philosophical model of theory-testing. This insight is along the same lines of Norwood Russell Hanson's view on how the theory-ladenness of observation plays a positive role by giving intelligibility to observations (see Lund 2010).

If we are lucky, the necessary auxiliary assumptions can be found in the form of facts or theories that have already become well-established elsewhere. In the case at hand, all the available auxiliary assumptions were quite new and shaky. When Nicholson and Carlisle first electrolyzed water, there was no settled theoretical view at all about what a current of electricity should do when it met water. Indeed, the very concept of a "current" of electricity was still unfamiliar, as steady flows of electric current were not available until Volta invented the pile. There was even a view, advanced by Georg Christoph Lichtenberg in 1799, that *electricity* would be decomposed into two parts in that situation (see Ostwald [1896] (1980), 24/24). And what was best known about the interaction of electricity and water before the Nicholson–Carlisle work was Cavendish's employment of electric sparks in exploding oxygen and hydrogen together to *synthesize* water. As the London surgeon Charles Hunnings Wilkinson (1763/1964–1850)[20] put it in his fascinating early compendium of electrochemistry: "It has appeared difficult to comprehend how an electrical explosion should effect a decomposition of water; and, when it has been decomposed, should occasion its re-composition." (Wilkinson 1804, 382)

The most basic question to be answered was the following: does the action of electricity tend to create simpler or more complex arrangements of matter? Ritter gave one answer: electricity was a material substance (though an imponderable one, like Lavoisier's caloric), and it would have a tendency to combine with ordinary matter and form compounds (like Lavoisier's caloric did, when it combined with solids to make liquids and with liquids to make gases). On that basis, one can infer from the hypothesis of elementary water (*E*) the prediction that water and electricity would form some sort of compounds if they met each other. If we add the further auxiliary assumption that positive and negative electricity are two distinct fluids (following the two-fluid theory of electricity, which was perhaps the most common view of the nature of electricity at the time), then we get a prediction of two different substances issuing separately from the two electrodes. As Ostwald pointed out in a passage quoted above (p. 79), there was nothing unusual or unreasonable about this reasoning. And the prediction is very nicely vindicated by observation. Therefore I think we are obliged to conclude that the observed facts of electrolysis provided positive empirical support to the hypothesis of elementary water, in the context of a set of auxiliary hypotheses that were both reasonable and widespread at the time.

---

[20] On Wilkinson's life and work, see Thornton (1967).

Since hypothesis *E* directly contradicts hypothesis *C*, the defenders of *C* felt the need to find reasons to reject *E*. At least, they thought that any alleged confirmation of *E* had to be dismantled. Strenuous efforts were made to discredit Ritter's reasoning, almost by any means possible. Did these efforts work? We do know for a historical fact that Ritter's view was rejected by almost everyone within a relatively short amount of time. But we need to take a closer look at how and why this rejection was made. It has to be admitted that intellectual and professional inertia played a very important role. As Cuvier and Trommsdorff put it quite frankly in the passages quoted in the first section (Sect. 2.1), accepting elementary water would have required the revision of far too much in the system of chemistry widely accepted by then. We also have to imagine some fatigue in the chemical community. Trommsdorff, for instance, had only in 1796 given in to the tide of Lavoisierian chemistry, not as a "blind adherent" but on the balance of probabilities, after a long and spirited resistance (Hufbauer 1982, 132–141). The prospect of now going back over to elementary water, and just because of one new phenomenon (though undoubtedly an important one), would have seemed incredibly tedious to him and many others in similar situations.

But were there actually any specific arguments against Ritter's view on water? There are surprisingly few that one can find in the literature of the time, despite some retrospective declarations that Ritter's view had been conclusively refuted at the time. For example Ostwald, while being quite sympathetic to Ritter in some ways, declared that "a scientific closure of the matter was brought about" as early as 1802, through an experiment by Paul Louis Simon (1767–1815), professor at the Civil Engineering Academy in Berlin. There were some disputes about Simon's results, but a similar experiment by J. F. Erdmann confirmed them, and "in this way, the matter was finally decided." (Ostwald [1896] 1980, 159–161/163–166) What Simon and Erdmann did was to verify that the total weight of the oxygen and hydrogen produced by the electrolysis of water was very nearly equal to the weight of the water that disappeared (4.61 grains and 4.60 grains, in Simon's experiment). This only showed that the electric fluids had no weight,[21] and electricity was widely assumed to be weightless, as with all imponderable fluids, so the demonstration of its weightlessness was little more than a red herring. In stressing the importance of the Simon–Erdmann result, Ostwald invoked a parallel to Rumford's nearly con-temporary experiment (1799) demonstrating that caloric had no weight. I would have imagined that Ostwald was a good enough historian to catch the irony here: the very same Lavoisierians who dismissed Ritter's ideas were also in the process of dismissing Rumford's anti-caloric argument by noting that caloric was *meant* to be weightless.[22] So they would have had considerable discomfort in giving much

---

[21] Or that it had so little weight as to be undetectable by the technology of that time (which could in fact be said very fairly about electrons, too).

[22] See Brown (1950), 372, and Brown (1979) on Rumford more generally. They also managed to put up sufficiently strong objections to Rumford's more powerful anti-caloric argument based on the more famous "cannon-boring" argument showing the indefinite production of heat by friction (Chang 2004, 171, and references therein).

credence to Simon's weight-based argument against Ritter. In fact I do not get a sense in the primary literature of Ritter's own time that Simon's experiment had such a decisive impact as Ostwald attributes to it.

The most convincing argument I have seen against Ritter's view was first articulated in 1803 by the young Jöns Jakob Berzelius (1779–1848) and his patron Wilhelm Hisinger (1766–1852), in Berzelius's first major publication: "the theory becomes very ineffective in all cases where the phenomena are due to the decomposition of other substances than water. For example, when [the sulphate of potash] is decomposed by a gold or lead wire, according to Mr. Ritter it must be said: [the sulphate of potash] is a simple substance which when combined with negative electricity makes [caustic potash] and when combined with positive electricity makes sulfuric acid."[23] (quoted in Ostwald [1896] 1980, 311/322) A similar argument was later articulated by Donovan (1816, 47), who thought that Ritter had arrived at his view of electrolysis in a hasty way ("without much reflection") by considering just the case of water. Hydrogen and oxygen can be produced by the electrolysis of other substances, too, and that certainly creates complications for the synthesis view. For example, Donovan indicated that the electrolysis of "nitrous acid" yielded oxygen and nitrogen and asked, rhetorically: "Would he [Ritter] have supposed that oxygen and azote [nitrogen] are nitrous acid in the positive and negative states of electricity? Would he have supposed that the air which we breathe is nitrous acid?" The answer to that could well be: yes, why not? But there is a more serious point to Donovan's question, which is about consistency. Did Ritter want to argue that *both* positively electrified water *and* positively electrified nitrous acid were oxygen? Although not mentioned by Donovan, the case of ammonia would have added force to his argument: is hydrogen both negatively electrified water and negatively electrified ammonia, and is nitrogen both positively electrified nitrous acid and positively electrified ammonia?

Even these arguments, however, do not constitute a knock-out blow against Ritter's position in my view. First of all, it is not logically impossible that two different compounds (e.g., water plus negative electricity, and ammonia plus negative electricity) should both have the same key properties and therefore be identified as one substance (e.g., hydrogen). More seriously, I think that Berzelius and Hisinger, and Donovan too, were giving an uncharitable interpretation of Ritter's view. Surely Ritter could not have meant that the effect of the Voltaic pile was *always* synthesis, or that anything subjected to such treatment was an element. On the contrary, it would have been perfectly reasonable to maintain that electricity would break up a compound (such as nitrous acid) but form compounds when it acted on an element.

---

[23] I have corrected the translation appearing in the English version of Ostwald's text, which has the terms in square brackets as "potassium sulfate" and "potassium hydroxide", which is anachronistic in a problematic way, as Berzelius and Hisinger were writing before Davy's work on the isolation of potassium, when potash was widely regarded as elementary with only an unfounded suspicion that it might be a compound. (Rendering "vitriolic acid" as "sulphuric acid" is not problematic in the same way.)

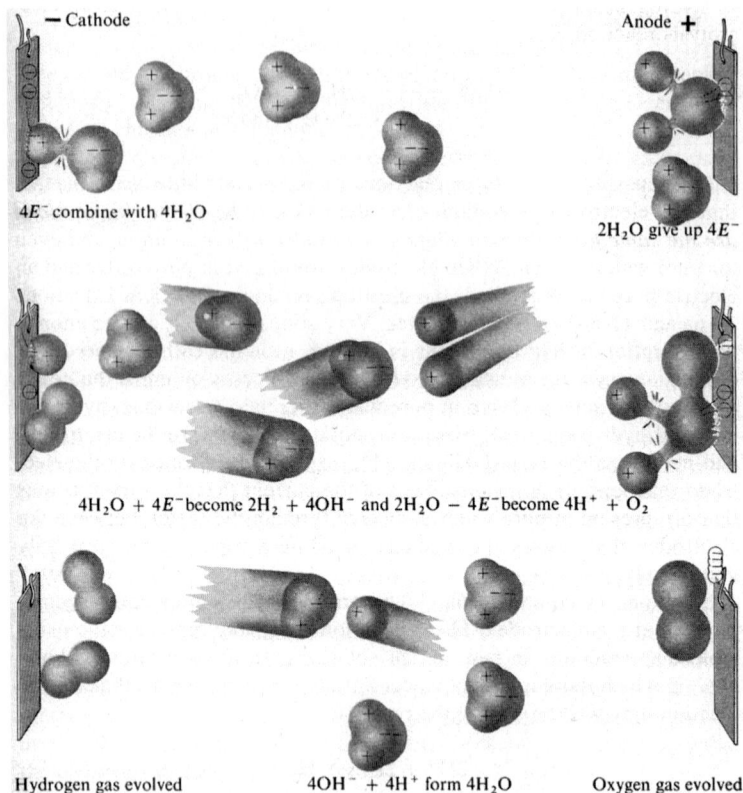

—Cathode   Anode +

$4E^-$ combine with $4H_2O$

$2H_2O$ give up $4E^-$

$4H_2O + 4E^-$ become $2H_2 + 4OH^-$ and $2H_2O - 4E^-$ become $4H^+ + O_2$

Hydrogen gas evolved $4OH^- + 4H^+$ form $4H_2O$ Oxygen gas evolved

Fig. 2.5 One modern view of the electrolysis of water

Even more reasonably, it may be supposed that electricity would trigger a two-step process in reacting with an already-compound substance, first combining with it and then causing a subsequent break-up of the resulting composite. In case this sounds absurd, consider one modern view of the electrolysis of water, from the great Linus Pauling (1901–1994) and his son Peter (1931–2003) in their textbook, which is reproduced in Fig. 2.5 (Pauling and Pauling 1975, 357). On the cathode side, negative electricity in the form of electrons enters the liquid and combines with water molecules, which then break up into hydrogen gas and hydroxyl (OH⁻) ions. This is not the most standard picture one gets in modern textbooks, but it is not in the realm of fantasy, either. It is perfectly reasonable to think that there is a synthesis of electricity and matter in electrolysis, at least in the first stage of the process.

Returning to the actual history in the early nineteenth century: Donovan also pointed out a fundamental uncertainty about the kind of chemical behavior of electricity presumed by Ritter: "And what grounds were there for supposing that electricity ever entered into permanent combinations?" But Donovan was equally skeptical of various "galvanic hypotheses" on all sides, and not particularly harsh on Ritter on this count. For all that anyone knew at the time, electricity was just as likely to form chemical combinations with other substances as not. Ritter was not the only person who had the notion of electricity as a chemical substance, nor was he the most extreme. Perhaps most instructive is the case of Luigi Valentino Brugnatelli (1761–1818), whom we met briefly in Sect. 1.2.1.3 of Chap. 1 in relation to his *ad hoc* concept of "thermoxygen" designed to defend Lavoisier's theory of combustion against the anomaly of combustion supported by oxygen in non-gaseous states. It was Brugnatelli's *Giornale Fisico–Medico* where Volta had first published his views of galvanism opposing Galvani's own views. Considering the phenomena revealed by Volta's pile, Brugnatelli concluded that electricity was an acid[24]:

> The electric acid is a liquid which closely resembles the heat substance and the light substance [Lavoisier's *calorique* and *lumière*]. It is expansible, has a characteristic unpleasant odor . . . and an acidic burning taste. It irritates and burns the skin. . . . It turns blue litmus tincture red. . . . When the electric acid is set in motion it dissolves metals as water dissolves salt. In doing so it carries along with it the dissolved metal to very great distances and in particular through several substances. The electric acid is soluble in water. In such a solution most of the metals get oxidized at the cost of water which in these cases is decomposed and produces hydrogen gas. . . . The salt formed by the electric acid with copper has a beautiful green color and is transparent . . . (quoted in Ostwald [1896] 1980, 208/215)

Brugnatelli's view was not widely accepted, but I have quoted it here to illustrate two important points: it was not just Ritter who was inclined to treat electricity as a substance capable of forming chemical combinations, and moreover this kind of view was also present even on the Lavoisierian side.

All in all, I do not think that Ritter was ever refuted specifically on his view of the electrolysis of water. In my view, explained further in Sect. 2.3.5, what caused the demise of Ritter's view of electrolysis was a failure of the whole program of science that he epitomized. I cannot resist reproducing here some key details in the poignant portrait of Ritter by Walter Wetzels (1978a, b, 1990). Mainstream scientific communities did not know what to do with Ritter's profuse output of experimental results, metaphysical convictions and speculations, and even prophecies. This is the flavor of Ritter's work: he predicted the existence of ultraviolet rays,[25] putting together William Herschel's discovery of infrared rays with his own conviction that the symmetry of the spectrum had to be preserved; and then he duly found them, through an ingenious experiment. Ritter's work became increasingly mystical and speculative, even going into the investigation of divining rods, yet during the same period he also opened up the new field of plant electrophysiology. He never

---

[24] Etienne Gaspard Robertson (1763–1837) in Paris independently advanced a similar view of a galvanic acid (Ostwald [1896] (1980), 209/216). Mottelay (1922, 350–351) explains that Robertson had a personal friendship with Volta, and he was one of the first in Paris to pay proper attention to Volta's work; curiously, it was through Brugnatelli's intervention in a lecture given by Robertson that the latter first began his interaction with Volta.

[25] Wollaston also found ultraviolet rays, independently of Ritter's work.

gained a university appointment, and the one academic post that he did eventually get (membership of the Royal Bavarian Academy of Arts and Sciences in 1805) ended in disputes, censorship and loss of support. He thought there was a correlation between the maximum inclination of the ecliptic and the occurrence of major discoveries in electrical science, and predicted another major event in 1820. This weird prophecy was vindicated in the form of his friend Ørsted's discovery of electromagnetism, but Ritter did not live to see this, as his life was cut short by consumption in 1810 at the age of 34, not helped by years of poverty and electrical self-experimentation. Six years later Donovan (1816, 107) paid his tribute to "the ingenious and extraordinary Ritter": "A premature death deprived the world of one whose constitutional singularity of opinion, ardency of research, and originality of invention, rendered him at once systematic in eccentricity, inexhaustible in discovery, and ingenious even in error."

Understandably, not many chemists and physicists were willing to follow Ritter's imaginative leaps and daring experiments to see which parts of his fantastical-sounding ideas and observations could be verified and built upon. Although he was very popular with scientists and philosophers sympathetic to Romantic *Naturphilosophie*, Ritter's standing in science declined as unfavorable reactions to *Naturphilosophie* set in among men of science. Even phlogistonists were on the whole quite sober-minded people who did not like Ritter's wildly brilliant style of science. It is telling that Priestley did not ally himself with Ritter. Ritter's view on the electrolysis of water was thrown out by mainstream science as part of a comprehensive rejection, and layers of it: the rejection of his views on electricity in general, the rejection of any theories of chemistry incorporating elementary water (including the phlogiston theory), and the rejection of *Naturphilosophie*. I will not enter here into the question of whether Ritter's general outlook should have been maintained in science. What is clear is that the idea of elementary water was logically independent of Ritter's other views, and it was perhaps unfortunate that the unfortunate Ritter was its leading advocate at the turn of the nineteenth century.

2.2.2   How the Lavoisierian Rescue-Hypotheses Fared

Let us accept the historical fact that the notion of elementary water ceased to be in serious play beyond the early years of the nineteenth century, rightly or not. That still leaves the question of what exactly the results of electrolysis implied about the constitution of water, assuming it was a hydrogen–oxygen compound. It would be wrong to see the persistence of the distance problem as a decisive refutation of the compound nature of water—scientific hypothesis-testing is more complex than that. But there is no denying that it created a certain degree of discomfort, and this is what the Lavoisierian rescue-hypotheses were trying to deal with.

Recall the positive role of auxiliary assumptions. For those wishing to take electrolysis as positive evidence for compound water, the task was to establish a credible

Fig. 2.6 Ritter's V-tube
arrangement

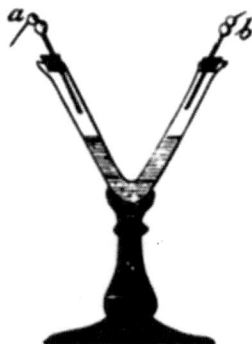

body of theory that would allow the deduction of the observed facts of the electrolysis of water from hypothesis C ("Water is composed of oxygen and hydrogen"). What kind of theory would this be? There were two background factors that reasonably restricted the theoretical imagination in electrochemistry until the 1830s. First, the standard mode of reasoning was based on electrostatics, inspired by the observation that certain substances (oxygen, acids, etc.) moved to the positive pole of the battery and certain others (hydrogen, metals, etc.) to the negative pole. Second, by this time, especially after the work of Charles-Augustin Coulomb (1736–1806) was widely accepted, electrostatic reasoning was predominantly corpuscular. Applying corpuscular reasoning to chemistry meant theorizing about microphysical particles; therefore, electrochemical theory was typically couched in atomic terms, although electrolysis by the Voltaic pile predated the publication of Dalton's chemical atomic theory by almost a decade. So, it was "natural" to think that if water was a compound it was an atomic combination of a positively charged particle of hydrogen and a negatively charged particle of oxygen. This prevalent electrostatic–atomic picture formed the basis of most attempts to solve the distance problem, at least until Faraday's work in the 1830s, and well beyond that for most people.

In the first section (Sect. 2.1) I briefly outlined three hypotheses offered for solving the distance problem by the advocates of compound water, two of which were widely considered viable: invisible transport and molecular chains. Since they were specifically designed to account for the basic aspects of the electrolysis of water, they were fit to serve as auxiliary hypotheses making it possible for electrolysis to provide positive evidence for compound water (hypothesis C), and they became sufficiently detailed to serve as reasonably complete models for the mechanism of electrolysis. They also became focal points of contention. As C ("Water is composed of oxygen and hydrogen") directly contradicted E ("Water is an element"), it makes perfect sense that advocates of E desired to produce experiments that invalidated any auxiliary hypotheses helping to support C.

Ritter made a strong attack on the auxiliary hypothesis of invisible transport. He put a liquid capable of capturing hydrogen or oxygen in between the two bodies of water, and demonstrated that the electrolysis of water proceeded as normal in each of the two separated bodies of water (see Fig. 2.6, from Ostwald [1896] 1980,

156/160). Although this result did not serve as a positive proof of Ritter's own hypothesis, it did create a serious problem for the invisible-transport hypothesis, according to which the production of the two gases from the two poles could continue only if either hydrogen or oxygen (or both) could get across the body of water. It is interesting to read Ostwald's assessment of this argument:

> For [capturing] hydrogen the sulfuric acid he [Ritter] used should do this work but this is somewhat doubtful. For oxygen he uses a solution of potassium sulfide against which no doubt can be raised. If after the atom of hydrogen is separated from it, one follows an atom of oxygen on its way to the other side, then at the instant at which it enters the solution of potassium sulfide it must be compounded by it and cannot appear on the other side. This, however, it does. In fact very little can be said against this argument. (Ostwald [1896] 1980, 158/163)

And very little can be said against Ostwald's assessment. So, even though Ritter did not manage to get his own hypothesis accepted, he played a large role in stopping one of the most promising alternatives in its tracks.

However, as it so often happens in science, there were ways of blunting the force of this apparent refutation. Singer (1814, 343 and 379) took some sort of invisible transport as the unavoidable implication of the distance problem, though he was not entirely satisfied by that idea, either. Accepting invisible transport as a *fact*, Singer tried to make sense of it (pp. 349–350). He reported: "So powerful are these means of decomposition and transfer, that the elements of compound bodies may be conveyed through chemical menstrua for which they have a strong attraction." For example, Singer reported that sulphuric acid could pass through ammonia, and acids and alkalis could pass through "delicate vegetable colors without affecting them". Singer reasoned: "The want of chemical action between the interposed menstrua and the transmitted bodies, appears to arise from some peculiar annihilation of energy during the process, which is perhaps also the cause of the invisible transmission of gas." That is certainly not a full-fledged theory, but it would have served as a humble stop-gap while waiting for a better theory.

So much for invisible transport. What about the idea of molecular chains? Some historians have suggested that Grotthuss's hypothesis was the clear front-runner in the race to solve the distance problem. For example, Williams (1965, 232) declares: "The Chemical Revolution was saved. Grotthus [sic] showed beautifully how it was possible to preserve the compound nature of water and also understand the transfer of two gases through a solution without any visible sign of them." K. M. Gorbunova et al. (1978, 232), who call Grotthuss "the Dalton of electrochemistry", say that Grotthuss's "chain mechanism of electroconductivity" was "generally accepted up to the middle of the nineteenth century". J. R. Partington (1964) says that Grotthuss's theory "lasted until about 1890", though he falls short of claiming general acceptance. Similarly, Florian Cajori (1929, 224) says that there were "several curious theories", but "the one which held its ground for over half a century" was Grotthuss's. Paul Fleury Mottelay (1922, 390) quotes Lardner and Fahie's assessments that it was the "most plausible of the many" hypotheses that were proposed in the early period.

But these assessments do not accord with my own sense of the primary literature, in several respects.

First of all, we have to recognize that there were different versions of the molecular-chain hypothesis, not just Grotthuss's. So it is difficult to claim that there was a consensus on a very specific idea that was attributable to Grotthuss. Even among those who followed the Grotthuss configuration of a single chain of partner-swapping molecules flipping around at the end of each step, there were subtle yet important differences. For Grotthuss himself, the electrical polarity in the water molecules was *induced* by the influence of the battery. Since "every element" of the Voltaic pile (that is, each pair of disks) "possesses its negative and positive pole", Grotthuss (1806, 353) thought that the action of the pile "might establish a similar polarity among the elementary molecules of the water". This idea, he confessed, gave him "a spark of light on the subject." He thought that the hydrogen and oxygen became electrically charged by "a separation of their natural electricity in such a manner that the former acquires the positive and the latter the negative state". In contrast, most others, especially during the dominance of electrochemical dualism (see Sect. 2.2.3.2), would have assumed that the atoms had inherent electric charges, and the polarity of molecules already existed independently of the application of the battery.

And not everyone followed Grotthuss's exact geometry. There was a double-chain version, as shown in Fig. 2.7, in which a chain of hydrogen atoms slides alongside a chain of oxygen atoms, going in opposite directions. This idea was advanced by Berzelius (1811, 278), and also by the Manchester-based chemist William Henry (1775–1836) in 1813. Figure 2.9 below reveals that Donovan had a similar geometry in mind, too, though the mechanism he postulated was different (see Sect. 2.2.3.2 below). And that was not the only variation. As we will see in Chap. 3, there was a growing opinion in other corners of chemistry that the atomic constitution of water was not HO, but H_2O. One of the reasons for this was that the volume-ratio of hydrogen and oxygen gases produced in the electrolysis of water was 2:1, and there were no obvious reasons to think that the same number of hydrogen and oxygen atoms would somehow make up such different volumes. If there are two atoms of hydrogen to each atom of oxygen, the simple geometry of Grotthuss's chain is seriously disturbed. Curiously, in a second paper he published on the subject Grotthuss (1810) gave an updated model in which he assumed water was HO_2 (see Fig. 2.8). Although it was possible to accommodate the extra oxygen (or hydrogen) in the way Grotthuss did, the lack of simple polarity in each water molecule destroyed his pleasing bar-magnet analogy. For the double-chain mechanisms it might have been easier to accommodate H_2O (or HO_2), as one would just have to conceive of an additional chain sliding alongside the originally imagined two. But the already-strong whiff of the *ad hoc* surrounding the chain hypotheses would have got stronger by any of these accommodations.

Can we at least say that there was a broad consensus on *some* kind of molecular chain? Now we meet those who were skeptical about the whole idea. A good example

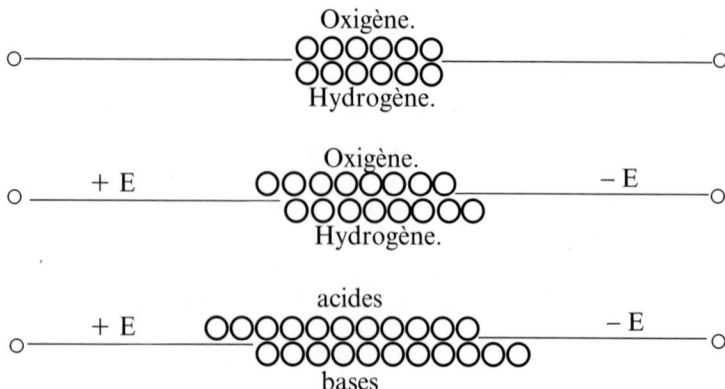

Fig. 2.7 A double molecular chain, by Berzelius

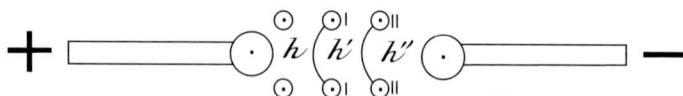

Fig. 2.8 The second Grotthuss chain, with water as HO_2 (from the original French version of his paper in *Annales de chimie*, vol. 63 (1807), plate facing p. 35)

is Singer again (1814, 380), who had the following complaint in relation to Berzelius's version of the idea:

> To me this supposition appears to increase the difficulty, for it infers a series of decompositions, and recompositions, of which we have no proof; and yet it does not seem probable that such phenomena could occur, without producing some apparent motion, or change in the interposed fluid.

So the horn of the dilemma that Singer preferred to take was an unexplained lack of chemical action, rather than undetectable chemical action. In terms of unobservability it may seem that invisible transport was just as bad as molecular chains, but the idea of a set of isolated atoms traveling through water undetected would have seemed less incredible than a chain of macroscopic length (and most likely a whole bundle of them) operating inside the liquid with no detectable effect.

If Singer was merely skeptical, there were others who claimed to have positive experimental evidence against the hypotheses of molecular chains. In fact, some of this evidence went against *both* molecular chains and invisible transport. For example, Ritter and Davy produced experiments in which the water being electrolyzed was put in two separate cups connected by non-liquid conductors. I have already shown a similar arrangement by Davy in which hydrogen and oxygen could be produced in separate cups, connected by asbestos (Fig. 2.3). Now, that is not so problematic,

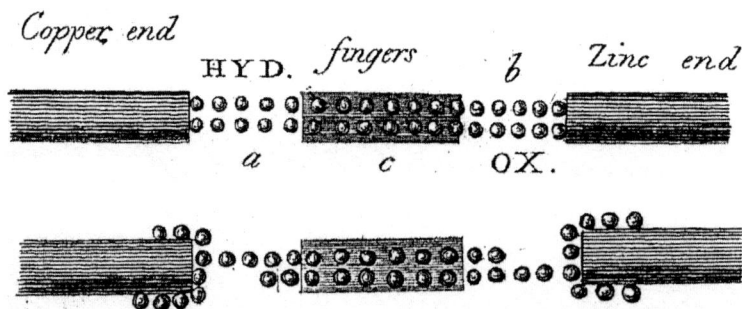

Fig. 2.9 Donovan's schematic representation of the electrolysis of two bodies of water connected by a human body (Donovan 1816, plate facing title page, figures 4 and 5)

since asbestos is absorbent and creates channels of water in its midst; in such watery channels one could have either invisible transport or molecular chains. But the connection between the two cups could also be made by metallic wires, or the human body—one hand dipped in each cup did nicely (Donovan 1816, 340–341; Ostwald [1896] (1980), 156/161). If it seemed unlikely that hydrogen and oxygen could pass unnoticed through a body of water or form an invisible chain within it, it seemed downright bizarre that it would do so through solid metal or a human body.

There were ways of getting around this argument, too, as explained by Donovan (1816, 341–348). In the case of the metallic-wire connection, there is no *new* problem since in each body of water there are two wires, from which hydrogen and oxygen are produced as normal; so it's just the same old distance problem in each cup, and the metal can be supposed to be merely conducting electricity between the two cups. The case of the human body is more difficult, but Donovan, following Grotthuss, pointed out that animal tissue was actually full of water, which is actually in contact with the outside through the pores of the skin. So the connection by the human body is essentially the same as the connection by wet asbestos (see Fig. 2.9), and in either case there is effectively only one body of water, and there is just the same old distance problem. In support of this argument Donovan brought up an extraordinary recollection: "I have often received severe burns by touching the conducting wires of 1,000 pairs of 4 in. plates [constituting an enormous Voltaic pile], and always observed that the parts thus seared [getting the pores closed up] were no longer capable of communicating a shock". Without the ability to pore microscopically into the human body, those who entered this dispute had to be satisfied with a truce.

To take an overview of the situation, let us choose a poignant moment in the history as a vantage point, though it is as arbitrary a moment as any other at which to rest: Grotthuss died in 1822 at the age of 37, driven to suicide by an incurable illness.[26] The situation at that point would not have seemed promising. Although

---

[26] For the circumstance of Grotthuss's death and his legacy, see Gorbunova et al. (1978), 233–234.

chemists working in the Lavoisierian tradition managed to do enough defensive work to prevent electrolysis from being used to resurrect the idea of elementary water, they were unable to agree on a way to solve the distance problem, which stood in the way of turning electrolysis into positive evidence for the compound nature of water. In the end, the distance problem was not solved, but dissolved. By the late nineteenth century a wholly new ontology of electrolytes came, which negated an important premise behind the formulation of the problem itself: "if what happens in electrolysis is the breakdown of each water molecule by the agency of electricity…" The centerpiece of the new ontology was free ionic dissociation: some molecules of water are already broken up into electrically charged ions, before any external source of electricity is applied; these pre-existing ions, diffused throughout the liquid, are picked up at the electrodes according to their electrical charges. The distance problem no longer existed in the new ontology, and there was no need for invisible transport, or for molecular chains. Now, all this, too, is an oversimplification, but I will not enter into that historical discussion fully since it would take us too far away from our story, into the twentieth century.

2.2.3 The Character of Compound-Water Electrochemistry

The discussion so far makes it clear that no one solved the distance problem concerning electrolysis to everyone's satisfaction. Early electrochemical theories were certainly too uncertain to support a positive experimental demonstration of the compound nature of water. Rather, what we have seen so far is that water was taken to be a compound on the basis of other reasons. What I will call "compound-water electro-chemistry",[27] which quickly became the main-stream of electrochemistry in the early nineteenth century, began by assuming without a clear justification that Ritter's synthesis view of electrolysis was wrong, and that water consisted of one component manifesting itself as hydrogen at the negative electrode, and one component manifesting itself as oxygen at the positive electrode. A great multitude of theoretical ideas and a growing body of experimental techniques and facts accumulated on the basis of these assumptions. In this section I want to make a more careful consideration of the character of compound-water electrochemistry in its early stages, and show how it was able to grow in productive ways despite what would seem like a pile of confusion at its foundation. The developments were guided by a sense of coherence within each particular system of practice, without a presumption of certainty in its foundations.

---

[27] By this phrase I don't mean just the electrochemistry of water, but the electrochemical system that took water as a compound.

2.2.3.1 The Stabilization of Experiment

One thing that will be evident even to fairly casual observers of nineteenth-century electrochemistry is that the experimental side of the enterprise was much more stable than its theoretical side.[28] Take Faraday's view of the situation, as he began to report on the initial results from his work on electrochemistry:

> What may be considered as the general facts of electro-chemical decomposition are agreed to by nearly all who have written on the subject. They consist in the separation of the decomposable substance . . . into its proximate or sometimes ultimate principles . . . ; in the evolution of these principles at distant points . . . ; and in the constant determination of the evolved elements or principles to particular poles according to certain well ascertained laws. But the views of men of science vary much as to the nature of the action by which these effects are produced; and as it is certain that we shall be better able to apply the power when we really understand the manner in which it operates, this difference of opinion is a strong inducement to further inquiry. (Faraday 1833, 683, §§ 478–479[29])

Half a century later, the theoretical field was no more unified. In this context it is interesting to revisit the well-known recollection by Arrhenius about the initial phase of his work on the ionic theory in the early 1880s:

> I came to my professor, [Per Cleve], whom I admired very much, and I said, "I have a new theory of electrical conductivity as a cause of chemical reactions." He said, "This is very interesting," and then he said "Good-bye." He explained to me later that he knew very well that there are so many different theories formed, and that they are almost all certain to be wrong, for after a short time they disappeared; and therefore by using the statistical manner of forming his ideas, he concluded that my theory would not exist long.[30] (quoted in Gray and Haight 1967, 90)

But the stability of the experimental side of electrochemistry also requires a closer examination, as it is not straightforward. If there is anything I have managed to demonstrate so far in this chapter, it is that what Faraday called "the general facts of electro-chemical decomposition" were open to serious dispute, almost as much as the high theory. Indeed, Ritter and his supporters would not have recommended electrolysis as a method of chemical decomposition, either for water or more generally; they would not even have agreed to speak about "electrolysis" (which was introduced by Faraday later), since the term already pre-judges the issue. And I have argued that there was no knock-down argument showing that Ritter was wrong. But it is also not the case that mainstream experimental electrochemistry simply rested on dogma in its interpretations. Rather, its stabilization came through the establishment of coherence. The presumption of the compound nature of water, for example, was

---

[28] That, too, will become less certain with deeper knowledge.

[29] As Faraday meticulously numbered the paragraphs in all of his papers on "Experimental Researches on Electricity" in one consecutive sequence, I will note the paragraph numbers in my citations.

[30] So we can see that Cleve was a whole century ahead of Laudan in making the pessimistic meta-induction from the history of science! Laudan's point is a stronger one, as his examples concern theories that were once well-established, as opposed to Cleve's ephemera. Cleve may have been wiser than his pupil, but Arrhenius was both the more typical and the more productive player in this game.

a central doctrine that initially formed the basis of experimental practice but later became buttressed (and also refined) by further coherent developments in whole electrochemical systems encompassing both theory and experiment.

A curious and encouraging fact about the development of compound-water electrochemistry is that it doubled back on its initial starting point and corrected and refined it. Much of early electrochemistry (and early atomic theory, too) forged ahead on the assumption that water was HO. By the time atomic chemistry reached its maturity and electrochemistry adopted the idea of free dissociation, the accepted formula for water had changed to H_2O, and the ionic composition of it into H^+ and OH^-: still a compound, but not a simple hydrogen–oxygen compound as initially imagined. This is the kind of iterative development that I characterized elsewhere as "progressive coherentism" (Chang 2004, ch. 5; 2007a).

The crucial starting point in the building of the system of compound-water electrochemistry was the match between the outcomes of electrolysis and pre-existing ideas of the composition of the substances that were electrolyzed. This rendered chemical analysis by means of the Voltaic battery a coherent activity which could serve as a core of various systems of electrochemistry (see Chap. 1, Sect. 1.2.1.1 for my definitions of "system", "activity" and "coherence"). For those subscribing to the Lavoisierian orthodoxy, the production of oxygen and hydrogen from the electrolysis of water was a key point to hold on to, despite the reasonable doubts raised by Ritter and others. It matters little here that the detailed picture of the process that we now have is really not at all like what the Lavoisierian electrochemists would have believed, especially about what happens at the anode as OH^- ions are converted into O_2 gas. The relevant point for coherence is that the mainstream-electrochemical and Lavoisierian accounts went well together.

There were other important cases of such calibration, too. Cruickshank subjected the solutions of metals in acids (and in ammonia) to electrolysis, and succeeded in recovering the metals in very pure form (1800a, 189–190; 1800b, 259–260). This was superior to the case of water as a vindication of electrolysis as an analytical tool, as the composition of the metallic solution was not itself in dispute, but on the contrary known with as much certainty as chemistry can ever attain, by preparing the solution oneself. So Cruickshank's electrochemical practice cohered strongly with some very basic chemistry of metals in solution. Many others made similar experiments to consolidate Cruickshank's observations. Similarly, the successful electrolysis of various salts into their acid and base parts enhanced confidence that electrolysis was decomposition, precisely matching the pre-existing affinity-based notions of the composition of salts, and the laboratory practices of the actual manufacture of those salts. Berzelius and Hisinger (1803) took the lead on that work (see also Lowry 1936, 273, 287). Again, it is immaterial that we now don't think of a salt as actually being made up of an acid and an alkali in their entirety[31]; what matters for present purposes is that there was a coherence among the

---

[31] T. M. Lowry (1936, 270) notes that it was only in 1840 that John Frederic Daniell advanced the general view that a salt was a binary compound of two radicals, not of an acid and an alkali in their entirety.

old practice of making a salt as a compound formed by reacting an acid and an alkali, and the new practice of decomposing it into the acid and the alkali by electrolysis. But in order to uphold this coherence, electrochemists had to remove one anomaly: the apparent production of acid and alkali from the electrolysis of pure water; it was Davy's (1807) achievement to identify hidden salts in the water as a source of these mysterious acids and bases.

Thanks to these successes, within 5–6 years electrolysis was sufficiently established at least as a promising method of analytical chemistry. On that basis came the next stage of development, namely attempts to use the battery to effect previously unknown decompositions. The spectacular success of this attempt in the form of Davy's isolation of the alkali metals potassium and sodium through the electrolysis of potash and soda is a very well-known story (Davy 1808a, b; Golinski 1992, chapter 7, and references therein). Now, Davy's claim to have decomposed potash, soda and other previously "undecompounded" substances could have been, and was, challenged. Jan Golinski gives a thorough account of the many factors that contributed to the victory of Davy's interpretation against, for example, Joseph-Louis Gay-Lussac and Louis-Jacques Thenard's contention that potassium/sodium was a compound of potash/soda and hydrogen, the latter being produced by the electrolysis of water mixed in with the potash/soda. Without denying the cogency of Golinski's points about the importance of the rhetorical tools and spaces of discourse that Davy effectively utilized, I wish to highlight how the coherence in Davy's practice in this episode contributed to the consolidation of experimental electrochemistry.

Building on the previous stages of development discussed above, Davy designed his experiments on potash and soda on the assumption that electricity would break down these substances, if they were compounds. Targeting the alkaline earths was a coherent strategy, since there had already been some suspicion, dating back to Lavoisier's time, that these substances might be compounds. Once Davy produced potassium and sodium, their spectacular properties (such as bursting into flames when thrown into water) drew much attention. For my story, however, even more important were some rather mundane properties. The comparison of such properties between potassium and potash, and between sodium and soda, fitted well into the familiar pattern of relationship between metals and their oxides, which cohered well with Davy's initial claim that potassium/sodium were metals and potash/soda their oxides. (Again, it is immaterial, for the coherences of practices at the time, that modern chemists do not identify potash and soda as simple oxides of potassium and sodium.[32]) Among other things, Davy thought he could show that potassium/sodium could be turned back into potash/soda by burning, which also fitted the old routine of decomposition-and-recomposition for demonstrating the constitution of a compound. For Davy there was also another aspect of coherence in the whole picture,

---

[32] Lowry (1936, 11, 62, 288) states that Davy used caustic potash and caustic soda, and identifies them as the hydroxides of the metals, KOH and NaOH. The non-caustic varieties are the carbonates: K_2CO_3 and Na_2CO_3. See Lowry's explanation (pp. 283–284) on how Davy and others gradually moved away from his initial view that potash and soda were simple oxides.

although not everyone shared this one. Davy (1809, 1810) was unwilling to accept that there were a great number of ultimate chemical elements, so he welcomed the apparent power of Voltaic electricity to break down substances that had been considered elements, in the hopes that in the end one could do with fewer ultimate elements.[33] This simplicity-driven inclination toward decomposition is what made Davy turn away from Ritter's synthesis view, even though he was severely critical of Lavoisier and did not share the Lavoisierian motivation for rejecting Ritter (see Chap. 1, Sect. 1.2.2).

In the end, the interpretations put forward by Davy came to be regarded as facts. As Faraday put it a quarter-century later (1833, 683, § 478): "What may be considered as the general facts of electro-chemical decomposition are agreed to by nearly all who have written on the subject. They consist in the separation of the decomposable substance acted upon into its proximate or sometimes ultimate principles". These "facts" provided a stump on which various theoretical systems were built, as I will discuss further in the next section. Taking electrolysis as decomposition amounted to a partial operational definition of electrolyzable substances as compounds, applicable even in the absence of any agreed understanding of the mechanisms underlying electrolysis. Any such substance was also assumed to have a dualistic constitution; the conceptual framework of electrochemical dualism was almost written into the physical shape of the battery having two poles.

Davy's triumph helped create great expectations for electrolysis as a decomposition method. Singer (1814, 347) reported that the early electrochemical work culminating in Davy's experiments "displayed the importance of the Voltaic battery as an instrument of analysis; for the elements of almost all the bodies subjected to its action, were separated and collected at the wires connected with its opposite surfaces." It seemed to many that ultimately *any* compound could be broken up by electrolysis (with the accompanying notion that all chemical combinations were electrical in their nature). And even though a great deal of disagreement remained about the theory of electrolysis, the experimental techniques and phenomenological interpretations became quite quickly standardized. Voltaic batteries were made in the stable and easy-to-handle "trough" configuration invented by Cruickshank (1800b, 258–259; Coutts 1959, 124), with a wooden trough containing pairs of metallic plates set into flat cells containing acids or salt solutions (see Fig. 2.10 for an example, from Wilkinson 1804, frontispiece). This standard configuration continued to serve as a reference point for a few decades, despite numerous variations and innovations. Various other types of batteries were also developed, and a few became standardized tools, for example the Daniell cell invented in 1836 by John Frederic Daniell (1790–1845),

---

[33] As Knight (1967, 21) explains, this was one reason for which he did not wholly embrace John Dalton's atomic theory, which postulated a distinct atom for each chemical element recognized as such at the time. While accepting the Lavoisierian operational definition of an element as a hitherto undecomposed substance, Davy focused his effort on effecting new decompositions. One of his motivations for entertaining the revival of phlogiston (see Chap. 1, Sect. 1.2.2) was to see if he could not reduce the number of chemical elements (see Siegfried 1964).

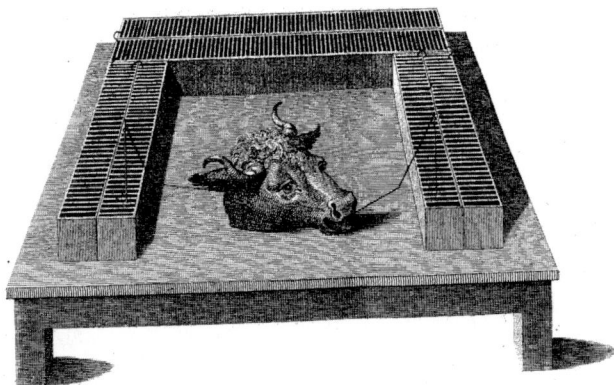

Fig. 2.10 A Voltaic battery in the trough configuration, being applied to a cow's head

Professor of Chemistry at King's College London.[34] For the performance of electrolysis, Singer (1814, 347) indicated that the arrangement with two separate cups connected by wet asbestos was standard by 1814 (see Fig. 2.3 above). And it was widely accepted that gold (or, better yet, platinum) should be used as electrodes so that the products of decomposition can be manifested on their own rather than becoming combined with the metal of the electrodes, as Nicholson and Carlisle already recognized in 1800.

Although the workings of the Voltaic battery remained shrouded in mystery and controversy (see Sect. 2.3.4), from early on there was a widely shared understanding that the number of metallic pairs determined the "intensity" of the electrical power (what we would now call voltage or electromotive force), and that the surface area of each metallic piece determined the rate at which the electricity flowed out (what came to be called "current"). A lack of theoretical sophistication probably delayed further developments, but the instrumental practice remained very robust and stable. In the realm of the phenomenological study of electrochemistry, Faraday's laws rightly stand as a landmark (see Williams 1965; James 1989). It is interesting to note that Faraday's own electrochemical thinking was rooted in his highly idiosyncratic view of forces, as I will discuss further in the next section. While hardly anyone followed Faraday in his theoretical and metaphysical speculations, that did not hinder the wide acceptance of Faraday's quantitative results. Partly with the help of assumptions such as the compound nature of water, electrolysis established itself as a standard (and standardized) method of effecting chemical decomposition, proving its value as an analytical tool long before there was any unified theory of it.

---

[34] In modern days the Daniell cell has also displaced Volta's as the paradigm of theoretical exposition in electrochemistry, as explained in Sect. 2.3.4 and further in Chang (2011c).

2.2.3.2   The Diversification of Theory

Let us now turn to the situation in early electrochemical theory. It is instructive to share the overview of history that Faraday made as he began his mature work on electrochemistry in the early 1830s. While acknowledging the "utmost value" of Davy's much-celebrated "facts" concerning "electro-chemical decomposition", Faraday felt dissatisfied by Davy's theoretical account of the effects: "The mode of action by which the effects take place is stated very generally, so generally, indeed, that probably a dozen precise schemes of electro-chemical action might be drawn up, differing essentially from each other, yet all agreeing with the statement there given" (Faraday 1833, 684, § 482). Two years later, when challenged by Davy's brother John on this statement, Faraday duly produced a long list (reprinted in Faraday 1844, 216). On the mode of decomposition, there were different theories by Grotthuss, Davy, Riffault and Chompré, Biot, De la Rive, and Faraday himself. Regarding the action of the battery itself, there were contact theories as opposed to chemical theories; Davy himself thought that the electricity was "excited by contact, but continued by chemical action." There were also differences as to whether the particles of matter had inherent electrical charges, or acquired the charges contingently through their mutual interactions. And so on. And it can hardly be said that Faraday's own subsequent work brought any greater unity to the field, as most people did not adopt his unique theoretical point of view.

In order to come to a full understanding of the dispute over electrolysis, we must place it properly in this complex theoretical field. It is important to keep in mind that the theoretical situation was one of coordinated diversification, not complete chaos. The consolidation of compound-water electrochemistry was mostly in the experimental realm, but it did involve an agreement on some fundamental theoretical interpretations of experiments, most importantly the view that electrolysis was decomposition and that substances prone to successful electrolysis (such as water and potash) were compounds. So, in the terminology developed in Chap. 1, Sect. 1.2.1.1, I would say that there were many electrochemical systems of practice, all of which shared some core experimental activities but differed significantly in their more theoretical activities. Another way of expressing the same idea is to say that nineteenth-century electrochemistry was an electrolytic system-type (see Chang 2011d for more on system-types). What the term "electrolytic" in that phrase refers to is the shared understanding of the action of the Voltaic battery as decomposition, subscribing to the intended literal meaning of "electro-*lysis*", excluding interpretations such as Ritter's but allowing a great deal of room for various theories about how exactly the decomposition happens by the agency of which powers.

It is interesting to consider how one would view nineteenth-century electrochemistry in Kuhnian terms. It does not seem possible to say that nineteenth-century electrochemistry had a clear paradigm. If the agreement on the more experimental side of the business provided a paradigm, then at least there were a great number of divergent articulations of that paradigm as scientists attempted to make further theoretical development. In the Kuhnian scheme divergent paradigm-articulation is a key symptom of a crisis, but that implies the existence of a previously established paradigm that is unraveling, which is not the situation here. Or was electrochemistry in this period a mere "pre-science", which occurs before the establishment of any paradigm? There would be some justice in that description, if we just looked at the heady first few years of the century. But after the establishment of compound-water electrochemistry, there was plenty of activity that was not only scientific in the broad sense of the word, but also resembled the puzzle-solving activity characteristic of Kuhnian normal science. What we see in nineteenth-century electrochemistry is neither pre-science nor a revolutionary spasm in between periods of normal science, but a long-lasting plurality in which multiple systems co-existed. The creators and advocates of the different systems communicated well with each other, in print and in person. The inter-systemic and inter-personal interactions were on the whole productive, as I will argue in the next section. Not unlike modern philosophy, nineteenth-century electrochemistry was a discipline that thrived on disagreement, dispute and debate. There was enough common ground of both concepts and customs to allow a continually disagreeing scholarly community. I think science operates on this kind of basis more often than we might imagine, as I will discuss in more detail in Chap. 5. To return to Kuhn: it may simply be that there is more of a continuum where Kuhn saw a rather sharp dichotomy, between normal science and extraordinary science.[35] In all situations that we recognize as scientific at all, there are disagreements among scientists but also sufficient common ground to allow productive debates; it is a matter of degree, how deep the disagreements go.

I will now briefly outline several distinct systems of electrochemistry that grew up in the first half of the nineteenth century. These were not the only important systems, but I am focusing on those that were sufficiently developed, at least reasonably well-known in their own time, and directly relevant to the understanding of electrolysis. Other systems I might have included here include those due to Biot, Ampère, de la Rive, Daniell, and Helmholtz. I will be covering some ground well-trodden by various historians of science. Yet, I think a succinct summary of several key systems will be helpful in giving a clear sense of the theoretical terrain and its plurality and complexity.

To begin with, this would be an appropriate place to pay some attention to Volta's own theoretical view of the battery and its actions, for which Giuliano Pancaldi's (2003) biography of Volta is the best up-to-date source. Volta's longstanding inclination was to reach a mechanical understanding of electrical phenomena. He liked to reason in terms of the movements and actions of the electrical fluid in the tradition of the one-fluid theory of electricity, though his ideas differed somewhat from Franklin's and Aepinus's. His early fame rested on his invention and interpretations of two instruments, the electrophorus and the *condensatore*, both of which arose from his work in the tradition of eighteenth-century physics of static electricity. The key concepts he employed in understanding the battery, as well as his earlier inventions, were the tension of the electrical fluid, and the capacity of bodies for holding

---

[35] Many historians and philosophers have criticized Kuhn on this point. For an early example see Toulmin (1970).

the electrical fluid. He saw the battery as an instrument that occasioned a continual flow of electricity by harnessing the tension that arose at the contact of two different substances. Much of his energy in electrochemistry was spent on keeping biology and then chemistry itself out of it, arguing against the existence of a separate animal electricity that was postulated by Galvani, and against the chemical explanations of the action of the battery.

After Volta, if we set aside Ritter, Davy was perhaps the first person who went beyond the initial fascination with electrochemical phenomena to create a comprehensive electrochemical theory. Although Davy was averse to making rigid and overarching theories, he did base his thinking on some sound theoretical ideas. A great deal has already been written about Davy's electrochemical works, so I will only highlight a few salient aspects (for further details, see Russell 1959; Golinski 1992, ch. 7). The core idea was that different chemical elements were made up of particles with different electrical inclinations, and that chemical combination was best understood through the electrical interaction of charged particles. Davy took Volta seriously on the electrification of bodies by mutual contact, and repeated and extended Volta's experiments. He envisaged the microscopic particles of different chemical elements developing electrical charges by contact in the same way, and linked chemical affinity to such electrification. Initially Davy was strongly tempted by a reductionistic thought, stating in a Royal Institution lecture course in 1808: "Is not what has always been called chemical affinity merely the union or coalescence of particles in naturally opposite electrical states?" By 1812 he retreated from such a simple idea: "Electrical effects are exhibited by the same bodies, when acting as masses, which produce chemical phenomena when acting by their particles; it is not therefore improbable that the primary cause of both may be the same" (quoted in Russell 1959, 16, 18). Davy's theoretical ideas never seem to have congealed into a definite shape; however, as Colin Russell puts it (1959, 24): "Davy was the first to link the well-known phenomena of chemistry with the new facts of Voltaic electricity. . . . the pioneer in attempting a union of the two branches of science was Davy."

Berzelius appreciated Davy's electro-chemical ideas, but developed them in a different direction (see Brock 1992, ch. 4; Russell 1963; Melhado 1980, ch. 4). As with Davy, the foundation of Berzelius's theoretical system was a classification of chemical substances into electropositive and electronegative, or rather, their placement on a spectrum of electropositivity/negativity ranging from oxygen to potassium. However, Berzelius saw the electrical charge of atoms as inherent, not merely developed by their contact with each other. He viewed chemical combination as a straightforward result of electrostatic attraction between corpuscular atoms positively and negatively charged; higher-order compounds formed between molecules with residual overall electric charge. In contrast to Volta and Davy both, Berzelius was heavily influenced by Lavoisier, and built the core of his chemical system on the basis of Lavoisierian ideas; the young Berzelius "came under the spell of the new antiphlogistic philosophy", which "affected all his thinking" (Russell 1963, 117). A pervasive consequence of this was Berzelius's steadfast emphasis on the importance of oxygen, to which he gave an electrochemical twist. It was commonly agreed that oxygen was a highly electronegative element, but for Berzelius it had "absolute electronegativity" (quoted in Russell 1963, 128). Davy and Berzelius agreed that most acids were oxides, but while Davy delighted in the discovery that muriatic (hydrochloric) acid had no oxygen in it, Berzelius accepted this fact with the greatest reluctance and regarded it as an anomaly that did not quite invalidate Lavoisier's oxygen theory of acidity (see Gray et al. 2007). When electrolysis revealed to Davy that most *alkalis* contained oxygen as well, he took pleasure in the mockery this made of Lavoisier's idea of oxygen as acid-generator. For Berzelius, on the contrary, this only increased the importance of oxygen: he now thought that all acids and alkalis were oxides, as well as all organic substances. For him the fundamental dualistic composition was electronegative oxygen plus an electropositive radical, a formulation obtained by generalizing Lavoisier's theory of acids.

For Faraday, a fundamental shortcoming in Berzelius's theory was its basis in simple-minded electrostatics. Faraday made various experiments designed to show that electrolytic action could not be caused by Coulomb-type action-at-a-distance attraction and repulsion exercised on electrolytic molecules by the poles of the battery (Arrhenius [1897] 1902, 111–113; Partington 1964, 115–116; Williams 1965, 241ff). The most striking of these included the demonstration that the intensity of electrolytic action was the same throughout the solution regardless of the distances from the poles, and that "a single ion, i.e. one not in combination with another, will have no tendency to pass to either of the electrodes, and will be perfectly indifferent to the passing current" (quoted in Williams 1965, 266). Faraday saw the action of electricity as modifying the inherent chemical affinities (see Faraday [1859] 1993; Sinclair 2009).[36] In an electrochemical circuit, Faraday imagined all actions linked up in a continuous loop of forces, whose balance determined how much electricity should pass in which direction. In resisting the temptation to read our modern notions into Faraday's electrochemical theory, it will be useful to remember that Faraday's ions were *not* electrostatically charged, at least in the usual sense. In fact, it was his objection to electrostatic chemical theories that compelled him to invent his new terminology with the help of William Whewell and others, to remove the unwanted theory-ladenness of terms like "poles" (Williams 257–269). So his electrolytes contained "ions", which meant travelers; "electrodes" were not poles of attraction, but merely the portals through which ions passed; the terms "cathode" and "anode", simply denoting "down" and "up" movement, were deliberated coined as to mean nothing specific about electricity or chemistry. Ionic movement was caused by somewhat mysterious field-theoretic forces, and even Faraday himself

---

[36] In this he had two important predecessors. One was Davy, whose conception of the relation between the electrical and chemical forces was more subtle, complex and vague that Berzelius's. Russell's view (1959, 12) is that "Faraday was the one to influence the world to look favourably on his master's theories. And he did this by enshrining them in his own." The other predecessor I want to highlight is Donovan (1816, 278), who published this insightful view 15 years before Faraday's work: "it was found that copper lost its affinity for oxygen, by contact with zinc; . . . the affinity of the zinc for oxygen was much increased by contact with copper. I think therefore there is nothing overstrained in the inference that one has gained what the other lost, or in other words that the copper has transferred a portion of its affinity for oxygen to the zinc."

confessed to be unclear about how "the power under consideration can appear at one time as associated with particles giving them their chemical attraction, and at another as free electricity" (Faraday 1834, 470, "Note" at the end of the paper). Faraday's electro*dynamic* ideas were later reconstructed by James Clerk Maxwell into a system that others could relate to; no one did a similar service regarding Faraday's electro*chemical* ideas.

A very different way of departing from the simple electrostatic picture originated with Rudolf Clausius (1822–1888) (see Arrhenius 1902, 114–116; Cajori 1929, 225–226). While Faraday brought sophistication to the notion of force in electrochemical action, Clausius brought in further considerations of corpuscular mechanics. Clausius approached the problems of electrochemistry from the viewpoint of the up-and-coming kinetic theory of matter, bringing in two basic ideas that were quite alien to traditional chemistry, and also to both Berzelius's and Faraday's systems of electrochemistry. First, Clausius saw molecules as not fixed in places but bouncing around in random motion, at least in gases and liquids. Second, he understood the progress of chemical reactions, like all natural processes, as a matter of probabilities. Clausius did not focus on why atoms combined with each other, and he thought that the great speed at which molecules were seen to be traveling according to the kinetic theory made it plausible that they would have to be breaking up every so often. Perhaps too casually for most chemists' liking, he allowed that atoms could combine in any number whatsoever, and that what came stuck could also come spontaneously unstuck with a definite probability. Clausius's picture of water, therefore, was a dizzying, dancing soup of hydrogen and oxygen atoms, in which HO was the most probable species but by no means the only one. *All* possible combinations of H and O would exist in the mix (H, O, HO, H_2O, HO_2, H_2O_2, down to any H_nO_m), though the more complex molecules would have very low probabilities of forming or remaining. So Clausius thought that there would be some free atoms of hydrogen and oxygen already present in water, ready to be coaxed out. All this is fantastic, in retrospect, as it anticipated free ionic dissociation (which is why Arrhenius paid attention to Clausius in his historical retrospective), and that was Clausius's main contribution in breaking the hold of Berzelian electrostatic thinking. But at the time his ideas would have seemed fantastic in the other sense as well—too speculative, and not sufficiently tied to any real problems of chemistry.

2.2.3.3 Pluralism: Benefits of Toleration and Interaction

All in all, the theoretical situation in electrochemistry remained pluralistic throughout the century, for good reasons. The persistent theoretical disagreements did not result in any widespread positivistic renunciation of theory, although it did foster a healthy degree of skepticism and humility about hypotheses. The theoreticians kept theorizing, and kept trying to come to a better understanding of what was going on at the atomic–molecular level in electrochemical phenomena. Recall how Faraday plunged into his own electrochemical theorizing after surveying the various theories of electrolysis up to that point and noting his dissatisfaction with all of them, and how he continued in his endeavor even after he frankly admitted the dead-end he had reached. The usual tendency among twentieth-century scientists and many historians of science has been to skim over the years of disagreement, zooming into details again only at the origins of the modern ionic theory with the work of Arrhenius and van't Hoff.[37] I want to throw a different light on the pre-Arrhenius phase of electrochemistry, to show the merits in the pluralistic mode of working. In Chap. 5 I will make a general argument in favor of pluralism in science; my comments now will constitute some threads that I will pick up again and weave together there.

There were two main ways in which the flourishing of various electrochemical systems was beneficial: encouragement of different strengths, and productive interactions. First, by tolerating different systems, science could benefit from their different strengths. Since there wasn't any one system of electrochemistry that was strong in every way, multiple systems were needed if electrochemistry as a whole were to gain and retain sufficient empirical adequacy and explanatory power. Electrochemists in the nineteenth century recognized this fact, and organized their science in a suitably pluralistic way. Even though there were some dominant figures and strong personalities in the field, they advocated their own systems without the book-burning, name-calling destructive hostility of the kind displayed by Lavoisier and some of his associates, which was geared to annihilate the opposition instead of admitting that they could learn something from it. The only major exception in that regard was the suppression of Ritter's synthesis view, which is linked to the Lavoisierian legacy.

Let us take in just some key details of the benefits of plurality. First of all there were some important phenomena that could not be accommodated very well by all systems. For example, Volta's system (and also Davy's, more vaguely) was designed to cope well with electrification by contact, and the others were not. And while some electrolytic phenomena were clearly amenable to electrostatic explanations, Faraday discovered others that were not (such as the undiminished strength of action in the middle of the solution far away from the poles of the battery). One may think that it would have been better to have one unified theory which explained all the facts; in Chap. 5 I am actually going to argue against a general preference for unity, but for now there is a more immediate and pragmatic point. Given that an all-conquering unified theory was not forthcoming, it was better to have multiple theories so that every important phenomenon could receive *some* theoretical account, which not only provided a measure of understanding but also facilitated further research. If we do not allow such *provisional* role to theory, then theoretical science would have to be suspended until we reach the perfect theory and know that we have reached it.

---

[37] As usual, the exhaustive treatments by Ostwald [1896] (1980), Mottelay (1922) and Partington (1964) provide very useful exceptions. Another notable exception, though very brief, is Harold Hartley's discussion of "Faraday's successors and the theory of electrolytic dissociation" (Hartley 1971, ch. 7); Hartley was not worried about avoiding historiographical whiggism, but his perspective in 1931, when he composed that piece, was that his own current situation resembled that of the rich and uncertain field that Faraday faced, rather than the over-clarity of Arrhenius's work.

Different systems also provided different modes of explanation, which different people found enlightening to different degrees. For example, Volta's system would have been the most satisfying for those partial to the explanations of electrical phenomena on the basis of the electrical fluid, and Clausius's for those excited about the new thermodynamic and kinetic theories of heat and matter in mid-century. Berzelius's system would have appealed to those attracted by particle-based electrostatic thinking, while Faraday's catered for those who found field-based continuum thinking more satisfactory. Unless we are willing to ignore completely the inherent and heuristic values of intuitive explanations, the provision of explanations to suit different tastes is an important aim of science.

Empirical adequacy and intuitive explanation were not the only epistemic aims at play here. For example, unity or systematicity served as an important driver of theorizing, and that was also satisfied in different ways in different electrochemical systems. For example, as Russell (1963, 127) points out, a central concern in Berzelius's systematization of chemistry was classification, and his dualistic system began with classification as "its immediate purpose" though it later expanded its remit greatly into explanation (p. 134). Using the behavior of substances (or their oxides) in electrolysis as the key operational criterion, Berzelius classified all elements as oxygen, metals and "metalloids", all of them laid out on a single spectrum of electronegativity–electropositivity (pp.125, 131–133). For Faraday, the key aspect of systematicity in theorizing was to display the unity of all forces of nature. For Clausius, the chief point of unity was in the reduction of chemistry to physics. For Davy, as well as the unity of forces, it was important that there were only a small number of ultimate constituents of matter. These were all very different versions of systematicity/unity pursued in different systems, giving epistemic satisfaction to different people.

The second type of benefit coming from plurality consisted in productive interactions taking place between the different systems. Even disagreements had useful functions. For example, Faraday was able to use Davy's work both as a stimulus to spur his own work in disagreement, and as a basis on which to build his ideas (see Russell 1959, 12). The long-running debate between the contact and the chemical theories of the Voltaic battery stimulated the development of a great number of new experiments, as each side tried to refute the other (see Sect. 2.3.4 for further details). I find it difficult to assume that no similar benefits would have emerged if Ritter's theory of electrolysis had been maintained and allowed to compete in reasonable ways with the various theories built on the presumption of compound water. Even with Ritter ruled out, the fact of persisting disagreement among respectable scientists also kept men humble about the merits of their own theories, and open about the possibility of new ideas.

The simultaneous maintenance of multiple systems also created and maintained more conceptual possibilities for productive syntheses. Arrhenius's ushering-in of twentieth-century electrochemistry rested on a productive interplay between the three systems of Berzelius, Faraday, and Clausius: Berzelius provided charged ions, Faraday broke the hold of electrostatic reasoning, and Clausius provided the idea of spontaneous dissociation arising from kinetic factors. Had there been a monopoly by one of these systems in the years leading up to Arrhenius's work, his breakthrough would not have been possible, at least not in the form and not by the path that it took. Even if what one ultimately wants is one system that is comprehensively superior, it may only be possible to get there through a properly supported pluralistic phase of development; a premature enforcement of consensus would create obstacles in this process.

In summary, what can we say about the electrolysis of water in relation to the complex theoretical field in nineteenth-century electrochemistry? There are two major points to note. First, electrolysis did not provide any conclusive additional argument for the compound nature of water; rather, the bulk of work in electrochemistry beyond the first few years of the nineteenth century was partly *defined* by its exclusion of elementary water as postulated in the phlogiston theory and in Ritter's theory of electrolysis. Second, within the tradition of what I have called compound-water electrochemistry, there was a frank recognition of many unsolved problems and deep underlying theoretical uncertainty. As a result, electrochemistry developed in a pluralistic manner despite having ruled out Ritter's synthesis view. Without reaching a grand theoretical consensus, electrochemistry based its theoretical debates on a reasonably stable and expanding body of experimental work. Disputes on the mechanism of electrolysis, which continued throughout the nineteenth century, did not disturb the assumption of compound water, which retained the axiomatic status that was assigned to it in the early years.

2.3 In the Depths of Electrolytic Solutions

2.3.1 The Value of Studying Messy Science

Having read the second section of this chapter (Sect. 2.2), you may not yet be convinced that I ought to have focused on such a messy and uncertain stage of science, while neglecting more glorious and productive stages such as the arrival of the modern ionic theory. There are several reasons for my peculiar focus, aside from the simple fact that the developments that I was following did have a direct relevance to the debate concerning the constitution of water.

The first point is precisely that the stage of development I am focusing on *has* been neglected, with much more attention paid to the more unified and confident phases. In the historiography of nineteenth-century electrochemistry, a great deal of attention has been paid to five points only: (1) Volta's invention of the battery; (2) novel decompositions of substances by electrolysis (Nicholson/Carlisle on water, Davy on alkalis, etc.); (3) Berzelius's dualistic theory of chemical combination; (4) Faraday's work on the proportionality between the amounts of electrical and chemical effects; (5) Arrhenius's theory of ionic dissociation. What I have already shown in this chapter is that a great deal of electrochemistry happened between points (3) and (5), of which (4) is only a very small aspect. And even on the points that are heavily reported, many aspects have been neglected, for example the distance problem in the account of point (2). It has to be of *some* historiographical interest to pay more attention to the so-far neglected phases and aspects of scientific development. That is an obvious and relatively shallow point, but there are deeper points as well.

To begin with, I would dispute the unspoken assumption that the messier phases of nineteenth-century electrochemistry were not productive, including the long stretch between the height of Davy's achievements (around 1807) and Faraday's work on the electrochemical equivalent (early 1830s), and the even-longer 50-year stretch between Faraday and Arrhenius. As I have discussed briefly in Sect. 2.2.3, there were certainly useful developments coming out in these phases, even out of apparently futile struggles such as the attempt to solve the distance problem. It is important to understand these developments, even if all we care about is a true understanding of the later triumphs.

I also dispute the implicit assumption, especially common among philosophers of science but also present among some historians and sociologists, that it is more important to understand consensus-points than to remain focused on more pluralistic phases of science. (In Chap. 5 I will give a critique of an obsession with closure in our commentary on science.) Much of the life of science is spent in the messier phases, rather than in moments of consensus. In this connection it is interesting to take note of the ideas of Joseph J. Schwab (1962), who published *The Teaching of Science as Enquiry* in the same year as the first edition of Kuhn's *Structure of Scientific Revolutions*. There is a close parallel between Kuhn's distinction of normal/ extraordinary science and Schwab's distinction of stable/fluid inquiry, but Schwab's view was that with the continuing development of science more and more research was devoted to fluid inquiry.[38] Kuhnian extraordinary science and Schwabian fluid inquiry have a strong tendency to be pluralistic. So if we want to understand the nature of science fully, it is necessary to understand how such phases of science function. Science, on the whole, is pretty much always going to have pluralistic parts to it; even in Kuhnian normal science, the frontier of research must have Schwabian fluidity to some extent. It is not likely that there will ever be a point of time at which all fundamental debates in all fields of science reach a nice closure (with God mercifully taking us all at that moment so we don't have to live on and mess up the beautiful state of affairs).

Moreover, it is not even clear that pluralistic phases of sciences are confused and uncertain, and therefore inferior, to the more unified phases. The bright shining points of unification and consensus are quite likely to be epiphany moments, which are essential for the basic level of insight but not very useful for doing concrete work. They are points of *over*-simplification and *over*-confidence, after which scientists typically take a more realistic and seasoned attitude and go back to dealing with complications, exceptions, problems, wrinkles, hidden conceptual absurdities,

---

[38] Schwab's focus was on science education. The increasing prevalence of fluid inquiry in science means that it becomes increasingly necessary to train science students for it—in other words, to equip them for critical thinking; see Siegel (1990), 99–102, for further reflections on this point.

paradoxes, failed predictions, and puzzling new phenomena. Molecular genetics could only mature by moving beyond Watson and Crick's "central dogma" with its oversimplified notion of information-flow from DNA to RNA to proteins. Elementary particle physics could not rest with the pleasure of having only electrons, neutrons and protons to deal with. Copernican astronomy would not have got anywhere if it has stayed with Copernicus's own enchantment with uniform circular motions. When scientists get down-and-dirty mired in the complexities of nature, some degree of plurality will probably arise as they explore various ways of solving various difficult problems. If historians and philosophers of science do not give due recognition to these common, difficult and worthy phases of scientific development, who will? It would be wrong for us to follow the customs of history of science as found in science textbooks, which ignore the *post*-triumphal mess as well as the pre-triumphal mess.

2.3.2   Was Priestley Deluded? A View from the Laboratory

In the first section (Sect. 2.1) I discussed briefly Joseph Priestley's intervention in the electrochemical debates in the twilight of his life. This episode deserves some further attention, as an intriguing case of apparently absurd results reported by an able and reputable scientist. Many primary sources from the past of science are full of observational reports that sound very wrong from the modern point of view. Some of them may turn out to be valid observations, only forgotten or considered implausible because of our current biases or narrowness of focus. This is reminiscent of a well-known and controversial thesis in the philosophy of science, which says that the progress of science results in some *loss* of knowledge as well as obvious gains; for Kuhn (1970, ch. 9), this was an inevitable consequence of revolutionary change. In previous works I have argued that the recovery of such lost knowledge is one of the main tasks of history and philosophy of science (HPS) conceived as "complementary science" (Chang 1999; 2004, ch. 6). Cases like Priestley's electrochemistry provide an excellent opportunity for a recovery attempt, with the help of experiments as appropriate (see Chang 2011c on this function of historical experiments, and Chang 2007b for details of another case, of which I have given an extensive treatment).

There are two main strange items in Priestley's electrochemical reports. First, he claimed that the electrolysis of water took place only in the presence of dissolved oxygen. Second, he reported that the anode (made of various metallic wires), instead of serving as a site of oxygen-production, dissolved in the water and formed various compounds. Is there any plausibility to these reports?

We know that pure water is very difficult to electrolyze and usually we put a little bit of acid or salt to help the process along, as indicated in Fig. 2.2 above. Assuming that Priestley was using pure enough water (he does not specify the provenance of the water he was using), it is possible that his battery was not able to decompose water to generate oxygen and hydrogen from it, but able to coax out the oxygen that was dissolved in it. In order to test this suspicion, I made some preliminary experimental tests.[39] For ease of variation and monitoring, I used a modern electrical power supply capable of giving a direct current with a voltage anywhere from 0 to 60 V. I used de-ionized water (of very low conductivity, with resistivity typically around 1 MΩ per centimeter), and graphite electrodes to minimize any chemical reactions that might occur at the electrodes. Electrolysis in de-ionized water happens with great difficulty, and in my trials typically there was no visible bubbling from either electrode until the voltage was taken to around 50 V, which would be in the neighborhood of what Priestley had with his pile, if it was functioning well. I base that estimate on Priestley's description of his apparatus (1802, 198) "consisting of 60 plates of copper coated with silver, and as many thin rolled plates of zinc", which was supplied to him by "Mr. Weatherby Phipson, a young man of Birmingham".

On 29 September 2010, I inserted graphite electrodes into a tube containing de-ionized water, and attempted an electrolysis using a voltage-adjustable direct-current source. I monitored the amount of current passing through the water with an ammeter (a commercial multimeter), and measured the current as only 90 μA when the voltage applied was 25 V, with no visible formation of gas on either electrode. At 60 V the current increased to around 300 μA, and after a few minutes this began to produce visible bubbles sticking on the anode (+), but none on the cathode (−). Figure 2.11 shows a snapshot of this experiment 45 minutes after the voltage was set and left at 60 V. If a small amount of hydrochloric acid is added to the de-ionized water, the electrolytic behavior changes completely: this experiment, carried out on the same day, showed bubbles beginning to form and come off both electrodes at just 3 V, with more bubbling from the cathode than from the anode (see Fig. 2.12). At 5 V the current was 7.4 mA, or 7,400 μA—in other words, more than 20 times the level of current that was passing with 60 V applied to the de-ionized water.

The phenomena observed with the acidulated water are just as one would expect in a normal electrolysis of water: there should be hydrogen from the negative electrode and oxygen from the positive electrode, with twice as much hydrogen as oxygen. But with purer water one does not get that familiar result: with the small amount of gas generated in that experiment I was not able to make direct tests of its identity, but it seems reasonable to assume that it is oxygen rather than hydrogen, as it collects on the positive electrode (and it is difficult to imagine any other gases than oxygen and hydrogen arising in this setting). So it seems that in pure water there is a pre-threshold reaction in which the external voltage applied to the water only serves to disengage dissolved oxygen gas, without breaking down the water molecules into hydrogen and oxygen.[40] (There would also be hydrogen dissolved in

[39] I carried out these experiments in the electrochemistry lab of Daren Caruana at the Department of Chemistry at University College London. I would like to thank Dr. Caruana and his colleagues most sincerely for the use of the laboratory facilities and all the friendly advice they gave me. I also would like to thank Rosemary Coates, who assisted me most congenially and ably in these and other experiments, and the Leverhulme Trust, whose research grant provided much-needed funds and an authoritative seal of approval.

[40] Or is it possible that the application of electricity generates oxygen by decomposing CO_2, which will be found dissolved in the water in relative abundance?

Fig. 2.11 The result of an
application of electricity at
60 V to de-ionized water,
after 45 min.; the *black rods*
are graphite electrodes; the
green clip is connected to the
positive end of the power
supply, and the *white clip* to
the negative end; bubbles of
gas are only seen on the
anode (photograph by Hasok
Chang, 29 September 2010)

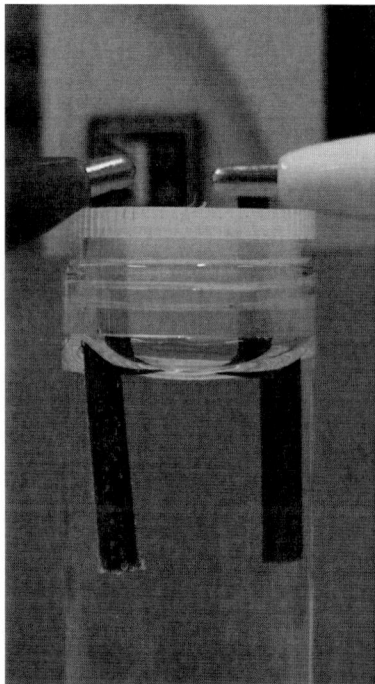

Fig. 2.12 The result of an
application of electricity at
3 V to water with a drop of
HCl added to it; bubbles of
gas are arising from both
electrodes, more of them
from the cathode (photograph
by Hasok Chang, 29
September 2010)

water, but it would only be a very small amount: the solubility of hydrogen is lower than that of oxygen and, more importantly, the atmosphere does not contain much hydrogen gas there to be dissolved.)

This is not quite a confirmation of the result that Priestley reported, but at least it is a vindication of the explanation that Priestley offered in order to make sense of his result. As for the result itself, I have not yet succeeded in observing the cessation of bubble-production on cutting off the contact between the water and the atmosphere, or exhibit the lack of action in sufficiently de-gassed water. I hope to carry out these experiments. Meanwhile, the rest of Priestley's observations are consistent with what I have seen so far. When he did get electrolysis going, Priestley got varying ratios of oxygen and hydrogen in different runs of the experiment, and also varying purities of the oxygen produced. That indicates to me that his samples of water had different amounts of dissolved air, and probably some other traces of impurities in them. But why would the application of electricity disengage dissolved oxygen from water? Someone of Berzelius's theoretical inclination would have easily agreed that oxygen, being electronegative, would be attracted to the positive pole of the battery and become disengaged there. More generally, we need to ask: how do dissolved gases in water interact with electricity passing through it? And that also raises a closely related question: what is the effect of dissolved gases in water or solutions that constitute the wet layers in the Voltaic pile? These are not typical questions that are treated in modern textbook electrochemistry.

On the last question, it is interesting to note Davy's report in his famous Bakerian Lecture of 1806 (Davy 1807, 46–47):

> Thus the Voltaic pile of 20 pairs of plates of copper and zinc exhibits no permanent electro-motive power when the connecting fluid is water free from air.... Concentrated sulphuric acid, which is a much more perfect conductor, is equally inefficient, for it has little action upon zinc, and is itself decomposed only by a very strong power.... Water containing loosely combined oxygene [sic] is more efficient than water containing common air, as it enables oxide of zinc to be formed more rapidly, and in larger quantities.... Diluted acids, which are themselves easily decomposed, or which assist the decomposition of water, are above all other substances powerful....[41]

These observations fitted well with Priestley's own phlogistonist explanation of the battery (Priestley 1802, 202): the operation of his pile (made of zinc and silver-plated copper) was dependent on the calcination (dephlogistication) of the zinc. The phlogiston leaving the zinc piece would need to go somewhere, and was taken up by the silver/copper, which thereby became "supersaturated" with phlogiston. This excess phlogiston flowed out, creating the electric current issuing from the battery; combining with water, it made phlogisticated water, in other words hydrogen.

Let us now turn to the second set of strange results reported by Priestley, namely the dissolution of the metallic wires used as the anode in the electrolysis of water, coupled with the production of hydrogen from the cathode. The "surest method of

---

[41] Davy notes that he had performed these experiments several years earlier and published the results in Nicholson's Journal in 1800 (volume 4); see also the summary in Donovan (1816), 43. Priestley would have read these papers.

Fig. 2.13 The dissolution
of copper anode in the
electrolysis of tap water, at
50 V (photograph by Hasok
Chang, 17 September 2010)

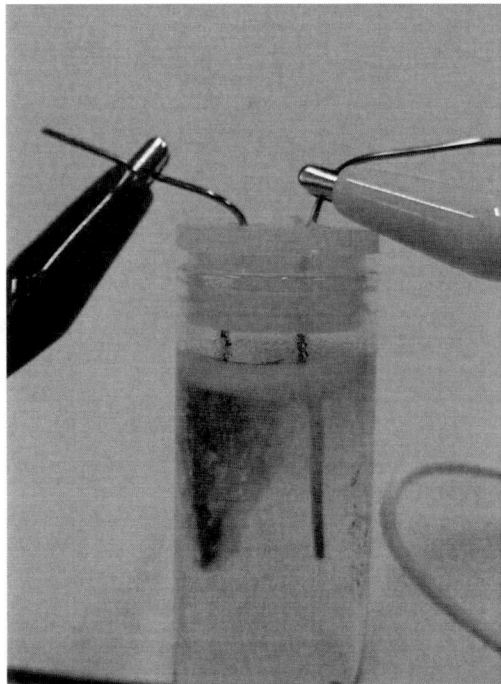

producing this solution of metal" was by using charcoal as the material for the
cathode. Priestley reported (p. 200): "I once dissolved pure gold in this manner, and
I preserve the solution as an evidence of it; but I could never do it a second time . . .
nor could I by this process dissolve platina." These results may be fairly easy to
explain in modern terms, if we assume that some of his water samples contained
impurities that would have combined with the metals to form soluble salts, particularly any chlorides. Using untreated tap water in London, I easily succeeded in
dissolving copper wires used as positive electrodes connected to 10 V, with the
production of a blue-green gunk. In trials I made on 17 September 2010, I detected
the solution of the copper anode with voltage as low as 4 V (with 3.3 mA of current
at that voltage). The amount of current increased steadily with increased voltage,
reaching 10 mA at 10 V and around 90 mA at 60 V, with correspondingly more
impressive destruction of the copper anode (see Figs. 2.13 and 2.14). As one might
expect, I failed to dissolve gold with the same arrangement, but this is not incon-
ceivable if Priestley would have had chloride impurities present in his water. In
some experiments carried out in a different context, I have easily dissolved gold
anodes in the electrolysis of saturated solutions of NaCl (common salt), using just
two ordinary batteries in series (3 V) (see Fig. 2.15). According to some further
tests that I have carried out in September 2011,[42] the dissolution of gold anode

---

[42] For hosting these experiments, I thank the Department of Chemistry at the University of
Cambridge, and Dr. Peter Wothers, Mr. Chris Brackstone, and Mr. Gary Herrington.

Fig. 2.14 The dissolution of
copper anode at 60 V
(photograph by Hasok
Chang, 17 September 2010)

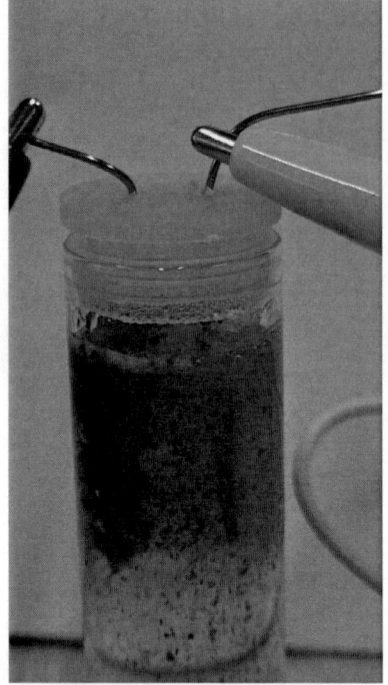

Fig. 2.15 The dissolution of
gold anode (*right-hand wire*)
at 3 V in NaCl solution;
hydrogen bubbles form on
the cathode (*left-hand wire*,
of copper), creating the fuzzy
appearance of the wire and in
the top part of the solution
(photograph by Hasok
Chang, 17 November 2009)

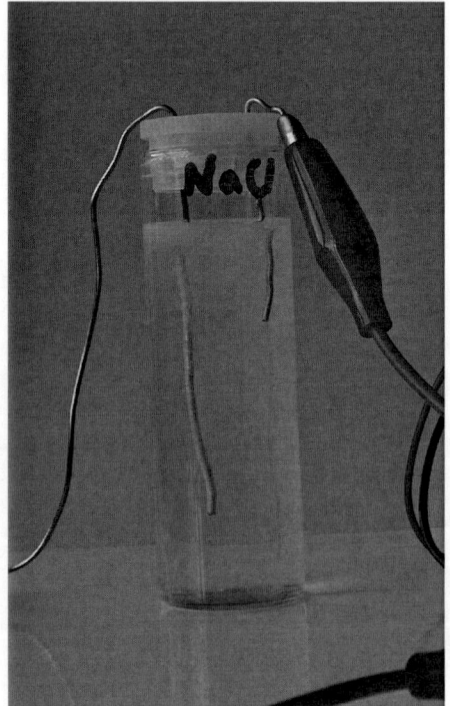

(with graphite cathode) can take place with NaCl concentration as low as 1/25 of saturation. But it happens only in a small window of voltages (roughly 2.2–3.0 V, depending on the concentration); beyond 3.0 V, the gold anode remains intact but produces chlorine gas. I plan to make further investigations into this phenomenon.

In fact it was not just Priestley who reported a battery-assisted dissolution of gold. Cruickshank had already reported (1800b, 256) that by electrolyzing a solution of the muriate of lime (calcium chloride) he obtained "a perfect solution of gold". And Donovan (1816, 83–84) recounted an intriguing experiment by Francesco Giuseppe Pacchiani of Pisa (see also Mottelay 1922, 392): "when water is decomposed by means of gold conducting wires, oxygen is continually emitted, and the water becomes acid: a smell of oxymuriatic acid [chlorine] is rendered perceptible; the gold corrodes, and an orange solution is obtained which tinges the skin of a rose colour." From this experiment reported in 1805, he inferred that "muriatic [hydrochloric] acid was produced by the abstraction of oxygen from water". Donovan hastened to add (p. 85): "subsequent researches proved that substances contained in the water, in the vessels, or in the connecting media, had been the real origin of these appearances." It does not seem unlikely to me that a chloride impurity present in the water would have been able to produce both gold chloride and hydrochloric acid.

All in all, both my tinkering in the laboratory and other textual evidence support the conclusion that at least most of Priestley's reports are pointing to some real phenomena—"real" enough to be exhibited across 200 years of time and very different material conditions. They also open up some interesting scientific questions worthy of modern consideration.

2.3.3   The Intricacies of Ion-Transport

In the first and the second sections of this chapter (Sects. 2.1 and 2.2) I presented, as and when needed, rather simplified views of the history of disputes concerning the mechanism of ion-transfer in electrolysis. I wish to make amends as much as I can. Here I will give a more synoptic view of the questions that were widely debated in the nineteenth century, and also try to point out some developments that go against the grain of my earlier narrative, though they do not contradict it. Modern historical treatments of this subject are rare: Olivier Darrigol, who provides a judicious overview of the mid-nineteenth century situation, remarks that "there is unfortunately no authoritative history of electrochemistry."[43] However, a good deal of further relevant material can be found in some of the old secondary literature, not to mention the primary literature (see especially Wilkinson 1804; Singer 1814; Donovan 1816; Ostwald [1896] 1980; Partington 1964).

(a) What is an ion? All those who agreed that electrolysis was decomposition were also agreed that some parts of the decomposed molecules travelled to the

---

[43] Darrigol (2000), 266–274, quotation on p. 266, footnote 1.

electrodes immersed in liquids. As noted earlier, Faraday's new terminology ("ion", "electrode", etc.) was explicitly designed to express just these agreed-upon facts without implying any further theoretical interpretations. Beyond those facts, as described in Sect. 2.2.3.2, there were rife theoretical disagreements about the nature of ions. Most fundamentally, Davy and Berzelius thought that ions were electrostatically charged particles, which Faraday denied. Since there was no agreed understanding of the *process* by which molecules got broken up into ions, it was also difficult to know the nature of the *product*.

(b) What is the chemical affinity of ions while they travel? This question was raised especially by Ritter's experiments in which the alleged products of electrolysis would have had to pass through substances to which they had a strong chemical affinity (see Sect. 2.2.2). As noted already, Singer (1814, 349–350), Ritter's results only meant that the galvanic agency was "so powerful" as to enable this unlikely transfer. But how would that enabling work? On the Berzelian view in which electrically-charged atomic particles move around by electrostatic attraction, they would be captured by other particles of opposite charge in the midst of the solution; the force of attraction or repulsion from the distant poles of the battery would be negligible, as electrostatic force is inversely proportional to the square of the distance. This kind of concern drove many people away from the electrostatic basis of Berzelius's system. Faraday wrote in his notebook in September 1832: "The effects of the decomposition would seem rather to depend upon a relief of the chemical affinity in one direction and an exaltation of it on the other rather than to direct attraction and repulsion of the poles." (quoted in Hartley 1971, 161). This is the kind of view he continued to hold to the end, but it never became articulated clearly enough for others to adopt and develop (see Faraday [1859] 1993; Sinclair 2009).

(c) What exactly drives the ions? For reasons just stated, it would have been difficult to think that the movement of ions was simply governed by the electrostatic attraction and repulsion from the poles. Grotthuss and Davy, while agreeing that the basic force involved was electrostatic, thought that the ionic movement was broken up into short-range movements dependent on the attraction and repulsion between parts of neighboring molecules. The Grotthuss mechanism has survived into modern physical chemistry, though only as a mechanism of proton-transfer rather than a general explanation of electrolysis. Meanwhile, those who followed the "invisible transport" view rather thought that ions moved only by virtue of being grabbed by the electrical fluid, which had its own reasons for rushing around in the electrochemical circuit. Dumas in 1837 declared that "the first sane view of the decomposition of water by the pile" was Fourcroy's, a version of invisible transport. And Auguste De la Rive had offered a new version of the same kind of theory as late as 1825, postulating a double current (Partington 1964, 23, 28). So the "invisible transport" theory was not killed off so easily as I might have seemed to imply earlier.

(d) How fast do ions move? This question does not seem to have been addressed seriously until the 1850s, when Wilhelm Hittorf and Friedrich Kohlrausch took it up (see Partington 1964, 665–672 for a concise summary). Hartley (1971, 174) notes that Hittorf's work began with an attempt to explain the changes in the concentration of ions in the vicinity of the electrodes. But this means that the imbalance view made a quiet return, though it seemed to have disappeared quickly in the early decades of the nineteenth century. To be more precise, the imbalance of concentrations within the electrolyte undergoing electrolysis was rejected as a satisfactory solution to the distance problem, but it slowly became established as an experimental fact, itself requiring an explanation and posing an interesting problem. Hittorf subscribed to the double-chain version of the Grotthuss mechanism, and he concluded that different chains moved at different speeds as they slid by each other. Building on Hittorf's work, Kohlrausch improved the measurements, paid special attention to dilute solutions, and generally helped establish the independent mobility of different species of ions.

(e) How do ions become converted to electrically neutral matter at the electrodes? For those who believed that ions were ionic only by virtue of an unnatural electrical state they assumed, it made sense that they would take the normal form again once their regained electrical neutrality by gaining or losing the requisite amount of electricity at the poles. But for those who believed that the electric charge was inherent to atoms (e.g. Berzelius) or that ions were not electrostatically charged (e.g. Faraday), a more elaborate story was required.

2.3.4   Disputes on How the Battery Works

One vast subject of which I could really only scratch the surface so far is the working of the Voltaic battery itself. Of course this is related to the mechanism of electrolysis, being just the converse of it in an important sense as Davy recognized early on. But there is much more to it, so much so that I have begun working on an entire book on this subject. Numerous intriguing questions present themselves: what exactly is the nature of the interaction between the two metals in the Voltaic battery, and between the electrolyte and each of the metals? How do the mechanisms of electricity-production differ, depending on whether the electrolyte is water, acid, salt, etc.?

What is the standard modern explanation of Volta's cell? Surprisingly, there isn't one readily available. What we get almost everywhere we turn is an explanation of the *Daniell cell*, in which the electrolyte consists of two different solutions, connected by a salt bridge or a porous barrier (e.g., Housecroft and Constable 2010, 638; Gilbert et al. 2009, 894–895; R. Chang 2010, 841). In this setup, each metal is dipped in its own solution, and the electrical activity is conveniently explained in terms of the imbalance of the redox potentials on the two sides. But the action of Volta's original cell, which has only one electrolyte, containing no ions of either metal to begin with, cannot be explained in this way. Consequently Volta's original cell has disappeared from basic electrochemical thinking; so has Volta's theory, which attributed the electrical action to the contact between two different metals, not to chemical reactions. Volta's notion of contact action survives in the form of the physicist's contact potential (linked to the work function of each metal), but this is not part of the standard chemical discourse today; in my admittedly limited survey, I have only seen one chemistry textbook in which the contact potential is mentioned (Levine 2002, 413), and even in that case it is not actually employed in giving an explanation of electrochemical cells.

Helge Kragh (2000) gives an insightful overview of the long and complex debate that raged throughout the nineteenth century between those who believed (following Volta) that the electrical action was caused by the contact between two different metals, and those who believed that the electricity was produced by chemical reactions. The relation between the contact and chemical theories of the Voltaic cell make a tidy case of incommensurability, as Kuhn himself points out (2000, 21–24).[44] And there were many, such as Davy, who attempted to make some sort of compromise between these two theories. Kragh concludes that the dispute was never really resolved; rather, it lost its urgency and fizzled out coming into the twentieth century. Today in chemistry textbooks the standard story is given in terms of chemical redox potentials, but physicists happily continue to talk about contact potentials between different metals; meanwhile, battery technology has moved on steadily, seemingly without much help from fundamental theory. The vexing questions driving the nineteenth-century debates and many of the various experiments invoked by the opposing camps of scientists are now mostly forgotten, and they certainly do not feature in standard textbooks of chemistry. Even among professional historians of science, the details of nineteenth-century electrochemical debates are no longer common knowledge. The most thorough treatments of this history are still to be found in the older secondary literature, such as the classic treatises by J. R. Partington (1964) and Wilhelm Ostwald [1895] (1980). A happy exception to the current dearth of interest is the set of papers published in the *Nuova Voltiana* volumes (Bevilacqua and Fregonese 2000–2003), especially those by Kragh (2000) and Nahum Kipnis (2001).

Let me briefly highlight some of the most interesting experiments and arguments offered from each side of the dispute. By the middle of the century it would have seemed that the contact theory was no longer viable, especially as Faraday established the quantitative identity of electrical and chemical action arising from the battery. Early on Davy (1807, 33) had confounded Volta by making a cell using just one metal, or even no metal at all but a piece of charcoal and two different liquids. Volta himself, fascinated by the thought that his "pile" was a realistic model of the torpedo (electric fish), made a battery using pieces of bone instead of metal (Pancaldi 2003, 205). In the cells not involving contact between two different metals, it seems clear (in modern terms) that the net flow of electrons is caused by the different rates of chemical electron-generation on the two sides, which creates an imbalance.

On the side of the contact theory, Sungook Hong (1994) gives a detailed account of one curious phase of this history, in which Kelvin revived Volta's contact theory

---

[44] Yet it seems evident that Kuhn had not taken an in-depth look, as he says that "both viewpoints were *briefly* in the field at once" (p. 23, emphasis added).

in the 1860s. Earlier, De Luc and others made various "dry piles" that used dry layers (paper, etc.) instead of electrolytes, which were subjects of great experimental and theoretical debate (Hackmann 2001; Ostwald [1895] 1980, 346–353/359–366; Partington 1964, 16–17). There is a long-surviving example of a dry pile in the Clarendon Laboratory at Oxford which, as of 1984, had been ringing a bell nearly continuously for 144 years! A. J. Croft (1984), who reports on this remarkable instrument, says that "what the piles are made of is not known with certainty", but that "a considerable number" of dry piles inspired by this instrument were made for military purposes during the Second World War by the Oxford physicist A. Elliott.[45] After much debate back in the nineteenth century an agreement was reached that the operation of the dry pile did rely on the presence of moisture in the air; however, there was never a conclusive agreement on whether the role of moisture was to make the dry layers conducting or to generate electricity by facilitating chemical reactions.

2.3.5   Ritter and Romanticism

I have invoked the general rejection of *Naturphilosophie* as one of the important factors leading to the rejection of Ritter's synthetic view of electrolysis. Ostwald lamented ([1895] 1980, 67/68): "J. W. Ritter was ruined by the natural philosophy of those times [*Naturphilosophie*]." To reinforce his own sense that *Naturphilosophie* was an utterly misleading and unproductive way of thinking, he quoted Liebig's recollection of his own life in the Romantic period: "Alas! I lived through this period so rich in words and ideas but so poor in real knowledge and solid study that it cost me two valuable years of my life; I cannot describe the horror and fright as I was awakened from this frenzy to consciousness." But a more careful discussion is needed, in order to get a full and accurate sense of Ritter's relation with the tradition of romanticist natural philosophy, and the linkage between the rejection of Ritterian electrochemistry and the general rejection of romanticism in science. The discussion here will have to be sadly sketchy and ill-informed, as this is a subject that goes well beyond the scope of this book and the limits of my own expertise, and in fact I have been able to find surprisingly little in the extant Anglophone literature that is directly relevant to my questions here.

Of Ritter's affinity with romanticism and his close personal relationship with the leading romanticist thinkers, there is little doubt, as Walter Wetzels (1990) explains. Here I will just summarize the ways in which Ritter's electrochemical science showed the influence of romanticism. The basic structure of electricity seemed to be a perfect instantiation of nature's polarity that is yet unified, so it is not a surprise that

---

[45] Partington (1964, 17) also mentions that the Oxford dry pile had continued to work for more than a century.

electricity (and similarly magnetism) was one of the romantics' favorite scientific subjects. And the study of galvanism seemed marvellous, since it showed the same forces of nature running through living beings as well as inert materials. Some experimental success in this line of work emboldened Ritter to ask: "Where then is the difference between the parts of an animal, of a plant, of a metal, and of a stone?— Are they not all members of the *cosmic-animal*, of *Nature*?" (quoted in Wetzels 1990, 203) Ritter's style of experimentation also reflected the romanticist notion that true scientific knowledge was gained by a direct intuitive communion with nature; Ritter added a bodily dimension to this work, engaging in an endless series of self-experimentation. If shocking himself with the pile was a mere expedient for Volta necessitated by the lack of suitable measuring instruments, for Ritter it was a heroic act of a romantic genius.

It may seem obvious, and it is too big a story to tell here, why modern science has rejected romanticism on the whole. But in fact it is not so straightforward to invoke the taint of romanticism as the reason for the rejection of a theoretical idea or the ostracization of a scientist. Some major scientific figures in Ritter's period and later, including Humboldt, Davy, Ørsted and Faraday won acclaim for their work even though they were visibly associated with romanticism. It has even been argued that romanticism served directly and positively as an inspiration for their scientific work, or at least that "the ideology and institutions of the new natural science owed much to Romanticism and Naturphilosophie." (Cunningham and Jardine 1990, 8)

The difficulty of using romanticist association as an explanatory factor in the rejection of a given idea is exacerbated by the fact that an idea may be separated out from its broader associations as it develops and travels between different contexts. For example, the polarity of electricity held great attraction for Ritter for its romanticist associations, but in Berzelius's hands it became something quite mundane and prosaic. The grand unity of nature was a key romanticist conception and motivated the work of Ritter, Ørsted and Faraday in the romanticist vein as they all sought the mirage of the grand unity of all forces of nature, and miraculously found it sometimes. Yet the unity of nature was also an idea that predated the Romantic period, and persisted long after it, right down to today's seekers of superstrings, the Higgs boson, and quantum gravity.

Even at the level of scientific methodology, the associations can be so fluid as to be analytically useless. For Ritter, empiricism was an integral part of romanticism, since sensory experience was on a par with intuitions of genius as avenues of direct communion with nature. One can sympathize with Ostwald to an extent as he finds contrary inclinations in one and the same work by Ritter, showing both sober scientific reason and irresponsible flights of fancy. But Ostwald merely projected the associations familiar to himself on to Ritter's remarks, and the result was utter incomprehension. Commenting on Ritter's paper of 1798 on galvanism, Ostwald found it "psychologically impossible" that "the same head" could have generated both the "masterly presentation" of "a plethora of experimental investigations" exhibiting "remarkable power and boldness in thinking", and the "last series of conclusions, which shorn of its flowery presentation stands there in its naked senselessness." (Ostwald [1895] 1980, 66/67–70/71)

Having learned to be cautious about sweeping explanations involving the rejection of romanticist science, let us return to the question about the reasons for the rejection of Ritter's ideas. I think Ritter's failure to be accepted by the communities of chemists and physicists had more to do with style and strategy of communication rather than the substance of his work, just as much as Lavoisier's success owed much to his vigorous and effective campaigning. It does not even seem that Ritter's work was strongly ruled by a desire to be accepted by the scientific communities. As factors hindering the scientific acceptance of Ritter's work, I note the following. (1) His style was profuse and diffuse, and not easily accessible to those who did not invest a great deal of time and attention. (2) He was not strategic in placing his publications in high-profile outlets. (3) He made no efforts to downplay certain thoughts and results, such as his work on water-divination, that were likely to put off the audiences he was addressing. (4) Placing himself against the solidifying Lavoisierian orthodoxy reduced his chances of being accepted in the German chemical community. There may have been other reasons, too. What seems certain, however, is that the rejection of Ritter's electrochemical theory was idiosyncratic and contingent, rather than principled and inevitable.

References

Arrhenius, Svante. 1902. *Text-book of electrochemistry* (trans: McCrae, J.). London: Longmans, Green, and Co.

Berzelius, Jöns Jakob. 1811. Essai sur la nomenclature chimique. *Journal de Physique* 73: 253–286.

Berzelius, Jöns Jakob, and Wilhelm Hisinger. 1803. Versuch, betreffend die Wirkung der elektrischen Säule auf Salze und auf einige von ihren Basen. *(Gehlen's) Neuen allgemeinen Journal der Chemie* 1: 115–149.

Bevilacqua, Fabio and Lucio Fregonese, eds. 2000–2003. *Nuova Voltiana: Studies on Volta and his times*, 5 vols. Milan: Hoepli.

Brock, William H. 1992. *The Fontana history of chemistry*. London: Fontana Press.

Brooke, John Hedley. 1980. Davy's chemical outlook: The acid test. In *Science and the sons of genius: Studies on Humphry Davy*, eds. Sophie Forgan, 121–175. London: Science Reviews Ltd.

Brown, Sanborn C. 1950. The caloric theory of heat. *American Journal of Physics* 18: 367–373.

Brown, Sanborn C. 1979. *Benjamin Thompson, Count Rumford*. Cambridge, MA: The MIT Press.

Cajori, Florian. 1929. *A history of physics*. New York: Macmillan.

Chang, Hasok. 1999. History and philosophy of science as a continuation of science by other means. *Science and Education* 8: 413–425.

Chang, Hasok. 2004. *Inventing temperature: Measurement and scientific progress*. New York: Oxford University Press.

Chang, Hasok. 2007a. Scientific progress: Beyond foundationalism and coherentism. In *Philosophy of science (Royal Institute of Philosophy Supplement 61)*, eds. Anthony O'Hear, 1–20. Cambridge: Cambridge University Press.

Chang, Hasok. 2007b. The myth of the boiling point. http://www.cam.ac.uk/hps/chang/boiling. First posted on 18 Oct 2007.

Chang, Raymond. 2010. *Chemistry*. Boston: McGraw-Hill.

Chang, Hasok. 2011c. How historical experiments can improve scientific knowledge and science education: The cases of boiling water and electrochemistry. *Science and Education* 20: 317–341.

Chang, Hasok. 2011d. Compositionism as a dominant way of knowing in modern chemistry. *History of Science* 49: 247–268.

Christensen, Dan Ch. 1995. The Ørsted–Ritter partnership and the birth of romantic natural philosophy. *Annals of Science* 52: 153–185.

Court, S. 1972. The *Annales de chimie* 1789–1815. *Ambix* 19: 113–128.

Coutts, A. 1959. William Cruickshank of Woolwich. *Annals of Science* 15: 121–133.

Croft, A.J. 1984. The oxford electric bell. *European Journal of Physics* 5: 193–194.

Crosland, Maurice. 1978. *Gay-Lussac: Scientist and bourgeois.* Cambridge: Cambridge University Press.

Crosland, Maurice. 1980. Davy and Gay-Lussac: Competition and contrast. In *Science and the sons of genius: Studies on Humphry Davy*, ed. Sophie Forgan, 95–120. London: Science Reviews Ltd.

Cruickshank, William. 1800a. Some experiments and observations on galvanic electricity. *(Nicholson's) Journal of Chemistry, Natural Philosophy, and the Arts* 4: 187–191.

Cruickshank, William. 1800b. Additional remarks on galvanic electricity. *(Nicholson's) Journal of Chemistry, Natural Philosophy, and the Arts* 4: 254–264.

Cunningham, Andrew, and Nicholas Jardine , eds. 1990. *Romanticism and the sciences.* Cambridge: Cambridge University Press.

Darrigol, Olivier. 2000. *Electrodynamics from Ampère to Einstein.* Oxford: Oxford University Press.

Davy, Humphry. 1800a. Account of some experiments made with the galvanic apparatus of Signor Volta. *(Nicholson's) Journal of Chemistry, Natural Philosophy, and the Arts* 4: 275–281.

Davy, Humphry. 1800b. An account of some additional experiments and observations on the galvanic phenomena. *(Nicholson's) Journal of Chemistry, Natural Philosophy, and the Arts* 4: 394–402.

Davy, Humphry. 1807. The Bakerian Lecture [for 1806]: On some chemical agencies of electricity. *Philosophical Transactions of the Royal Society* 97: 1–56.

Davy, Humphry. 1808a. The Bakerian Lecture [for 1807]: On some new phenomena of chemical changes produced by electricity, particularly the decomposition of the fixed alkalies, and the exhibition of the new substances which constitute their bases; and on the general nature of alkaline bodies. *Philosophical Transactions of the Royal Society* 98: 1–44.

Davy, Humphry. 1808b. Electro-chemical researches, on the decomposition of the earths; with observations on the metals obtained from the alkaline earths, and on the amalgam procured from ammonia. *Philosophical Transactions of the Royal Society* 98: 333–370.

Davy, Humphry. 1809. The Bakerian Lecture [for 1808]: An account of some new analytical researches on the nature of certain bodies, particularly the alkalies, phosphorus, sulphur, carbonaceous matter, and the acids hitherto undecompounded; with some general observations on chemical theory. *Philosophical Transactions of the Royal Society* 99: 39–104.

Davy, Humphry. 1810. The Bakerian Lecture for 1809: On some new electrochemical researches, on various objects, particularly the metallic bodies, from the alkalies, and earths, and on some combinations of hydrogene [sic]. *Philosophical Transactions of the Royal Society* 100: 16–74.

Davy, Humphry. 1812. *Elements of chemical philosophy.* London: J. Johnson and Co.

Donovan, Michael. 1816. *Essay on the origin, progress and present state of galvanism, etc.* Dublin: Hodges and McArthur.

Faraday, Michael. 1833. Experimental researches in electricity, fifth series. *Philosophical Transactions of the Royal Society* 123: 675–710.

Faraday, Michael. 1834. Experimental researches in electricity, seventh series. *Philosophical Transactions of the Royal Society* 124: 77–122.

Faraday, Michael. 1844. *Experimental researches in electricity, 2 vols, reprinted from the Philosophical Transactions of 1838–1843, with other electrical papers from the Quarterly Journal of Science and Philosophical Magazine.* London: Richard and John Edward Taylor.

Faraday, Michael. 1993. *The forces of matter (Royal Institution Christmas Lectures).* Amherst, NY: Prometheus Books.

Gilbert, T.R., R.V. Kirss, N. Foster, and G. Davies. 2009. *Chemistry: The science in context*. New York: Norton.

Golinski, Jan. 1992. *Science as public culture: Chemistry and enlightenment in Britain 1760–1820*. Cambridge: Cambridge University Press.

Gorbunova, K.M., L.J. Antropov, Yu.I. Solov'ev, and J.P. Stradins. 1978. Early electrochemistry in the USSR. In *Proceedings of the Symposium on Selected Topics in the History of Electrochemistry*, eds. George Dubpernell and J.H. Westbrook, 226–256. Princeton: The Electrochemical Society.

Gray, Harry B., and Gilbert P. Haight Jr. 1967. *Basic principles of chemistry*. New York: W. A. Benjamin.

Gray, Tamsin, Rosemary Coates, and Mårten Åkesson. 2007. The elementary nature of chlorine. In *An element of controversy: The life of chlorine in science, medicine, technology and war*, eds. Hasok Chang and Catherine Jackson, 41–72. London: British Society for the History of Science.

Grotthuss, Christian Johann Dietrich (Theodor). 1806. Memoir upon the decomposition of water, and of the bodies which it holds in solution, by means of galvanic electricity. *Philosophical Magazine* 25: 330–339.

Grotthuss, Christian Johann Dietrich (Theodor). 1810. On the influence of galvanic electricity in metallic arborizations. *(Nicholson's) Journal of Chemistry, Natural Philosophy, and the Arts* 28: 112–125.

Hackmann, Willem. 2001. The enigma of Volta's "contact tension" and the development of the "Dry pile". In *Nuova Voltiana: Studies on Volta and his times*, vol. 3, eds. F. Bevilacqua and L. Fregonese, 103–119. Milan: Hoepli.

Hartley, Harold. 1971. *Studies in the history of chemistry*. Oxford: Clarendon Press.

Haüy, René-Just. 1806. *Traité élémentaire de physique*. Paris: Courcier.

Henry, William. 1813. On the theories of the excitement of galvanic electricity. *(Nicholson's) Journal of Chemistry, Natural Philosophy, and the Arts* 35: 259–271.

Hong, Sungook. 1994. Controversy over Voltaic contact phenomena. *Archive for History of Exact Sciences* 47: 233–289.

Housecroft, C.E., and E.C. Constable. 2010. *Chemistry: An introduction to organic, inorganic and physical chemistry*. Harlow: Pearson.

Hufbauer, Karl. 1982. *The formation of the German chemical community (1720–1795)*. Berkeley/ Los Angeles: University of California Press.

James, Frank A.J.L. 1989. Michael Faraday's first law of electrochemistry – How context develops new knowledge. In *Electrochemistry, past and present*, eds. John T. Stock and Mary Virginia Orna, 32–49. Washington, DC: American Chemical Society.

Kipnis, Nahum. 2001. Debating the nature of Voltaic electricity. In *Nuova Voltiana: Studies on Volta and his times*, vol. 3, eds. F. Bevilacqua and L. Fregonese, 121–151. Milan: Hoepli.

Knight, David. 1967. *Atoms and elements*. London: Hutchinson.

Knight, David. 1978. *The transcendental part of chemistry*. Folkestone: Dawson.

Kragh, Helge. 2000. Confusion and controversy: Nineteenth century theories of the Voltaic pile. In *Nuova Voltiana: Studies on Volta and his times*, vol. 1, eds. F. Bevilacqua and L. Fregonese, 133–157. Milan: Hoepli.

Kuhn, Thomas S. 1970. *The structure of scientific revolutions*, 2nd ed. Chicago: University of Chicago Press.

Kuhn, Thomas S. 2000. *The road since Structure: Philosophical essays, 1970–1993, with an autobiographical interview*. Chicago: University of Chicago Press.

Levine, I.N. 2002. *Physical chemistry*. Boston: McGraw-Hill.

Lilley, Samuel. 1948. 'Nicholson's Journal' (1797–1813). *Annals of Science* 6: 78–101.

Lowry, T.M. 1936. *Historical introduction to chemistry*, revised ed. London: Macmillan.

Lund, Matthew. 2010. *N. R. Hanson: Observation, discovery, and scientific change*. Amherst, NY: Prometheus Books.

Melhado, Evan M. 1980. *Jacob Berzelius: The emergence of his chemical system*. Stockholm: Almqvist & Wiksell International.

Mottelay, Paul Fleury. 1922. *Bibliographical history of electricity and magnetism: Chronologically arranged*. London: Charles Griffin & Company Limited.

Nicholson, William. 1800. Account of the new electrical or galvanic apparatus of Sig. Alessandro Volta, and experiments performed with the same. *(Nicholson's) Journal of Chemistry, Natural Philosophy, and the Arts* 4: 179–187.

Ostwald, Wilhelm. 1980. *Electrochemistry: History and theory*, 2 vols (trans: Date, N. P.). New Delhi/Bombay/Calcutta/New York: Amerind Publishing Co. Pvt. Lt.

Pancaldi, Giuliano. 2003. *Volta: Science and culture in the age of enlightenment*. Princeton: Princeton University Press.

Partington, J.R. 1964. *A history of chemistry*, vol. 4. London: Macmillan.

Pauling, Linus, and Peter Pauling. 1975. *Chemistry*. San Francisco: W. H. Freeman and Company.

Priestley, Joseph. 1788. Experiments and observations relating to the principle of acidity, the composition of water, and phlogiston. *Philosophical Transactions of the Royal Society* 78: 147–157.

Priestley, Joseph. [1796] 1969. *Considerations on the doctrine of phlogiston, and the decomposition of water* (and two lectures on combustion, etc. by John MacLean). New York: Kraus Reprint Co.

Priestley, Joseph. 1802. Observations and experiments relating to the pile of Volta. *(Nicholson's) Journal of Chemistry, Natural Philosophy, and the Arts, new series* 1: 198–204.

Rumford [Benjamin Thompson, Count]. 1799. An inquiry concerning the weight ascribed to heat. *Philosophical Transactions of the Royal Society* 89: 179–194.

Russell, Colin A. 1959. The electrochemical theory of Sir Humphry Davy (in 2 parts). *Annals of Science* 15: 1–25.

Russell, Colin A. 1963. The electrochemical theory of Berzelius (in 2 parts). *Annals of Science* 19: 117–145.

Schofield, Robert E. 2004. *The enlightened Joseph Priestley: A study of his life and work from 1773 to 1804*. University Park: Pennsylvania State University Press.

Schwab, J.J. 1962. *The teaching of science as enquiry*. Cambridge, MA: Harvard University Press.

Siegel, Harvey. 1990. *Educating reason: Rationality, critical thinking and education*. New York: Routledge.

Sinclair, Alexandra. 2009. *Beyond the law(s): Michael Faraday's experimental researches, series 8*. London: University College London.

Siegfried, Robert. 1964. The phlogistic conjectures of Humphry Davy. *Chymia* 9: 117–124.

Singer, George John. 1814. *Elements of electricity and electro-chemistry*. London: Longman, Hurst, Rees, Orme, and Brown; R. Triphook.

Snelders, H.A.M. 1979. The Amsterdam experiment on the analysis and synthesis of water 1789. *Ambix* 26: 116–133.

Snelders, H.A.M. 1988. The new chemistry in the Netherlands. *Osiris (2nd Series)* 4: 121–145.

Stevenson, W.F. 1849. *The composition of hydrogen, and the non-decomposition of water incontrovertibly established*. London: James Ridgway.

Thornton, John L. 1967. Charles Hunnings Wilkinson (1763 or 64–1850). *Annals of Science* 23: 277–286.

Toulmin, Stephen. 1970. Does the distinction between normal and revolutionary science hold water? In *Criticism and the growth of knowledge*, eds. Imre Lakatos and Alan Musgrave, 39–47. Cambridge: Cambridge University Press.

Volta, Alessandro. 1800. On the electricity excited by the mere contact of conducting substances of different kinds. In a letter from Mr. Alexander Volta, F.R.S. Professor of Natural Philosophy in the University of Pavia, to the Rt. Hon. Sir Joseph Banks, Bart. K.B. P.R.S. *Philosophical Transactions of the Royal Society* 90: 403–431.

Wetzels, Walter J. 1978a. J. W. Ritter: The beginnings of electrochemistry in Germany (with commentary by George Dubpernell). In *Proceedings of the Symposium on Selected Topics in the History of Electrochemistry*, eds. George Dubpernell and J.H. Westbrook, 68–76. Princeton: The Electrochemical Society.

Wetzels, Walter J. 1978b. J. W. Ritter: Electrolysis with the Volta-pile (with commentary by George Dubpernell). In *Proceedings of the Symposium on Selected Topics in the History of*

Electrochemistry, eds. George Dubpernell and J. H. Westbrook, 77–87. Princeton: The Electrochemical Society.

Wetzels, Walter J. 1990. Johann Wilhelm Ritter: Romantic physics in Germany. In *Romanticism and the sciences*, eds. Andrew Cunningham and Nicholas Jardine, 199–212. Cambridge: Cambridge University Press.

Wilkinson, Charles Hunnings. 1804. *Elements of galvanism, in theory and practice, with a comprehensive view of its history, from the first experiments of Galvani to the present time, etc.*, 2 vols. London: John Murray.

Williams, L. Pearce. 1965. *Michael Faraday*. London: Chapman and Hall.

Chapter 3
HO or H$_2$O? How Chemists Learned to Count Atoms

Abstract Water served as an emblematic locus for debates on the atomic constitution of matter. Today it is taken as common sense that water is H$_2$O, but this was a highly disputed hypothesis for the first half-century of atomic chemistry. In Dalton's original formulation of the atomic theory published in 1808 water was presented as HO, and consensus on the H$_2$O formula (first proposed by Avogadro) was not reached until after the mid-century establishment of organic structural theory based on the concept of valency. The main epistemic difficulty was unobservability: molecular formulas could be ascertained only on the basis of the knowledge of atomic weights, and vice versa. There were multiple self-consistent sets of molecular formulas and atomic weights, which were employed in at least five different systems of atomic chemistry that flourished in the nineteenth century, each with its distinctive set of aims and methods and in productive mutual interaction. At the heart of the distinctive systems of atomic chemistry were different ways of operationalizing the concept of the atom (weighing, counting, and sorting atoms). It was operationalization that enabled atomic theories to become more than mere hypotheses that may or may not be consistent with observed phenomena. If we examine the crucial phase of development in which the consensus on H$_2$O was achieved, the key was not the revival of Avogadro's ideas by Cannizzaro, but the establishment of good atom-counting methods in substitution reactions. This, too, was a triumph of operationalization. We also need to keep in mind that the H$_2$O consensus was not a straightforward unification of all systems of atomic chemistry; rather, it was a reconfiguration of the field which resulted in a new pluralistic phase of development.

3.1 How Do We Count What We Can't See?

"Water is H$_2$O." That is common sense among scientifically educated people today. What is not commonly known is that the initial molecular formula adopted for water in atomic chemistry was HO (one atom each of hydrogen and oxygen). Even less well-known is the fact that it took the leading chemists of Europe more than 50 years to reach a consensus that water was H_2O. In the mid-1860s there was even an author identifying himself as a certified lunatic at the Hanwell Asylum outside London who sent two satirical letters to *Chemical News* mocking chemists' inability to agree on the formula for something as simple as water, or even an agreed notation in which to express their disagreement (Anonymous 1864, 1865). One can ponder the bewildering variety of formulas for water, he lamented, "until the brain first becomes confused, then swims, and finally softens."[1] It may be some comfort to know that the lunatic was slightly out of date, as he was writing just as leading chemists were finally coming to a considerable degree of consensus over the matter. But it is difficult for us in the twenty-first century to imagine all the impassioned debates and intricate developments, many of them in organic chemistry, that were to be required before H_2O could be safely installed as the molecular formula for water.[2]

In this chapter I want to make sense of that turbulent first half-century of atomic chemistry. On the one hand, why did it take so long to agree on that simple formula H_2O? On the other hand, how was it actually possible at all to decide such a thing? There were no direct means of observing individual atoms at the time (or even now, in the case of atoms inside a water molecule). How can you count the number of atoms in a molecule, if you can't observe them individually? As late as 1851, more than 40 years into the history of atomic chemistry, the great German chemist Justus Liebig (1803–1873) declared: "We possess no means of ascertaining the *number* of atoms, even in the most simple compound, since for this purpose it would be necessary that we should be able to see and to count them. . . ." (Liebig 1851, 103; emphasis original) By the end of this chapter I hope you will agree with me that atoms *could* be counted, even though they couldn't be *seen*, or observed in any other very direct way.

In earlier chapters we have seen how water eventually came to be recognized as a compound by most people, despite some serious difficulties in reaching agreement on that conclusion. Shortly after the electrolysis of water (discussed in Chap. 2), the atomic theory arrived. There is common consent that the chemical atomic theory was the creation of John Dalton (1766–1844), a reticent schoolmaster in the north of England. Like Priestley and Davy, Dalton came from a humble background. Born to a working-class Quaker family in a small village in Cumbria in northwest England, Dalton clearly belonged to that venerable tradition of British amateur science.[3] With only several years of primary education in his own background, Dalton made his living by teaching—both privately and in various dissenting (non-Anglican) academies,

---

[1] Bill Brock has unearthed these letters, first mentioned in Brock (1992, 152), and discussed further in Brock (2011, 286–289), also with a conjecture regarding the real identity of the author.

[2] As prophecy goes, this is the best I have found, from Berzelius (1813, 449): "It is in the study of the composition of organic bodies that our knowledge of the laws of chemical proportions, and of the electrochemical theory, will one day reach that degree of perfection which the human mind is capable of giving it."

[3] For a collection of informative articles on Dalton's life and work, see Cardwell (1968).

including the Manchester New College, which had started its life as the Warrington Academy, where Priestley taught in the 1760s (Brock 1992, 134).

Dalton's main contribution was to marry the familiar old (even ancient) idea of atoms and the eighteenth-century chemistry of compositions,[4] creating the essential nexus of nineteenth-century atomic chemistry. He realized that some striking regularities in the proportions by which various chemical substances combined with each other could be explained nicely if one assumed that chemical combination was the grouping together of atoms possessed of definite weights. For example, examining the five oxygen–nitrogen compounds known to him, he discerned compounds that would be written in modern notation as NO, N_2O, NO_2, NO_3, and N_2O_3, as shown in Fig. 3.1 (diagrams 41–45).[5] Dalton published his atomic ideas in a book titled *A New System of Chemical Philosophy*, the first part of which appeared in 1808. It is gratifying to know that the ideas of this unknown country schoolteacher received proper attention from the scientific community. Although many chemists were reluctant to believe in Daltonian atoms in a fully literal sense, it soon became common practice to conceptualize chemical reactions in terms of the grouping and re-grouping of some atomic units of elementary substances. Dalton remained in the provinces and kept to a modest way of life, but his work in chemistry and physics was widely acclaimed. He became a Fellow of the Royal Society of London (though he almost never attended its meetings), had an audience with the King, and served as the President of the Manchester Literary and Philosophical Society. When he died he was given a state funeral. Not bad for a scientist who never even went to secondary school!

3.1.1   Unobservability and Circularity

So we can honor Dalton as "the father of chemical atomism" if we like, but it would be a serious mistake to imagine that he was thinking of anything like atoms as we now know them. He was convinced that atoms were spherical bodies made up of a small hard core surrounded by an "atmosphere" of caloric (the fluid of heat). The core determined the weight of the atom, and the caloric atmosphere determined its size. Most atomic chemists in the nineteenth century quickly abandoned Dalton's concern with the sizes of atoms, focusing on weights instead.[6] But Dalton was very wide of the mark on atomic weights, too, as shown in Table 3.1. Of the 20 elements

---

[4] See Chap. 1, Sect. 1.2.3, on "compositionism" in eighteenth-century chemistry.

[5] See Dalton (1808, 215) and Dalton (1810, 316–368). Figure 3.1 is a reproduction of Dalton's Plate 5, opposite p. 560; note that he was using the Lavoisierian French term "azote" for nitrogen. Modern formulas for these compounds match Dalton's, except that his NO_3 would be our N_2O_5, nitric anhydride (see Lowry 1936, 209).

[6] One interesting exception was Dalton's early supporter Thomas Thomson, who still gave an account of atomic volumes even in 1831 in the 7th edition of his *System of Chemistry*, ranging from 1 for carbon to 28 for potassium (Thomson 1831, vol. 1, 14).

ELEMENTS .
Simple

Plate 5.

Compound
Oxygen with Hydrogen

Oxygen with Azole

Oxygen with Carbone and Sulphur

Oxygen with phosph.

Hydrogen with azote & Carbone.

Hyd . with Sulph.& phosph

Sulphur with phosph

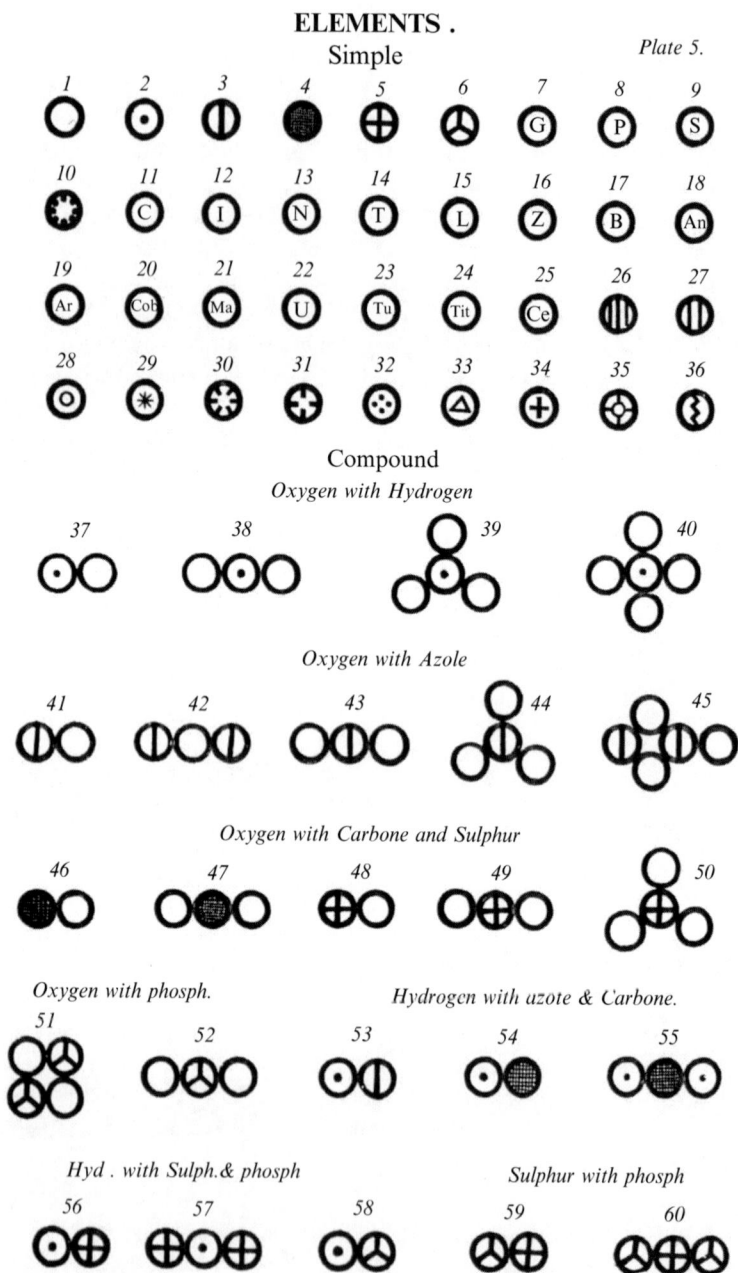

Fig. 3.1 Dalton's atoms and molecules ("compound atoms")

Table 3.1 Dalton's atomic weights given in 1808, and later updates by Dalton, in comparison to modern weights

	Dalton (1808)	Dalton (1810)	Dalton (1827)	Modern
Hydrogen	1	1	1	1.008
Azote (Nitrogen)	5	5	5±, or 10?	14.007
Carbon	5	5.4	5.4	12.011
Oxygen	7	7	7	15.999
Phosphorus	9	9	9	30.974
Sulphur	13	13	13, or 14	32.064
Magnesia	20	17	17 (compound?)	
Lime	23	24	24 (compound?)	
Soda	28	28	28 (compound)	
Potash	42	42	Not listed	
Strontites	46	46	46 (compound)	
Barytes	68	68	68 (compound)	
Iron	38	50	25	55.847
Zinc	56	56	29	65.37
Copper	56	56	56, or 28?	63.54
Lead	95	95	90	207.19
Silver	100	100	90	107.87
Platina	100	100?	73	195.09
Gold	140	140?	60±	196.97
Mercury	167	167	167 or 84	200.59

Sources: Dalton (1808, 219); (1810, 546–547); (1827, 352–353). For a more extensive tabulation, see Gjertsen (1984), 277, Table 11.2

whose atomic weights he determined by 1808, five were later judged not to be elements at all (lime, soda, etc.). Of the remaining 15, only one (silver) is within 10% of the modern value (excepting hydrogen, which was given the unit value of 1 by definition).

Some of the divergence may be attributed to the inaccuracy of Dalton's laboratory techniques, as in the case of zinc or copper. In other cases, however, Dalton's values are roughly half of the modern values (as when he gives 5 for nitrogen and carbon, and 7 for oxygen), or even a third (as with phosphorus and sulphur). These wildly pre-modern atomic weight values are directly related to the pre-modern molecular formulas he gave, the prime example of which is HO for water (Diagram 37 in Fig. 3.1). In the case of nitrogen, copper and mercury, we can witness Dalton becoming unsure about the relevant molecular formulas by 1827, as their atomic weights are listed as 5 or 10, 56 or 28, and 167 or 84, respectively. Figure 3.2 exhibits the un-modern character of Dalton's atoms very nicely: the atoms and molecules of gases are shown as stationary particles in a regular array; each atom/molecule looks *hairy*, which is Dalton's way of representing the "atmosphere of caloric" that surrounds each atom and determines its size; and the hydrogen molecules are single atoms, not the modern-day H_2 (Dalton 1810, Plate 7, with explanations on p. 548). What kind of "father of chemical atomism" does Dalton make, if he could hardly

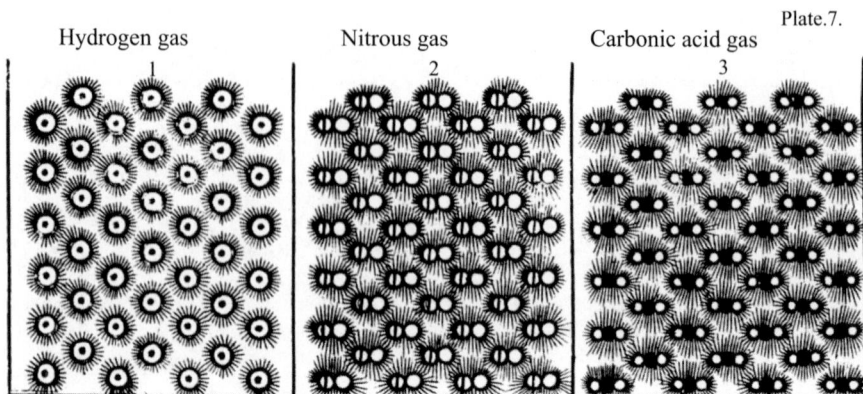

Fig. 3.2 Dalton's picture of gases

get an atomic weight right, didn't even know the molecular formula of water, and had all sorts of strange ideas about the shapes, sizes and structures of atoms?

Still, it is important not to lose our sympathy. Admitting that his atoms were not directly observable, Dalton arrived at HO by the assumption of simplicity. He knew of only one compound of hydrogen and oxygen. Why should he have assumed that the composition of water was not the simplest possible, namely a one-to-one combination of hydrogen and oxygen atoms? Yes, one could assume instead that a water molecule consists of 24 atoms of hydrogen and 37 of oxygen, but why? (Why would God have made the world in such a frivolous way?) As Alan Rocke points out (1984, 36), Dalton actually had a physical reason for favoring simplicity in the combination of atoms: he thought that atoms of the same element would repel each other, because they were not attracted to each other by chemical affinity (as atoms of different elements would be) which would counterbalance the repulsion of the caloric contained in each of them. If a molecule contained a larger number of like atoms, it would be less stable. HO, it must be.

And how do *we* know that the correct formula is H$_2$O? Give this problem to a bright schoolchild of today, and this is the sort of answer we will receive: the atomic weights of hydrogen and oxygen are 1 and 16 (hydrogen being taken as the unit); when we break down water in the lab, we get 1g of hydrogen for 8g of oxygen; so there must be two atoms of hydrogen that combine with each atom of oxygen to make a water molecule. What the bright student usually can't say, having just memorized these values from the textbooks, is how we know that the atomic weights of hydrogen and oxygen are 1:16. A clever answer would be that we know the molecular formula of water to be H$_2$O, so if the gross (macroscopic) combining weights of hydrogen and oxygen are 1:8, then the ratio of their atomic weights must be 1:16. But then we have to ask how we know the formula of water is H$_2$O, which is exactly where we started! Now this is the circularity that plagued Dalton and all of his contemporaries. All we can observe directly is the gross combining weights. If we know the molecular formula, we can infer the atomic weights from the combining weights; if we know the atomic weights, we can infer the molecular formula. But observation by itself gives us neither the atomic weights nor the molecular formula. We can make up any self-consistent system of atomic weights and molecular formulas, and observation cannot refute our system. Without breaking this circularity, atomic chemistry could not really get off the ground. This circularity is only noted in the most philosophical of science textbooks today (e.g., Langford and Beebe 1969, 18–20; Rogers 1960, 592–593; Holton and Brush 2001, 280–281).

This is why Dalton needed his "rules of greatest simplicity", which gave him a method for breaking the circularity; if molecular formulas could be fixed through simplicity considerations, then atomic weights could be determined on the basis of molecular formulas (Dalton 1808, 213–214). When only one compound of two elements was known, Dalton assumed that it was a one-to-one atomic combination. So water is HO, and the atomic weights are H = 1 and O = 8 (in approximate modern numbers).[7] If there are multiple compounds, then other combinations will form, in the following order of readiness: 1-to-2, 2-to-1, 1-to-3, 3-to-1, etc. Dalton himself was satisfied with this way of breaking the circularity, and he proceeded to build his "New System" of chemistry on that basis. In fact, when Thomas Thomson, Dalton's greatest early advocate, made his first presentation of Dalton's atomic ideas, he stated that the simplicity rules *were* Dalton's new theory, the idea of atom itself being old and widely accepted (Thomson 1807, 425).[8] From her early twentieth-century vantage point the Cambridge chemist and historian Ida Freund (1904, 284) gave a similar view: "Dalton dealt from the outset with the atom as a conception generally known." A look at Dalton's own assessment of the situation confirms Thomson's and Freund's views:

> In all chemical investigations, it has justly been considered an important object to ascertain the relative weights of the simples which constitute a compound. But unfortunately the enquiry has terminated here. . . . Now it is one great object of this work, to shew the importance and advantage of ascertaining the relative weights of the ultimate particles, both of simple and compound bodies, the number of simple elementary particles which constitute one compound particle, and the number of less compound particles which enter into the formation of one more compound particle. (Dalton 1808, 212–213)

And then he immediately proceeded to give the simplicity rules as the way to determine these relative weights and numbers of atoms.

Most others disagreed with Dalton on the simplicity rules, and it is easy to see why he could not satisfy everyone. Not only was his doctrine of simplicity ultimately unjustified, but it was actually not even sufficient for its intended purpose of fixing atomic weights and molecular formulas, whenever the situation was more

---

[7] Dalton (1808, 215) and Dalton (1810, 275). But he does briefly acknowledge that it is possible that water may be H_2O (Dalton 1810, 276). In my exposition in the current section (Sect. 3.1), I will sacrifice historical accuracy and use modernized atomic-weight numbers (rather than Dalton's own), in order to avoid confusing the modern reader.

[8] Thomson took the idea of physical atoms as prevalent common sense, and in fact even used the term "atom" freely in his text *before* the section where he introduced Dalton's ideas; in fact the same is done in Dalton's own text (e.g., Dalton 1808, 125).

complex than the water-like cases involving just one compound made from two elements. As an example of the simplest case of trouble, take the two oxides of carbon: the old "fixed air" (or "carbonic acid"), and a more recent discovery, the "heavy inflammable air" of Joseph Priestley. William Cruickshank (?–1810/1811), whom we met in Chap. 2 as one of the early pioneers of electrolysis, seems to have convinced everyone that the latter gas was a compound of carbon and oxygen,[9] and Dalton called this gas "carbonic oxide". The modern names for these gases (carbon dioxide and carbon monoxide) already embody our idea of their atomic composition, so let's stick with the old names for the time being. Analysis revealed that the combining weights were as follows (Thomson 1831, 166–169):

Carbonic acid = 0.75 parts carbon + 2 parts oxygen
Carbonic oxide = 0.75 parts carbon + 1 part oxygen

Ah, don't the formulas and CO_2 and CO just leap out of the page now, with the atomic-weight ratios of 3:4 between carbon and oxygen? *Not so fast.* The combining-weight ratios can be just as easily represented as 3:8 for carbonic acid and 6:8 for carbonic oxide, which would suggest CO and C_2O, with the atomic-weight ratio of 3:8 (see the atomic weights used by Dumas, in Table 3.2 below). Both options satisfy Dalton's rules of simplicity equally well, each giving one binary compound (CO) and one tertiary compound (CO_2 or C_2O). As William Hyde Wollaston (1814, 7), another great early support of Dalton, said quite plainly: "it is impossible in several instances, where only two combinations of the same ingredients are known, to discover which of the compounds is to be regarded as consisting of a pair of single atoms", and "the decision of these questions is purely theoretical".

The case of nitrogen oxides was similar, and even more complicated as there were as many as five known compounds. For decades chemists debated whether their correct formulas were NO, NO_2, NO_3, NO_4, NO_5, or the equally self-consistent N_2O, N_2O_2 (=NO), N_2O_3, N_2O_4 (=NO_2), N_2O_5, which is obtained by halving the presumed atomic weight of nitrogen or doubling the presumed atomic weight of oxygen.[10] The first series comes from taking the atomic weights as $N = 14$ and $O = 8$, and the second series from setting $N = 7$ and $O = 8$ (or, $N = 14$ and $O = 16$). Soon enough, the situation with water itself fell into this uncertain territory, when hydrogen peroxide was discovered by Louis-Jacques Thénard in 1818. It was called "*per*oxide" because it was shown to contain more oxygen than water did, in proportion to hydrogen. We now say the peroxide is H_2O_2 and water is H_2O, but it would have been entirely consistent to keep water at HO and the atomic weight of oxygen at 8, understanding the new compound as HO_2. Initially Dalton was blissfully unaware of the peroxide, but Thomson later considered the question carefully and concluded that the new compound was HO_2, calling it "hydrogen deutoxide", keeping water as HO (Thomson 1831, 11).

---

[9] Priestley never agreed, and maintained that it was a phlogiston-rich inflammable air (see, e.g., Priestley [1796] 1969, 37–38).

[10] The second series is similar to Dalton's formulas.

Table 3.2 Competing sets of atomic weights (rounded to whole numbers) and molecular formulas in the nineteenth century

	Atomic weights				Molecular formulas		
	H (hydrogen)	O (oxygen)	C (carbon)	Ag (silver)	Water	Silver oxide	Hydrochloric acid
Gmelin (also Liebig, Thomson, Wollaston)	1	8	6	108	HO	AgO	HCl
Dumas (1828)	1	16^a	6	216	H_2O	AgO	HCl
Berzelius, as of 1826	1	16	12	216	H_2O	AgO	H_2Cl_2
Gerhardt (4-volume formulas[b])	1	16	12	108	H_4O_2	Ag_2O	H_2Cl_2
Laurent (2-volume formulas)	1	16	12	108	H_2O	Ag_2O	HCl

Sources: Brock (1992, 214); Dumas (1828), Introduction, L; Ihde (1984, 153); Odling ([1855] 1963, 41); Thomson (1831, 12); and Wollaston (1814)
[a]Freund (1904, 600) records, in the context of the discussion of Prout's hypothesis, that Dumas (by 1859), like Liebig, had reverted to 8 for the atomic weight of oxygen.
[b]Ihde (1984, 206) explains: four-volume formulas "represented the volume of vapor that occupied the same space as four volumes of hydrogen, i.e., alcohol, $C_4H_{12}O_2=H_4$." That is still mysterious to the modern eye, so let me spell it out anachronistically: if we take four atoms of hydrogen, that would make two volumes of hydrogen gas ($2H_2$); to make the same volume of water vapor ($2H_2O$) would require four atoms of hydrogen and two atoms of oxygen, this is written as the four-volume formula for water H_4O_2. Similarly for two-volume formulas. It may seem like H_4O_2 and H_2O are just the same thing; on the contrary, constitutional reasoning will go very differently depending on which of these formulas one adopts.

Agreeing on H_2O was a matter of agreeing on an entire set of atomic weights and molecular formulas. The atomic-weight assignment of O = 16 had to be maintained in all the other reactions that oxygen participated in. Every other decision also had its own implications about other reactions and compounds, and they all had to be made consistent with each other. Still, reaching some version of consistency was in fact much easier than deciding which method of reaching consistency was the best one. Competing sets of atomic weights soon emerged, and there were in principle indefinitely many possible theoretical systems consistent with known observations.[11] As I will recount in much greater detail in Sect. 3.2, chemists did not come to a general consensus about atomic weights and molecular formulas until the 1850s. That is to say, the first half-century of atomic chemistry progressed without that consensus.

---

[11] The following simple exercise demonstrates one way in which the latter statement is true: take the system of atomic weights and molecular formulas that you accept; pick any element in that system and halve its atomic weight, and double the number of that atom in every molecular formula; then we have a whole new system that is self-consistent. For example, if we said that the atomic weight of oxygen was 8 instead of 16, we would end up with water as H_2O_2, carbon dioxide as CO_4, etc. This can be done to any element we like, as often as we like.

3.1.2   The Avogadro–Cannizzaro Myth

A coterie of historians have duly told complex in-depth stories about the first half-century of atomic chemistry, and I will be drawing from their works in the remainder of this chapter.[12] But many of these authoritative accounts are inaccessible except to the most determined research students and experts. Meanwhile, many other commentators seem to have felt that the story *shouldn't* have been so complex, and have given accounts that blame the irrationality or ignorance of mainstream chemists for the mess. Perhaps the most common story in that vein is that an eccentric Italian who had seen the truth got ignored, until a more methodical and determined Italian unearthed his work 50 years later and convinced everyone about it. This is the story of Avogadro and Cannizzaro, whose spirit is illustrated very nicely in a statement by the great Oxford chemist Harold Hartley in a paper of 1966: "Cannizzaro [in 1858] saw so clearly that this confused state of chemical theory was due to the reluctance of chemists to accept whole-heartedly the logical conclusions from the work of Gay-Lussac and Avogadro owing to their preconceived ideas on one aspect or another." (Hartley 1971, 186). As Joshua Gregory (1931, 109) put it, Avogadro's ideas were "coldly regarded at the time, repeated variously at intervals, and finally established by Cannizzaro."

  As hinted in Hartley's statement, this story of two Italians actually begins with a Frenchman, Joseph-Louis Gay-Lussac (1778–1850), who published a crucial paper immediately in the wake of the publication of Dalton's atomic ideas. Gay-Lussac and Dalton had both established their initial scientific reputations with their work on the thermal behavior of gases. Turning his attention to chemical combination, Gay-Lussac [1809] (1923) kept his focus on gases and paid particular attention to *volumes* rather than weights, arriving at a striking generalization: when gases reacted chemically with each other, they did so in very simple ratios of volume (under the same pressure and temperature). For instance, 2 volumes of carbonic oxide combined with 1 volume of oxygen to make 2 volumes of carbonic acid; 1 volume of nitrogen combined with 3 volumes of hydrogen to make 2 volumes of ammonia. Water comes on the stage again: Cavendish (1784) had already noted a 2:1 ratio between the volumes of hydrogen and oxygen combining to make water. Doesn't that just spell out H$_2$O, as schematically indicated in Fig. 3.3? Yes, but only if one makes an assumption that is implicit in that drawing, namely that equal volumes of all gases contain equal numbers of particles; I will follow Alan Rocke (1984, 24 onward) in his catchy designation of this assumption as "EVEN" (equal volumes—equal numbers). Gay-Lussac came close to accepting EVEN and its implications: "The numerous results I have brought forward in this Memoir are also very favourable to [Dalton's] theory." But citing his mentor Claude-Louis Berthollet's doubts about the universal truth of the

---

[12] These include (in roughly chronological order) Ida Freund, T. M. Lowry, Joshua Gregory, J. R. Partington, Colin Russell, David Knight, Aaron Ihde, William H. Brock, John Hedley Brooke, Evan Melhado, Arnold Thackray, Mary Jo Nye, Trevor Levere, Alan Rocke, Christoph Meinel, Ursula Klein, Joseph Fruton, Peter Ramberg, and Alan Chalmers.

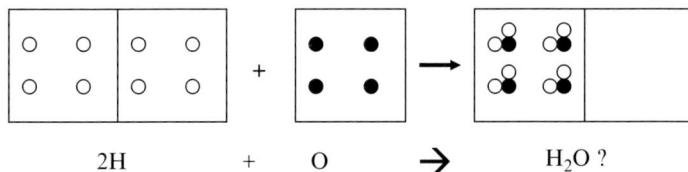

2H + O → H₂O ?

Fig. 3.3 The implication of EVEN for water

law of fixed proportions in chemical combinations, he shrank from making a definite theoretical commitment here (Gay-Lussac [1809] 1923, 23–24).

You might imagine that Dalton would have embraced Gay-Lussac's results as an attractive way of determining molecular formulas, using it to break the circularity between atomic weights and molecular formulas. However, he had physical reasons to resist EVEN. In fact, Dalton (1810, 556) pointed out that he had already considered and rejected a version of EVEN, even before Gay-Lussac's work.[13] Among other things, if EVEN were true, then when gases combine there should inevitably be a reduction of total volume, as the atoms would combine with each other and the overall number of particles would decrease. But this is not always the case. For the case of water, Fig. 3.3 again illustrates the problem: from EVEN and a 2-to-1 combination of hydrogen and oxygen, we would expect only one volume of water vapor to form, but experiment shows that two volumes are formed. Some cases are even worse, as in the formation of 2 volumes of nitrous gas by a 1-to-1 combination of nitrogen and oxygen (Dalton 1808, 70–71). Two years later, when Dalton published part 2 of his *New System*, he reiterated his objection: "In fact, his [Gay-Lussac's] notion of measures is analogous to mine of atoms; and if it could be proved that all elastic fluids have the same number of atoms in the same volume . . . the two hypotheses would be the same." But Dalton was so averse to EVEN that he even tried to argue that Gay-Lussac's observed volume-relations were only approximate and did not reflect anything fundamental about chemical combinations (Dalton 1810, 556–559).

It was not Dalton, but Amedeo Avogadro (1776–1856) who accepted EVEN and worked out its consequences for the atomic theory, venturing deeply into the realm of the unobservable. Initially trained as a lawyer, Avogadro was at this time Professor of Natural Philosophy at the Royal College at Vercelli, in Piedmont.[14] In his paper of 1811, published in the *Journal de Physique* in Paris, Avogadro showed how to reconcile Dalton and Gay-Lussac perfectly. Figure 3.4 represents his basic ideas. First he stated EVEN (Avogadro [1811] 1923, 29). Then he argued that the molecular

---

[13] For Dalton, who envisaged the atoms and molecules of gases stacked up without unnecessary gaps between each other, EVEN amounted to the same thing as what Dalton said he had rejected: "At the time I formed the theory of mixed gases, I had a confused idea, as many have, I suppose, at this time, that the particles of elastic fluids are all of the same size" (Dalton 1808, 188).

[14] For extensive details on Avogadro's life and work, see Morselli (1984).

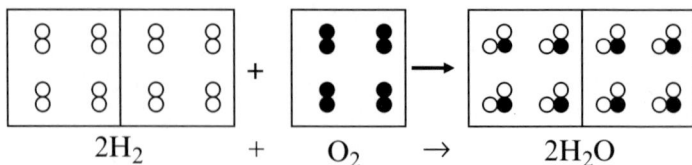

Fig. 3.4 Avogadro's hypotheses applied to the case of water

formula of a compound should be determined from the combining volumes, chiding Dalton for using "arbitrary suppositions as to the most likely relative number of molecules in compounds" (p. 33). EVEN gave the formula H_2O but also demanded two volumes of it, so Avogadro proposed that the water molecule had to split in two, in order to uphold EVEN. But that would have had the consequence of splitting the oxygen atom in half, which was impossible (consider how the H_2O boxes would have to be done, in Fig. 3.3). So Avogadro supposed that the oxygen particle must have been a double-atom molecule to begin with; hydrogen must have been, too, so that the correct ratio was maintained between the numbers of oxygen and hydrogen atoms (pp. 31–32). If we may write all this in modern notation, we have: $2H_2 + O_2 = 2H_2O$.

So it looks like Avogadro had the whole modern story, and many people have wondered why his insights were apparently neglected for 50 years.[15] John Hedley Brooke has quipped (1981, 235): "If there is one matter that has not been neglected it is the neglect of Avogadro's hypothesis." In some badly informed accounts it has been said that Avogadro's work was simply unknown, coming from the periphery of the main European scientific communities. But this is not likely, as his main paper on this subject was published in a major French journal. Besides, very similar ideas were also advanced by André-Marie Ampère (1775–1836), who was clearly part of the Parisian scientific establishment—by 1809 Professor of Mathematics at the *Ecole Polytechnique*, and Napoleon's Inspector General for the new French university system. And Ampère's work was followed and developed energetically by Marc-Antoine Gaudin (1804–1880).[16] It has also been hinted that Avogadro's exposition was so unclear that hardly anyone understood him. This, too, is unlikely. True, he wasn't the clearest of writers, but it is not the case that his terminology was so idiosyncratic and obscure as to confuse everyone. For example, his now-quaint phrase "integrant molecule" was a term in common use in crystallography, which was always connected with atomic chemistry, most of all in France under the influence of René-Just Haüy.[17] The incomprehension is much more on the part of modern chemists and historians, than it would have been on the part of Avogadro's contemporaries.

---

[15] It seems that some authors did ignore Avogadro, whether they were aware of his ideas or not; Joseph Fruton (2002, 56) notes that Berzelius chose not to discuss Avogadro in his annual reviews, and that Hermann Kopp's history of chemistry (1843–1847) makes no mention of Avogadro's name.

[16] See Mauskopf (1969) on Ampère and Gaudin.

[17] See Mauskopf (1970) on Haüy's work and its connection with atomism.

The real story as I see it, which is told in fulsome detail by Brooke (1981) and Nicholas Fisher (1982), is that Avogadro's microphysical hypotheses were noticed, discussed, and rejected by most, for fair enough reasons. They were too hypothetical and blatantly *ad hoc*, not grounded in any experiments providing independent empirical evidence. In addition, Dalton had specific physical arguments against them, and so did the promulgators of electrochemical dualism, including Berzelius. For the latter, the story was simple: two atoms of the same kind have the same electric charge, so they would repel each other. As for Dalton, in connection with his rules of simplicity I have noted above his view that all atoms were full of self-repellent caloric, and chemical combination could happen only between different types of atoms which exerted an attractive force of chemical affinity on each other, enough to overcome the self-repulsion of caloric. Avogadro does not seem to have given a convincing account of why two atoms of the same kind should stick together, and if they do, why the lumping would stop at two atoms. Jean-Baptiste Dumas (1800–1884) made a serious attempt to develop Avogadro's ideas, but he gave up after his study of vapor densities revealed apparent contradictions or at least arbitrariness; for example, the elementary molecule of mercury, phosphorus and sulphur had to be regarded as Hg, P_4 and S_6, not binary like hydrogen, oxygen and nitrogen (see Nye 1976, 248).

An essential complement to the popular story about Avogadro the neglected Italian is the story of Cannizzaro the passionate yet clear-headed Italian who eventually made everyone see the light. Stanislao Cannizzaro (1826–1910), a native of Sicily, was a revolutionary politician as well as a chemist; in chemistry he was better known as a consummate pedagogue than a researcher. A standard reference on scientific biography declares that Cannizzaro, "through his revival of Avogadro's hypothesis, laid the foundations of modern atomic theory."[18] The story goes: Cannizzaro in the 1850s reached a perfect synthesis of all recent chemical research, with Avogadro's ideas as the foundation for everything. He taught this system to his students, but did not have the opportunity to communicate it to the leading chemists of Europe. When the famous Karlsruhe Congress was called in 1860 with the aim of bringing uniformity to chemical notation and atomic weights, Cannizzaro seized this opportunity to expound his system. At the congress itself he failed to make a decisive impact despite his best efforts, but he distributed a pamphlet there, which eventually had the effect of persuading everyone.[19] There is a famous report from the German chemist (and much-neglected pioneer of the periodic system of elements) Julius Lothar Meyer (1830–1895), who recounted reading Cannizzaro's pamphlet on the train back from Karlsruhe, and how "scales fell from his eyes" and suddenly everything made sense (quoted by many historians, e.g., Hartley 1971, 185).

---

[18] Anonymous (2000), vol. 1, 210. For a very clear exposition of this view, see Bradley (1992), which does not pretend to be a work of history.

[19] Hartley (1971), 188–192, provides a lively account of Cannizzaro's interventions at the Karlsruhe Congress.

I follow Rocke (1984, 295–296; 1992) in noting that the crucial battles had already been fought and mostly won by the advocates of reform before the Karlsruhe Congress. Karlsruhe did not generate much additional consensus, and what Rocke calls the "Quiet Revolution" continued to spread on its own momentum, with or without Karlsruhe. This is not to deny that some people may have been convinced by Cannizzaro's speeches or his pamphlet. But even so, the main service rendered by Cannizzaro was only a particularly perspicacious presentation of the new point of view, and a marshalling of various considerations that were already in widespread circulation. He advanced no significant new ideas, and he did not present the powerful organic side of the story. The cogency and independence of his work is undeniable, but it wasn't quite the decisive stroke in the revolution that it is made out to be by some commentators.

3.1.3   *Operationalism and Pragmatism in Atomic Chemistry*

If the Avogadro–Cannizzaro story is a myth, or at least an exaggerated or one-sided account, what is a better story? There are already historical accounts (e.g., Rocke 1984, 1992, 1993; Russell 1971; Partington 1964) whose thoroughness and sophistication I cannot hope to surpass in this book (or ever). My aim is to give a perspicacious account of this complex development that combines tolerable historical accuracy with useful philosophical insights. I also hope that my distinct interpretation of events will have some freshness even to the expert historians, and much of the content presented in the rest of the chapter will probably be news to the majority of professional historians of science, due to the recent neglect of this area of study.

My full account will be given in the second section (Sect. 3.2). For now, I will only give an illustration of the kind of new interpretation I'm trying to offer, through a brief consideration of the role of volumes in atomic chemistry. Volumetric reasoning did have crucial importance in the development of atomic chemistry, though not so much through Avogadro or Cannizzaro. It is in fact difficult to find any textbook of chemistry after 1809 that does not discuss Gay-Lussac's volume-relations, and does not at least entertain the idea that volumes may be indicative of the numbers of atoms involved in chemical reactions.[20] EVEN was such a natural idea for anyone thinking about volumes in the context of atomism, that it is in fact difficult to credit Avogadro specifically for it. And it remained in circulation long after people stopped talking about Avogadro. It is instructive to have a quick preview of Berzelius's take on the issue:

> there is no other difference between the theory of atoms and that of volumes, than that the one represents bodies in a solid form, the other in a gaseous form. It is clear, that what in the one theory is called an *atom*, is in the other theory a *volume*. In the present state of our

---

[20] Liebig (1851, Letters 6 and 7) is an interesting exception, though it is not really a textbook.

knowledge the theory of volume has the advantage of being founded upon a well constituted fact, while the other has only a supposition for its foundation. (Berzelius 1813, 450; emphases original)

It may seem like Berzelius was rashly taking EVEN for granted without having sufficient evidence for it, but the last part of the passage quoted above should give a pause to that interpretation. As Cannizzaro recognized very clearly ([1858] 1910, 3), Berzelius's view certainly differed from Avogadro's, in rejecting the application of EVEN to compound gases and also rejecting the idea of polyatomic molecules of elementary gases (see also Russell 1968, 268–269). He even made it clear, on the very same page as the passage quoted above, that he still considered atoms hypothetical.

Rather than being an unquestioning believer of EVEN as meant by Avogadro, I think Berzelius was taking volumes as a *measure* of atoms, independently of combining weights, leaving the relation between the two measures as an open question.[21] Pause for a moment here to consider what a chemical atom meant in practice. As Rocke puts it (1984, 12), it was "a chemically indivisible unit, that enters into combination with similar units of other elements in small integral multiples." Now, in giving this definition Rocke was specifically concerned to distinguish atomic weights from equivalent weights, but weight is not the only property by means of which chemical atoms can be parsed. Chemical units can be conceptualized in terms of the volumes they occupy, as well as their weights. For the moment, get the picture of point-like particles out of your head, and imagine little boxes containing gases. There is a 2:1 volume ratio between hydrogen and oxygen combining to make water. That ratio holds, however small the absolute amounts of the two gases involved are; as Ursula Klein (2001, 15) puts it, we are dealing with "scale-independent portions of elements and compounds" here. Extrapolating into the smallest possible units, you get the idea of *volumetric chemical atoms*. The thinking process here is exactly the same as what is involved in arriving at the idea of gravimetric chemical atoms, which are the most minimal weights of substances participating in chemical reactions. In practice one does not have to know what the absolute values of the unit-amounts are, in either weight or volume, as long as one can discern the relative numbers of different units involved in a reaction. Berzelius surely knew that there were unanswered questions about the relation between volumetric and gravimetric atoms, but he was not going to let those problems get in the way of making initial progress in working out some basic facts about atomic combinations. (In Table 3.2 above, it is striking that Berzelius had the modern atomic weights, very early on and without making unjustifiable metaphysical commitments as Avogadro did. Part of the outcome of the "Quiet Revolution" of the 1850s was a return to Berzelius's atomic weights, only on a more secure basis.)

The real power of volumetric thinking was manifested through the concept of valency.[22] A most beautiful rational reconstruction of the working-out of valency is

---

[21] In the second installment of the same paper (Berzelius 1814, section IV) he presented considerations on the "weight of elementary volumes compared with that of oxygen gas".

[22] In modern American usage the standard term to use is "valence", but "valency" is more faithful to the usage closer to the time of the events discussed here.

presented in August Wilhelm Hofmann's *Introduction to Modern Chemistry* published in 1865. Hofmann (1818–1892), one of Liebig's star pupils in Giessen, had come to London as the first Director of the new Royal College of Chemistry, and the textbook records his lectures given there. Hofmann began by noting that hydrogen and chlorine gases combined in a 1:1 volume ratio, to make hydrochloric acid; hydrogen and oxygen combined in a 2:1 volume ratio, to make water; and hydrogen and nitrogen combined in a 3:1 volume ratio, to make ammonia. These relations yielded the molecular formulas HCl for hydrochloric acid, H$_2$O for water, and H$_3$N for ammonia, which indicated *confirmed* relative numbers of *volumetric* atoms, not hypothetical numbers of atomic particles (as in Avogadro's formulas). Staying with volumetric atoms, it seemed clear to Hofmann that chlorine, oxygen and nitrogen had different "atom-binding powers"—namely, the ability to combine with 1, 2, and 3 volumetric atoms of hydrogen. These molecular formulas, and the implied "valency" of 1, 1, 2 and 3 for H, Cl, O and N respectively (and later 4 for carbon), formed the core around which a whole system of molecular formulas and atomic weights could be built. (After the dust settled, Avogadro's more metaphysical ideas could be vindicated, too, though still with no understanding of how they were physically possible.)

This momentous development was a triumph of *operationalism* (more on that in Sects. 3.2.1 and 3.3.1). Note that the kind of thinking represented in Hofmann's exposition is directly rooted in easily and clearly reproduced laboratory operations. This is typical of the majority of atomic chemists in the middle of the nineteenth century: they took most seriously those theoretical ideas that could be directly linked to laboratory operations, and they sought to find, invent, and secure more and more of those links.[23] There is a very subtle balance that they achieved here, which is often lacking in philosophers' characterizations of scientists. These chemists were willing to theorize, but they also knew which aspects of their theories had an operational basis, and they also sought to extend that operational basis. They were neither positivists eschewing all talk of unobservable atoms and molecules, nor naïve realists with a firm belief in their pictures of atoms and molecules. Their positions with respect to scientific realism are very difficult to put into the standard philosophical pigeonholes.

This philosophical subtlety was rooted in a pragmatism born of humility—not a resignation to the uncertainty of the unobservable, but an active pursuit of knowledge while accepting one's own limitations. The use of volumetric reasoning is, again, a good case in point: all concerned acknowledged that there was "something to it"—nobody was sure for a long time what that something was exactly, but that did not stop efforts to learn. There was a pragmatist spirit of going on with inquiry in the most productive ways possible (see Sect. 3.3.3 on pragmatism), while freely acknowledging the limitations of what was possible. In nineteenth-century atomic chemistry the most valuable progress was made by perseverance with uncertain and down-to-earth theoretical and empirical inquiry, not by dogmatic clarity or adherence to axiomatic first principles.

---

[23] This gives a clear pointer toward pluralism, as I will discuss further in Chap. 5.

3.1.4 From Underdetermination to Pluralism

There is one important consequence of proceeding without certainties: there are many possible ways to go. Various competing theories about the atomic–molecular constitution of basic substances were all compatible with the commonly accepted body of evidence. This constituted a case of the problem of the *underdetermination of theory by evidence* much-debated in the philosophy of science, and the circularity between atomic weights and molecular formulas was at the root of it in this case. That philosophical problem also gives rise to a very clear and worthwhile historical question: how did scientists, in this particular instance, handle the underdetermination problem? The traditional philosophical framework is inadequate for a full historical understanding of this episode (and many others). As I will argue in the second section (Sect. 3.2), early atomic chemistry encompassed not merely a set of competing theories, but competing systems of practice based on different operationalizations of the concept of the atom (refer back to Chap. 1, Sect. 1.2.1.1 for a fuller definition of "system of practice").

A telling manifestation of the underdetermination in question was the simultaneous existence of multiple sets of atomic weights in widespread use up to the middle of the nineteenth century. Rocke (2001, 2) says that already by 1816 at least nine systems of atomic chemistry had appeared. The profusion of systems showed no sign of abating till the middle of the century, and Table 3.2 provides a brief glimpse of the diversity. Disagreements about atomic weights were directly tied to disagreements about molecular formulas, but they were also linked to deeper divisions and plurality in the field. As explained fully in Sect. 3.2.2, I can distinguish five major systems of practice in operation in nineteenth-century atomic chemistry, developing in competition and interaction with each other. (1) The *weight-only system* focused on determining and using atomic-weight values inferred from the macroscopic combining weights of substances (on the basis of some assumptions about molecular formulas). The practitioners of this system concentrated on chemical analysis rather than theoretical explanations of phenomena. (2) In contrast, there was a strong emphasis on explanations in the *electrochemical dualistic system*. A key operation in this system was the electrolysis of various substances using the Voltaic battery, and chemical reactions were understood as consequences of the electrostatic attractions and repulsions of atoms.[24] (3) In the *physical volume–weight system*, first constructed by Avogadro, chemists took not just weights but also volumes as measurable properties of physical atoms. The focus here was realist: to find out the real properties of the atoms and molecules of various substances. EVEN was a cornerstone of this system; while many chemists took EVEN for granted in a rather casual way, Avogadro's program was to defend it, whatever the consequence. (4) Disillusionment with the physical volume–weight system led to the rise of the

---

[24] In Chap. 2 it was noted that not everyone doing electrochemistry with the Voltaic pile shared this electrostatic view; however, the dualists within atomic chemistry all seem to have thought electrostatically.

substitution–type system. Instead of speculating about the real properties of atoms and molecules, an influential group of organic chemists took *classification* as their main aim and activity. Dumas led the way with his idea of "types", namely structural templates given by certain simple substances such as water and ammonia. (5) Many early type-theorists denied that type-formulas were meant to represent the actual geometry of molecular structures. Those who did concern themselves with getting at the structures practiced what I call the *geometric–structural system*. Often inspired by the crystallographic tradition, these chemists attempted to get directly at the geometry of molecular structures.

Philosophers who argue for underdetermination are correct, at least in this case: there *were* indefinitely many possible theoretical systems consistent with known observations. On the other hand, it is also the case that at least initially there wasn't even one system that was perfect if we consider other desiderata in addition to bare consistency with observations. For example, the weight-only system suffered from an arbitrariness in settling molecular formulas. The physical volume–weight system had to contend with the physical difficulties in Avogadro's hypotheses. The electro-chemical dualistic system suffered a serious discomfort when it was discovered that very different atoms could substitute each other, for example the highly electronegative chlorine and the highly electropositive hydrogen. And so on.

It is very interesting to see how the nineteenth-century chemists made progress in this doubly frustrating situation: with lots of alternatives, none of them perfect. (And it seems to me that this sort of situation is actually quite typical in science, as in the rest of life.) What we know is that, somehow, by the 1860s, consensus was reached on basically the same system of atomic weights and molecular formulas as what we now accept. As many historians have noted, this underdetermined situation was resolved with the help of developments in organic chemistry. Results in struc-tural theory, such as the fixing of the valency of carbon at 4, enabled the unique determination of molecular formulas for simpler compounds, which in turn led to the fixing of atomic weights. But how exactly was this achieved?

Initial progress was made by having a plurality of systems, each "zooming in" on what it could handle particularly well. For example, the weight-only system focused on gravimetric analytical chemistry, flourishing into the middle of the nineteenth century. The electrochemical dualistic system focused on substances that were amenable to clean electrolysis. And so on. Each system delivered a different set of new facts and insights, and contributed to the progress of chemical knowledge in ways that other systems could not easily manage.

After much development, it was possible to "zoom out" to make a synthesis of some of the competing systems. Most crucially, the concept of valency enabled the synthesis of the last three systems mentioned above (with some suitable modification of each system). When the substitution–type system improved its operational success, this gradually encouraged chemists to attribute reality to the models of molecular structure that they had invented for the purpose of classification. So they began to think that an oxygen atom really did bind two hydrogen atoms together in a water molecule, that a carbon atom held four hydrogen atoms to make a molecule of marsh gas (methane), and so on. Happily for them, the molecular formulas worked out in that way matched up well enough with the formulas used in the physical volume–weight system. And the increasing confidence and realism in that synthesis also allowed a further synthesis with the geometric-structural system. The key there was to take the carbon atom as a tetrahedral structure in three dimensions. With the synthesis of these three systems, nearly everyone came to agree that water was H_2O by the 1860s.

On the other hand, this synthesis was only possible through the renunciation of certain aims. Organic structural chemistry entirely neglected the need to explain how and why chemical bonds were made and broken. Many electrochemists retained the idea of electrostatic attractions and repulsions between atoms, and nothing better than that for explaining the chemical bond came along until electrons and quantum mechanics arrived. Meanwhile, the new discipline of physical chemistry began to make attempts to understand chemical reactions on the basis of thermodynamic and molecular–kinetic principles, developing in a very different direction from organic chemistry.[25] We should also note that it was quickly recognized that the valency of an element was actually not fixed, although the assumption of fixed valency did so much initially to establish organic structural chemistry. (This reflects a common pattern of development in science, which I have dubbed "epistemic iteration".[26]) What we have here is, again, a system of practice "zooming in" on what it can do best, leaving aside well-known anomalies. So the general picture of chemistry after the grand synthesis of organic structural theory is not one of a perfect unified science living happily ever after, but another pluralistic configuration of multiple imperfect systems competing and interacting with each other.

Let me conclude with a brief look back at the underdetermination of theory by evidence. Philosophers have had two diametrically opposite reactions to this problem. Some, joined by many sociologists and historians, celebrate the prospect of multiple theories and the lack of certainty in any of them. Others have an intense discomfort with underdetermination, and try to argue that when all the evidence is in, one and only one theory in each domain will emerge as the true or best one. A great deal of philosophical ink has been spilled on this debate. My study of the history of chemical atomism in the nineteenth century illustrates how both sides in this debate miss the point. Scientists develop various systems of practice containing different theories, which are suited for the achievement of different epistemic aims. In this way the great achievements of science come from *cultivating* underdetermination, not by getting rid of it.

3.2 Variety and Convergence in Atomic Chemistry

The aim of this section of the chapter is to give a *systematic philosophical account* of the development of our knowledge of chemical atoms in the earlier parts of the nineteenth century. My account will go up to the point where a reasonably firm consensus was reached on basic items such as H_2O as the molecular formula of

---

[25] See Servos (1990), ch. 1, for an exposition of the motivations that gave rise to physical chemistry.
[26] See Chang (2004, ch. 5, 2007a).

water; however, I hasten to add that where I have chosen to terminate my narrative does not imply the achievement of a simple and lasting unity at that point. The main interpretive innovation I offer is to analyze the development of chemical atomism in terms of various systems of practice in this domain, and the interactions between them. This is a vast area of study already with an extensive historical literature. I will generally rely on the existing histories, making my own study of primary scientific sources only occasionally as necessary.[27]

There is one thing to make clear at the outset: my focus is on the development of the knowledge of *chemical* atoms, not *physical* atoms with all their properties including shape, size, solid volume, internal structure, and motion. The establishment of credible knowledge of physical atoms only began in the late nineteenth century, and lies beyond the thematic and chronological scope of this book.[28] Recall Rocke's definition of chemical atomism, which was "universally (if implicitly and often unknowingly) accepted throughout the course of the nineteenth century" according to him: "there exists for each element a unique 'atomic *weight*', a chemically indivisible *unit*, that enters into combination with similar units of other elements in small integral multiples." (Rocke 1984, 10–12; emphases added) There is only one modification, or rather shift of emphasis, that I wish to make to Rocke's definition: chemical atoms are not simply weight-units; they are meant to be minimal units participating in chemical reactions, and such units can be discerned in various ways. I think this is close to what Rocke actually intends, when he says that "there is no single chemical operational definition" for chemical atoms. My story begins with how chemists learned to recognize the various properties of chemical atoms.

3.2.1   Operationalizing the Concept of the Chemical Atom

Working out atomic weights and molecular formulas was a painstaking achievement of nineteenth-century chemistry, made with no help from any direct means of observing the atomic–molecular level of reality. As we try to understand how this achievement was possible, I emphasize again that most nineteenth-century chemists were neither speculative metaphysicians theorizing about the ultimate atomic–molecular reality with little empirical basis, nor inflexible positivists categorically avoiding all theories and hypotheses that went beyond the observable realm. Rather, they learned how to integrate their atomic–molecular concepts directly with concrete empirical practices, and they were wary of concepts that could not be so integrated. This explains, for example, why Avogadro's ideas were not taken up by most chemists

---

[27] Otherwise, what is the point of secondary literature? For a list of authors I have found most helpful, even if I don't cite them extensively, see footnote 12.

[28] There are many instructive studies of physical atomism, with a strong focus on the debates about their reality. Knight (1967), Nye (1972) and the latter parts of Gardner (1979) are good places to start.

until they could get a more empirical handle on it. In general terms, my take on this history is in the same spirit as Alan Chalmers's on "how science succeeded and philosophy failed to gain knowledge of atoms". As he puts it, "nineteenth-century chemistry paved the way, rather than constituted a case, for an experimentally based and testable version of atomism." (Chalmers 2009, 188)

Here is a guiding idea, to help us understand the spirit of nineteenth-century atomic chemistry: in order to learn something real about atoms, *do* something with them. More generally, Ian Hacking (1983) has probably done more than anyone in recent decades to promote this sort of point of view in the philosophy of science. My own inspiration is the operationalist philosophy of science by the American physicist–philosopher Percy Williams Bridgman (1882–1961). My interpretation of Bridgman avoids taking operationalism as a narrow semantic doctrine reducing the meaning of a concept completely to its method(s) of measurement. Rather, I take operationalism as a philosophical outlook that emphasizes doings as the source of knowledge (see Sect. 3.3.1 and Chang 2009a for further discussion). There are certain resonances here with American pragmatism (see Sect. 3.3.3) and also with the ideas of Michael Polanyi and the later Ludwig Wittgenstein. In this broadly operationalist spirit, let us ask: what kind of concrete operations were chemists able to perform on chemical atoms, and what kind of things were they able to learn from the outcomes of these operations?

Here it is important not to get mired in a black-and-white debate about whether atoms are observable or not. Admitting that they of course remain much less observable than cats and dogs, we need to consider the various ways in which they have become more accessible to empirical investigations. Although atoms were, and still are, unobservable in the sense of not being directly accessible to human senses, it is not at all the case that atoms were purely theoretical entities in nineteenth-century chemistry. On the contrary, there were so many different ways of empirically engaging with chemical atoms, as I will explain shortly. Instead of focusing on what exactly "observability" should mean,[29] I want to consider the issue of *operationalization*, through which we gain a partial specification of the meaning of a concept by reference to a well-defined and clearly performable activity. For *physical* atoms, operationalization mostly involved some advanced technology or high-powered statistical reasoning, neither of which became available until the late nineteenth century. For *chemical* atoms, operationalization came much sooner and much more easily. By operationalizing the concept of atoms, chemists learned to do a pragmatic sort of metaphysics. This was a slow process of methodological learning, in parallel with the more empirical fact-learning that drove the whole enterprise onward. It is important that we reach a philosophical understanding of the mundane chemical operations, because it is these operations that truly formed the foundation of atomic chemistry.

Before delving into the details of how the atom concept was operationalized, I should make a couple of general remarks here about the nature of operationalization

---

[29] I do have a view on that issue, which is expressed in Chang (2005).

(see Sect. 3.3.1 for further elaborations). Those wishing to assimilate operational-ism to positivism tended to view operations as providing theory-free definitions of concepts. Bridgman himself may have shared that impulse at least sometimes, but it is a mistake all the same. Even the simplest of operational definitions embody cer-tain assumptions, if they are to make any sense to anyone. For instance, measuring length using a ruler is based on the assumption that the ruler itself does not change its length, with "length" understood in an intuitive and vaguely defined way. If my ruler seems to shrink and expand visibly in front of my eyes, I would begin to have doubts about its suitability as a ruler. At the same time, I would also come to realize that I had some pre-operational notion of what length is, which the operational definition builds on and refines (this, again, is an iterative process of development). So, an operationalization is not theory-free, and it can be disputed.

However, methods of operationalization do form a part of the very meaning of concepts. If the operationalization in question serves as an operational definition, then prior to that operationalization the meaning of the concept is not sufficiently determinate to allow definite judgments on the truth of all of the statements involv-ing the concept.[30] That is to say, the key assumptions involved in the operationaliza-tion of a concept do not have determinate truth-values, unless there is a prior definition of a concept capable of strictly regulating its meaning. Therefore, it is impossible to have an immediate and straightforward justification of a method of operationalization. This can be considered a generalization of what I have elsewhere termed "the problem of nomic measurement" (Chang 2004, ch. 2). These consider-ations have a direct bearing on the epistemic attitude that we ought to take about assumptions like Dalton's rules of simplicity, or EVEN (the equal volume—equal number hypothesis regarding gases). As central assumptions in methods of opera-tionalizing the concept of the atom, they inhabit an interesting twilight-zone of testability. At least in relation to the earlier stages of the science, we should not regard these assumptions as ordinary hypotheses subject to empirical testing. Rather, assumptions used in an operational definition of the chemical atom will be tauto-logically true, *within* a system that employs that definition. Or rather, I think of them as metaphysical principles that enable a particular kind of epistemic activity.[31]

3.2.1.1 Weighing by Equivalence

As noted in the first section of this chapter (Sect. 3.1), Dalton's great innovation was to assign weights to atoms. But here we need to ask how exactly he was able to achieve that, in operational terms. Think of it: Dalton and his followers somehow "measured" the weights of atoms using no more than an ordinary balance! Actually the initial groundwork for the operationalization of atomic weight was laid before Dalton's work, without much thought to atoms. It began most clearly when

---

[30] See Sect. 3.3.1 C1 for a further discussion of the different between definition and meaning.

[31] This sense of "metaphysical principle" or "ontological principle" is explained in Chang (2008, 2009c).

Jeremias B. Richter (1762–1807) in the 1790s recognized that there were interesting fixed relationships between the amounts of acids and bases that neutralized each other. For example, 793 parts (by weight) of lime became neutralized by 577 parts of carbonic acid, by 712 parts of muriatic acid, and by 1,000 parts of sulphuric acid. As these various amounts of the three acids all performed the same chemical function, Richter considered them to be "equivalent" to each other in this context. In fact Cavendish had already described such substances as "equivalent" to each other as early as 1766. Berzelius in 1807 was inspired by Richter's work to measure the equivalents of all known acids and bases, which is what initially set him off on his massive experimental project in analytical chemistry.[32]

All of this was before Dalton's publication of the atomic theory, and it was only a short step from equivalents to chemical atoms—so much so, that Justus Liebig (1851, 96) and some others have traced the origin of atomic chemistry to Richter rather than Dalton, perhaps with a tinge of Germanic nationalism. If one asked *why* Richter's equivalence relations should have obtained, one very plausible answer was that there were chemical atoms, namely discrete units of the substances in question, whose weights were in the same ratio as that of the observed macroscopic equivalents. However, hidden in this atomistic rendition of Richter's equivalent is a very significant step, namely the assumption that the combinations in question are one-to-one between the atoms[33] of the two substances involved in each compound. This one-to-one assumption is what gives operational meaning to "atom" in this context, and it seems to me that this was missing in Richter's thinking.

Another method of tracing chemical equivalents was provided by replacement reactions. There were certain chemical reactions in which one substance replaced another in a compound: schematically, $AB + C \rightarrow AC + B$. For example, as Pierre Duhem ([1902] 2002, 57) points out, this type of replacement reactions taking place in solutions of metals in acids were well-known for many centuries, and these reactions did give chemists access to chemical atoms. In such reactions, the substances B and C could be considered *chemically equivalent* to each other, and from their relative weights, atomic weights could be deduced. Again, this was based on the crucial assumption that B and C replaced each other one-to-one atomically, which would come to be questioned later. As I will explain further in Sect. 3.2.2.4, from the 1830s onward another kind of replacement reactions caught the excited attention of chemists. These were the substitutions of substances by other substances very unlike themselves, starting with the hydrogen–chlorine substitutions. These substitutions raised a chemical puzzle—what was it that chlorine and hydrogen had in common, so that they were able to take each other's place within a molecule? Setting that explanatory question aside, what was important in

---

[32] See Lowry (1936), 310–311, for these brief descriptions.

[33] In acid–base neutralization one is dealing with "compound atoms" rather than elementary atoms, but the conceptual structure is the same. In fact, from Dalton until the middle of the century it was perfectly routine for chemists to speak of the "atoms" of radicals and other compound units; see, for instance, the work in organic chemistry discussed in Klein (2001). The modern usage of "molecule" did not take universal hold till later.

terms of operationalization is that these substitutions unearthed previously unnoticed equivalents, to be added to the network of relations grounding the overall system of atomic weights.

3.2.1.2 Weighing by Combination

Not all determinations of atomic weights were carried out by means of equivalence, for two main reasons. First of all, there are not all that many sets of reactions displaying direct equivalence of chemical roles. Even more crucially, there were many sets of reactions with ambiguities that exploded the complacency of one-to-one atomic combinations that could be maintained in the equivalence-settings. Schematically: if one has $AB + C \rightarrow AC + B$, there is an equivalence relation between B and C, and one might easily presume that one atom of C has replaced one atom of B, since C and B are chemically analogous (e.g., A is an acid, B and C different metals or bases). On the other hand, if one only has $A + B \rightarrow AB$, then there is no equivalence relation to exploit there (e.g., A is nitrogen and B oxygen, with no chemical equivalence between them except in the fact that they combine with each other). Worse, if one has multiple compounds made up of A and B (e.g., Dalton's five nitrogen oxides), then it is not clear how many atoms of A and B are involved in all the different compounds. In such cases, more conceptual structure needed to be provided in order to enable atomic-weight determinations from gravimetric chemical analysis.[34] Then the key assumption in the operationalization has to be made explicit; any assumptions fixing the relevant molecular formulas could serve, by allowing the determination of atomic weights from the combining weights in any chemical reactions involving elementary substances.

That is precisely what Dalton's rules of simplicity were designed to achieve. And the first rule was relatively uncontroversial: if there is only one combination of two elements, treat the resulting compound as a one-to-one atomic combination; this can be regarded as a rather unthinking extension of the equivalence-based thinking. Dalton does not seem to have entertained much doubt about this assumption, and it was also accepted by many others including the two leading advocates of Dalton's theory in the early years, namely Thomson and Wollaston. This assumption of one-to-one combination, at this stage of inquiry, was not really an empirical hypothesis, but part of a scheme of operationalization; it was an important part of what "chemical atoms" meant for these chemists. Now don't go protesting: "But a water molecule *is* either a clumping-together of one oxygen atom and one hydrogen atom, or it *is not*!" Although Dalton himself did have a picture of atoms as tiny little balls, that was in the realm of unoperationalized theory, and the initial operationalization of the chemical atom was not directly linked to any particular picture of the physical atom. Dalton, Thomson and Wollaston were all agreed about water being HO, and

---

[34] This is just the kind of productive theory-ladenness of observation that Norwood Russell Hanson spoke of (see Chap. 2, Sect. 2.2.1).

the atomic weight of oxygen being roughly 8 rather than 16, at the operational level. But two elements forming only one compound is rather the exception than the rule (and water was a false case of it), so further operational rules were needed in assigning weights to most other elements.

Consider the case of carbon oxides, which I have used in the first section (Sect. 3.1) in order to illustrate the underdetermination problem. Here is Dalton's verdict (1808, 215): "carbonic oxide is a binary compound, consisting of one atom of charcoal, and one of oxygen, together weighing nearly 12"; "carbonic acid is a ternary compound, (but sometimes binary) consisting of one atom of charcoal, and two of oxygen, weighing 19".[35] But how did he reach the conclusion that carbonic oxide was binary and carbonic acid was ternary? A clue lies in Dalton's puzzling statement that carbonic acid was ternary "but sometimes binary"; I think he meant that, operationally, carbonic acid could be made directly from carbon and oxygen (i.e., by the combustion of carbon), but also by a combination of carbonic oxide and oxygen (i.e., by the combustion of carbon monoxide).[36] Now, if carbonic acid can be formed by adding an oxygen atom to a carbonic oxide atom, it is pretty clear which of the two carbon–oxygen compounds should be regarded as CO and which one as CO_2. So Dalton seems to have taken the actual laboratory methods of synthesis as a guideline for inferring molecular constitution (this comes to treating synthesis as a method of operationalizing molecular formulas). I think my interpretation is corroborated by what he says in the case of nitrogen oxides: "nitric acid is a binary or ternary compound *according as it is derived*, and consists of one atom of azote and two of oxygen, together weighing 19" (*ibid.*, emphasis added).[37]

Another instance of what I will call "weighing by combination" can be seen in the work of William Hyde Wollaston (1766–1828), London-based physician-turned-chemist who made a lasting contribution to chemical atomism with his paper of 1814. Rocke (1984, 12) notes that Wollaston's atomic-weight determinations were "dictated by an assumed formula for the lowest oxide of the element in question." It is useful to examine the starting point of Wollaston's enterprise in full. Stressing the great utility of the carbonate of lime (in today's terms, calcium carbonate, $CaCO_3$) in analytical chemistry, he stated:

> The first question, consequently, to be resolved is, by what numbers are we to express the relative weight of carbonic acid [as a key constituent of the carbonate of lime], if oxygen be fixed at 10. It seems to be very well ascertained, that a given quantity of oxygen yields exactly an equal measure of carbonic acid by union with carbon; and since the specific gravities of these gases are 10–13.77, or as 20–27.54, the weight of carbon may be justly represented by 7.54, which, in this instance, is combined with 2 of oxygen forming the deutoxide, and carbonic oxide being the protoxide will be duly represented by 17.54. (Wollaston 1814, 8)

---

[35] Modern measurements would give these weights as 28:44, instead of Dalton's 12:19.

[36] Dalton does not seem to have been entirely consistent in usage; in another context he would only call "binary" what is strictly made up of only two elementary atoms.

[37] It should be noted that Dalton's reasoning takes it for granted that after clumping together the composite atom made up of carbonic oxide and oxygen does not sub-divide; interestingly, that sort of post-combination division was precisely what Avogadro felt compelled to assume.

This is a bit mysterious, but it does make sense. Like Dalton, Wollaston started from the knowledge that carbonic acid contained more oxygen than carbonic oxide, and on that basis he inferred that carbonic oxide was a one-to-one combination of carbon and oxygen. Wollaston's general principle, or rule-of-thumb at least, must have been that the "lowest" known oxide of an element was a one-to-one atomic combination with oxygen. A simple calculation based on that principle gave Wollaston a ratio of 7.54:10 for the weights of carbon and oxygen, which was crucial for building up the rest of the system.

Some caution is necessary as we try to place Wollaston's work in the history of chemical atomism. The 1814 paper is where he gave his "synoptic scale of chemical equivalents", presented with such a strong theory-free gloss that it is difficult to spot his additional assumptions, and many commentators have said that Wollaston made no commitment to atoms, sticking only to observed regularities of combining weights. On the contrary, I think Rocke (1984, 12) is correct in his argument that Wollaston's equivalents were "operationally identical with chemical atomic weights". And there is a misleading aspect to Wollaston's use of the term "equivalent". I prefer to reserve the term to situations in which there was an operational *equivalence* between a set of substances (e.g., via neutralization or replacement), as discussed above in Sect. 3.2.1.1. Wollaston meant something more general, incorporating the relative weights by which substances combined with each other in any kind of circumstance. Wollaston's lumping of the different meanings here lends further credence to Rocke's argument that his "equivalent" was quite a theoretical concept, not a strictly operational one.

3.2.1.3   Counting by Volumes

With the conception of atoms as the smallest unit-amounts involved in chemical reactions, a fundamental aspect of operationalization was telling how many portions of those units were present in a sample. Although direct one-by-one atom-counting was not possible in the early nineteenth century, it was possible to determine the *relative numbers* of atoms participating in chemical reactions. Atom-counting could be done via combining weights if atomic weights were already known, but the lingering uncertainty about atomic weights stood in the way of that. The use of combining *volumes* was the most compelling method of atom-counting (and molecule-counting) in the early days. As hinted already, chemists only began to make real progress on this front when they realized that they could set aside weights and go directly to volumes as an independent measure of the relative number of atoms. I propose that many nineteenth-century chemists used volumetric atoms on a par with gravimetric atoms as complementary operationalizations of chemical atoms. Berzelius's attitude about volumes (quoted above on p. 146) then ceases to seem mysterious or cavalier. When Berzelius said that volumetric reasoning was "founded upon a well constituted fact" while (gravimetric) atomic theory had "only a supposition for its foundation", atom-counting must have been what he had in mind. The operationalization of the atom by volume was not a matter of regarding EVEN as a correct hypothesis. No one was able to test EVEN directly at this stage, nor offer a firm theoretical justification for it. Chemical atomists who were engaged in volumetric atom-counting *presumed* EVEN as a metaphysical principle (usually without even mentioning Avogadro's name). EVEN enabled them do volumetric atomic chemistry; they were going to *live* it, and test it only indirectly by the success of that life. Physical objections to EVEN offered by Dalton and others had no immediate impact on this volumetric operationalization of the number of atoms.

As discussed in the first section (Sect. 3.1), in the hands of Hofmann and others volumetric atom-counting contributed directly to the establishment of valency in mid-century. But volumetric reasoning was in use long before then, and an interesting example can be found in Thomas Thomson's textbook (1831, 166–169). Thomson, though primarily focused on weights, seems to have been more liberal than Wollaston and Dalton in making heuristic use of various types of operational clues, including volumes. He resolved the ambiguity in the atomic constitution of carbon oxides as follows. First he noted that 1 volume of carbonic oxide combined with ½ volume of oxygen to make 1 volume of carbonic acid. Then he noted that "carbonic acid gas contains its own volume of oxygen gas." In other words, 1 volume of carbonic acid was made from 1 volume of oxygen combining with some solid carbon. We might write the volume-equations as follows:

(i) 1 carbonic oxide + 0.5 oxygen = 1 carbonic acid
(ii) carbon (unknown volume) + 1 oxygen = 1 carbonic acid

The left-hand sides of the two equations must be equal to each other, so there must be 1 volume of oxygen in the left-hand side of (i) as well as (ii), which means that there must be 0.5 volume of oxygen contained in 1 volume of carbonic oxide (while there is 1 volume of oxygen contained in 1 volume of carbonic acid). So Thomson was able to conclude that carbonic acid contained twice as much oxygen as did carbonic oxide, which gave him the formulas CO_2 and CO.

Volumetric reasoning provided a plausible starting point in atom-counting, though it had clear practical limitations as there were numerous elements and compounds that chemists at the time could not volatilize (i.e., turn into gaseous form). The volumes occupied by atoms in their liquid or solid state were not operationalized until much later, so it is a matter of serendipity that some of the most crucial elements and compounds in nineteenth-century chemistry did exist naturally in gaseous forms under standard earthly conditions. It is also serendipitous that most of these elements exhibited straightforward volume-relations, unlike phosphorus, arsenic and sulphur for instance (see Sect. 3.2.2.3 on the latter).

3.2.1.4 Counting by Specific Heat

Although most chemists did not accept at face value Dalton's idea of atoms surrounded by the atmosphere of caloric, the measurement of heat had quite an important role in atomic chemistry. This was through the law of "atomic heat" advanced by Pierre Dulong (1785–1838) and Alexis-Thérèse Petit (1791–1820) around 1820, which stated the constancy of the product of atomic weight and specific heat for all elementary substances.[38] Since the specific heat of a substance is measured as the quantity of heat needed per unit weight to raise its temperature by a unit amount, Dulong and Petit's law amounted to saying that an atom of any substance had the same capacity for heat ("atomic heat"). This may actually be easier to see when set out in formulas:

Specific heat (by weight) $= \Delta H/W$
Atomic weight $= W/N$
Specific heat $\times$ Atomic weight $= (\Delta H/W) \times (W/N) = \Delta H/N$

In the formulas above ΔH is heat input required for raising the temperature of a body by $1°$, W is the weight of the body and N is the number of atoms in the body. If $\Delta H/N$ is constant regardless of the nature of the substance, that would imply that each elementary atom, of any kind, takes up the same amount of heat as the temperature goes up by the same amount.

Dulong and Petit's law was freely admitted to be only approximate, and there never was a good theoretical explanation for it, either. But it was difficult to deny that there was *something* behind it, and its judicious use helped chemists decide whether atomic weights had been estimated as multiples or fractions of the real value on the basis of mistaken molecular formulas. In other words, Dulong and Petit's law gave chemists another method of atom-counting, and therefore helped them fix molecular formulas and atomic weights. It was impossible to assign a precise value to atomic heat itself, but that is not what the nineteenth-century atomic chemists were after. Rather, the approximate constancy of atomic heat was a suffi-cient basis for telling how many atoms were involved in a given reaction. Given the fundamental atomist assumption that there could only be an integer number of each type of atom involved in a reaction, it didn't matter if an operational method of atom-counting was not precise. In the game of distinguishing 2 from 3, a value of 2.13 (or whatever) counted as 2, not 3, and that was good enough.

Had Dalton helped himself to this method, he could have resolved his uncer-tainty (expressed in Fig. 3.1, the column of numbers from 1827) about whether the atomic weight of nitrogen was 5 or 10, and whether the atomic weight of copper was 28 or 56. The Dulong–Petit law was particularly useful for determining the atomic weights of various metals. In the weight-based operationalization of metal-lic atoms the most common and reliable reactions were oxidations, but there was always a great deal of uncertainty about how many atoms of oxygen a metallic oxide contained, and it was common to have multiple oxides of a given metal. Petit and Dulong themselves used their law in order to correct some of the atomic weights that Berzelius had just published in 1818 (see Freund 1904, 363–365 for details). Thomson, again, provides some good examples. He used the Dulong–Petit law to fix the atomic weight of mercury and the formulas of mercury oxides; this also helped him to fix the formulas for copper oxides, through chemical analogy.

---

[38] See Freund (1904), chapter 14 and Fox (1968) for a detailed treatment.

In the case of water and hydrogen peroxide he found that chemical analogy and volume-considerations gave opposite verdicts, so he had recourse to specific heats—which actually ruled in favor of water being HO. Thomson says (1831, 9–12): "According to the experiments of Delaroche and Bérard, the specific heat of hydrogen gas referred to water is 3.2936. Now 0.376/3.2936=0.114 [0.376 being Thomson's value for the specific heat of oxygen]—a number much nearer 0.125, than to 0.0625. The specific heat then naturally leads us to determine in favour of 0.125 as the true atomic weight of hydrogen." Thomson was taking the atomic weight of oxygen as the unit, so 0.125 for hydrogen corresponds to oxygen's atomic weight being 8 if that of hydrogen is taken as the unit, and 0.0625 for hydrogen corresponds to oxygen being 16. Ironically, the modern values give an oxygen/hydrogen specific-heat ratio of 0.0645, which is very close to 0.0625!

3.2.1.5 Sorting by Electric Charge

Another clear method for operationalizing the concept of the atom was electrolysis. It is striking that electrochemical–atomic reasoning and techniques were already in place even before Dalton published his atomic theory. As I have discussed in Chap. 2, when one witnessed how electrolysis consistently produced different types of products at the two poles of the battery (oxygen and acids at the plus side, and hydrogen, metals and alkalis at the minus side), it was nearly impossible to deny that different substances had different electrical properties. It was plain that atoms were somehow inherently electrified, or at least liable to be electrified in particular ways. That much was granted by nearly everyone, even Ritter. Despite the deep mystery still surrounding the very nature of electricity and the exact mechanism of electrolysis, at the operational level there were definite and stable meanings to be assigned to the notion of electropositivity and electronegativity—these operational meanings were agreed even by Faraday despite his idiosyncratic ontology. All the elements could be placed on a spectrum, from oxygen as the most electronegative to potassium as the most electropositive. That already was a significant piece of operationalization of chemical elements, which was easily extended to atoms. The electrochemical operationalization of atoms had different versions (see Chap. 2, Sect. 2.2.3.2), but all of them could connect with the old notion of affinity: the farther away two elements were from each other in the electrochemical series, the stronger was the attraction between them. The electrochemical series also mapped nicely onto the old displacement series of metals (a classic expression of chemical affinity), and it has remained robustly in chemistry ever since then, in one form or another.

That is, however, as far as easy consensus went. In order to use electrolysis to operationalize the atomic *constitution* of compound substances, additional assumptions were required about what it was that the application of Voltaic electricity did to chemical substances. Anyone treating electrolysis as decomposition could not avoid a commitment to the dualist ontology in the atomic realm. If one engages in the activity of electrolysis-as-decomposition, the dualist ontology of chemical substances is already written in the bi-polar structure of the battery; dualism is a metaphysical principle inherent in the activity of electro-analysis. John Frederic Daniell (1790–1845), Professor of Chemistry at King's College London, took a further step in 1840 when he proposed to re-conceive salts as composed of "metallic" and "non-metallic" radicals, rather than an acid and a base as traditionally held. According to Lowry (1936, 270), Daniell's argument was driven by an electrolytic operationalization, on the ground that the metallic/non-metallic decomposition was what the action of electricity first effected before the acids and bases appeared as secondary products. I will not follow the reception of Daniell's work here, but would like to note that Daniell's argument serves as a reminder that the electrolytic operationalization of chemical composition rested on the assumption that the action of electricity broke up a molecule at the *natural* dividing line within it. This assumption was necessary for the standard electrolytic operationalization of atoms and molecules, which could not be subjected to an empirical test within any system of practice resting on that operationalization. Daniell was only taking that principle to its logical conclusion. But all of that is not to say that the traditional acid–base conception of the composition of salts was simply wrong; rather, it was the product of a different operationalization, based on the more traditional methods of synthesis and analysis.

3.2.2   Competing Systems of Atomic Chemistry

The discussion in the previous section should make clear that there were a number of different ways of getting at the atom empirically. Such operationalizations form the nuclei around which empirical systems of practice grow.[39] If we look at the actual history of atomic chemistry we find that there was a bewildering profusion of competing systems, almost one for each important chemist. In an attempt to create a clear framework of analysis, I will engage in an oversimplification and identify five idealized systems; these were already named in the first section (Sect. 3.1), and here they will receive a fuller characterization. In real history, many individual chemists had systems that drew from more than one of these idealized ones; I will try to give some indication of the shapes of these hybrid systems too, as I proceed with my description of the idealized systems. I will discuss the main aims and activities of each of the idealized system, as well as the hypotheses, beliefs and assumptions that were involved in those activities. If what I am doing seems too simplified even if I am identifying as many as five parallel systems in play, that is precisely one of the main points that I want to make about the first half-century of atomic chemistry: it was a complicated field, and for good reasons.

Even accepting the simple point that the historical situation was complicated, and accepting that philosophers may need idealized frameworks, one might still

---

[39] The image of the nucleus here is a new progressive coherentist metaphor, to supplement that of building on the round earth, which I have used in Chang (2004, ch. 4, 2007a).

question whether there is any historical reality to my idealized systems. And if not: what is the point in discussing non-existent entities in history? What I would argue is that my idealized systems are not entirely fictional. First of all, there were some chemists who did operate systems very much like the idealized versions that I describe, as I will point out below; there were also cases of closely related systems, all of which shared the core activities and assumptions of the idealized system. And regardless of the degree of actual realization, I think the idealized systems did serve as templates, or as coherent visions of how to do atomic chemistry, for the scientists at the time. So the idealizations are not *retrospective* but *contemporary* fictions. Such idealizations do serve crucial roles as ideologies, which may never be actualized in full detail but still guide people's thinking and actions. An analogy: it would be fair to say that there has never been a purely Marxist economy, nor a completely free-market capitalist economy. Still, such idealizations have served as extremely strong ideals and frameworks for economic policy and practice; often people mistake the imperfect versions as the real thing, or complain about the imperfections that they recognize, or conceptualize what they have as hybrids of the idealized types. I am putting forward, as a historiographical hypothesis, a claim that the five systems of atomic chemistry that I have identified have served as such contemporary fictions shaping the work of nineteenth-century chemists. I hope that my account in the rest of this chapter will give this hypothesis sufficient plausibility to induce others more competent than myself to put it to a serious test.

3.2.2.1   The Weight-Only System

Of the five systems that I will describe, the first one to become established was what I will call the *weight-only system*. In this system the chemical atom was conceptualized using only the first two of the operationalizations that I've discussed. It would be fair to say that the weight-only system arose from chemists' disappointment with the full-fledged version of Dalton's theory. This was a sort of modest and sanitized version of Daltonianism,[40] an attempt to deal in weights only without the distractions of other atomic parameters or properties that were considered more uncertain or less essential, renouncing Dalton's attempts to explain chemical and physical phenomena by reference to the shapes and sizes of atoms as well as their weights. The atomic weights employed were determined through both equivalence and combination (see Sects. 3.2.1.1 and 3.2.1.1). As discussed earlier, both of those operationalizations embodied certain assumptions about molecular formulas. The weight-only system has often been mistaken as an expression of positivism; however, as Rocke (1984, 10 and *passim*) has shown, this system still relied on a notion of chemical atoms if not physical atoms, and what it embodied was not positivism but a pared-down ontology of atoms only possessed of weights.

---

[40] And not "Daltonism", which refers to the red–green color-blindness, which Dalton suffered from and published a paper about.

Perhaps the best early exponent of the weight-only system was Wollaston—or rather, the Wollaston of the 1814 "synoptic scale of chemical equivalents", which enjoyed great popularity; Rocke says that in Britain Wollaston's weights were used "almost exclusively until the 1860s."[41] William Odling (1858a, 41; 1858b, 108) also noted that this system, with HO, was "ordinarily made use of in this country" (Great Britain). Thomson's practice also came close to the weight-only system, but he did not refrain from using volumetric and thermal operationalizations when convenient, so his system was somewhat hybrid. Although Germany was later to become the center of other, less restrictive systems of atomic chemistry, there was also a robust weight-only tradition there. Leopold Gmelin (1788–1853), longtime Professor of Medicine and Chemistry at Heidelberg (1817–1851), used it in successive editions of his influential *Handbuch der Chemie* (1843, etc.). Rocke (2010, 12) reports that Gmelin in 1838 convinced Liebig, Wöhler and two other leading German scientists that they should abandon unfounded theorizing and revert to Wollaston's system. More than a decade later, Liebig was still expounding a nearly pure version of the weight-only system in the third edition of his *Familiar Letters on Chemistry* (Liebig 1851, Letter VI, esp. p. 89), giving the equivalent weights of 1, 6, 8 for hydrogen, carbon and oxygen. The molecular formula of water in most variants of this system was given as HO. It is not logically necessary that a weight-only system should have water as HO; however, in the absence of volumetric operationalization, which provided the most convincing motivation to go for H_2O, simplicity considerations usually prevailed in favor of HO.

What did one *do* in this weight-only system of atomic chemistry? What were its core aims, and its chief activities? It is interesting to examine Wollaston's paper again in this light. He stated that in resolving ambiguities of atomic weights in his system, he had "endeavoured to make practical convenience [his] sole guide" (1814, 7). Practical convenience for what? From the rest of the paper it is quite clear that this was about the practice of chemical analysis. Scanning the body of Wollaston's research in chemistry more generally, we can also recognize that what he meant by chemical analysis included both the determination of the composition of various compounds in terms of their constituents, and the characterization of new or relatively unfamiliar elements. The most typical activity of the quantitative analytical chemist at that time was to tease out information about the substance to be studied by making it react with another substance, which was presumed to carry off a constituent part of the target substance and left the rest; often it was necessary to make various reactions and get all the different reaction products weighed up, before a substance could be characterized with sufficient certainty and accuracy. Wollaston (1814, 1–2) opened his paper by reminding the reader that at least 20 different weight-determinations were required in a full analysis of a substance as simple as "common blue vitriol" (copper sulphate). In "this field of patient investigation" it

---

[41] As Rocke (1984, 64–66) notes, Wollaston maintained a more realist side to his atomic chemistry as well. His 1812 Bakerian Lecture attempted 3D models (Wollaston 1813); see also Wollaston (1822) on the finite extent of the atmosphere.

was of great help to be in possession of *known* fixed ratios of combination, to cut down on the number of necessary fresh determinations and along with it the scope for uncertainty and error. On the other hand, the use of atomic weights also transformed the aim of analysis into the determination of the numbers of various atoms constituting each compound, in addition to the mere weight-ratios of the elements entering into combination.

There was certainly a lot of valuable work there to be done in the analysis of the myriad of chemical substances, often with great practical and economic utility as well. Analytical chemistry was an important enterprise, touching on many facets of human civilization including mineralogy and pharmacology, right down to the much-demanded analysis of spa waters. Weight-only atomic chemistry was perfectly suited for this analytical enterprise. Thomson's research was in a similar vein as Wollaston's, and more extensive. In the next generation, Liebig's mastery of the combustion-method of organic analysis spawned a major research program.[42] This was very demanding work, both because of the growing number of new substances to be tackled, and because of the high level of accuracy required due to organic molecules being composed of numerous atoms of only a few elements. Liebig's program generated a great deal of excitement and activity, but after that the weight-only system seemed to run out of steam. It did not support a broad enough range of research activity to be able to sustain the attention of ambitious chemists in the second half of the nineteenth century. But even so, it remained a useful system for supporting teaching and practical applications. This is a reminder that good systems of practice do not always make productive *research* programs.

3.2.2.2   The Electrochemical Dualistic System

Associated most often with the name of Berzelius (and also Davy), the *electrochemical dualistic system* of atomic chemistry grew up around the electrolytic operationalization of atoms (see Sect. 3.2.1.5), in addition to the two weight-based operationalizations (the latter were in fact used in all systems, though it was in principle possible to do atomic chemistry without them). Berzelius was fully committed to the electrical nature of atoms; this is perhaps not surprising, given that his first important work in chemistry was in electrolysis, as we have seen in Chap. 2. In the dualistic system, the most sensible thing to do with water was to regard it as a binary combination of hydrogen and oxygen; we have seen in Chap. 2 that this was how most electrochemists except Ritter and his fellow-travelers viewed it, and that trend continued in the dualistic system. (Curiously, Berzelius himself created the glaring deviation here, as he went with H_2O with the help of volumetric thinking and many others followed him.)

For Berzelius and his followers, electrochemical dualism was also integrated with the Lavoisierian legacy. How nice it was that oxygen, the linchpin of Lavoisier's

---

[42] On Liebig's school of organic analysis, see Morrell (1972), Brock (1997), and Jackson (2009).

chemistry, turned out to be the most electronegative of all known elements! Having retained Lavoisier's emphasis on the key role of oxygen, and having added a layer of electrochemical significance to it, Berzelius now incorporated these ideas into atomic chemistry. What he created was the mainstream of Lavoisierian chemistry in the age of electrochemical atomism. The electrochemical dualistic system enjoyed an impressive degree of dominance for a time, though that dominance was relatively brief and never total even at its height.

What were the main aims and activities of this system of atomic chemistry? Electrolysis gave chemists a splendid operational and conceptual tool with which they could probe into the constitution of substances. So, the most immediately important dualistic activity was analysis, but with a different mode of operation and a shift of emphasis compared to analysis in the weight-only system. Since Voltaic electricity was a new power at the disposal of chemists, it was most impressively used to effect reactions that were not known before, rather than demonstrating compositions that were already known. Again setting Ritter aside, these reactions were mostly understood as decompositions of tightly bound molecules that had resisted decomposition by other methods. Davy isolated several new elements by this means, starting with potassium and sodium in 1807, followed by strontium, barium, calcium and magnesium (see Lowry 1936, 281). All this was an outcome of a clearly conceived activity of electrolytic decomposition.

Another mainstay of the dualistic program was the activity of classification. I have already mentioned the ordering of elements along the electropositive–electronegative spectrum. When it came to the classification of compounds, the most important idea was the concept of "radical" (sometimes written as "radicle"; from Latin *radix*, root). In Lavoisier's theory, an acid was a compound of oxygen and a radical; oxygen made it acidic, and the type of radical determined the type of acid. Berzelius took up this notion and "electrified" it by identifying radicals as electropositive bodies attracted to the electronegative oxygen, and then extended this scheme into organic chemistry. Soon the notion became more generalized so that a radical could combine with various atoms and other radicals, not just with oxygen. At that point a "radical" came to mean any group of atoms that behaved as a stable chemical unit. The first radical to be discovered in this new vein was cyanogen radical (CN), which formed prussic acid by combination with hydrogen (not with oxygen); this occurred in the course of Gay-Lussac's work on prussic acid published in 1815 (see Ihde 1984, 185). Radicals were very handy classificatory tools as they could be used to put organic substances into *families*, rather than leaving them in a vast unordered set of compounds containing many and varied numbers of oxygen, carbon, hydrogen and nitrogen atoms. In the organic realm it would not have been much use to apply the familiar inorganic categories of "oxides", "hydrides", etc.

With growing confidence in the reality of radicals as entities of atomic–molecular chemistry, chemists began to use radicals as active tools of constitutional research as well. Klein (2001, 2003) argues that Berzelian atomic symbols in general worked as "paper tools" facilitating chemists' constitutional thinking, bits of formula serving as abstract chemical building-blocks to be used in paper-and-pencil operations. Radicals were the best embodiments of this practice, and they had clear potential to serve as links to laboratory practice, too. It became a treasured activity of ambitious chemists to identify a nice organic radical on paper through a comparative examination of the molecular formulas of various compounds, and then go hunting for it in the lab. Some high-profile successes made this an exciting cutting-edge of research, with a great deal of innovative work on the etherin, ethyl and benzoyl radicals (see Ihde 1984, 184–189). Hermann Kolbe (1818–1884) was the best die-hard cultivator of this practice, starting with his successful work in the electrolysis of organic acids (see Rocke 1993).

Another great advantage of dualism was that it provided a clear understanding of chemical bonding.[43] Explaining various instances of chemical bonding by reference to the inherent electric charges of atoms was a key activity in the electrochemical dualistic system. Berzelius developed a rather intricate view here involving polar atoms, but for those who followed him vaguely the key theoretical practice was to analyze a compound into two parts, one with positive and the other with negative electric charge; then each part would either be an element, or itself analyzed further dualistically into electropositive and electronegative parts, until elements are reached. At each step the combination was explained straightforwardly in terms of the electrostatic attraction between particles of opposite charge. Berzelius also explained the heat and light produced in chemical combination as a consequence of the neutralization of opposite electricities. Because the electropositivity and electronegativity of atoms had such a firm operational basis rooted in electrolysis (see Chap. 2, Sect. 2.2.3.1), these explanations were quite convincing to many people.

3.2.2.3 The Physical Volume–Weight System

The heart of what I call the *physical volume–weight system* was to take *both* gravimetric and volumetric operationalizations of atoms, and make them consistent with each other, *at any cost*.[44] One clear way of doing so was Avogadro's method, as explained in the first section (Sect. 3.1), which rested crucially on EVEN. It may not be obvious at first glance how EVEN related to the weight-based operationalization of the atom. The key here is to recall that the assignment of atomic weights through equivalence or combination required assumptions regarding molecular formulas; EVEN, as a method of atom-counting, provided those molecular formulas. What that enabled was a joint gravimetric–volumetric operationalization of the atom. Avogadro's program was to take EVEN for granted and embrace all of its consequences. This was a system of chemical *and* physical atomism, in which the

---

[43] There is much historical literature on Berzelian dualism, but see Brock (1992), 147–159, and also Ihde (1984), chapter 5, for concise and accessible introductions. A very extensive and detailed treatment can be found in Melhado (1980).

[44] This, in my view, constituted premature unification at the start, which we should not necessarily praise just because it turned out well in the end. But that is certainly not to deny that it did not make a promising and productive avenue of inquiry.

minimum units participating in chemical reactions were particles possessed of weight and volume (and presumable other physical properties as well). As noted above, Ampère had a similar approach, and Gaudin made a serious attempt to develop his system (see Mauskopf 1969). Cannizzaro's success was based on an *unflinching* adherence to the core tenets of the physical volume–weight system. In all instances of the physical volume–weight system, water was unequivocally H_2O.

There were two very important consequences of EVEN, which were closely linked with the main aims and activities of the physical volume–weight system. First, EVEN implied that certain elementary substances were seen as consisting of bi-atomic molecules, as explained in the first section (Sect. 3.1). This assumption helped remove the main difficulties in adopting the H_2O formula. The acceptance of such physical consequences is indicative of the realist and reductionist aims of the physical volume–weight system: to get at a literal physical picture of the microscopic realm, and to offer explanations of chemical phenomena on the basis of that picture. This realist agenda created additional pressure for the verification of the key assumptions in the physical volume–weight system, and unfortunately physical justification for bi-atomic elementary molecules was not forthcoming for a very long time (until covalent bonding was explained by quantum mechanics). Both Dalton and Berzelius had serious objections to the idea, and their objections were detrimental to the credibility of the physical volume–weight system. Although Berzelius accepted and used EVEN, he only applied it to elementary substances (see Freund 1904, 333); therefore, he had no need for Avogadro's hypothesis of bi-atomic molecules for elementary gases.

Secondly, EVEN implied that vapor densities were straightforward measures of molecular weights. Consequently, the determination of vapor densities became a crucial activity in this system, and it also became very important to vaporize substances that had only been known in the solid and liquid states before. But the vapor-density work also revealed difficulties. In 1826 Jean-Baptiste Dumas (1800–1884) started enthusiastically on this program, naming Avogadro and Ampère explicitly and declaring EVEN to be "a hypothesis admitted by all physicists".[45] By the mid-1830s he was so bewildered by the results he obtained that he abandoned EVEN altogether: "You can accept if you like that equal volumes of all gases contain equal numbers of some sort of atomic or molecular groupings; no one will disagree with you. But so far this has been of no use to anybody."[46] This experience drove Dumas away from physical atomism altogether, though he remained implicitly committed to chemical atomism like most chemists (see Ihde 1984, 150–153). Later Cannizzaro ([1858] 1910), Hofmann (1865) and others would obtain tidier results; they also bit the bullet that Dumas had dodged, accepting that, for whatever reason, the elementary molecule of phosphorus was P_4 and that of sulphur S_6 for instance.

How effective was the physical volume–weight system as a research program? After the vapor-density work hit a dead-end there was not very much one could do,

---

[45] See Partington (1964), 218f.

[46] Quoted in Fisher (1982), 88, from Dumas, *Leçons de philosophie chimique* (1837).

although any new technical advance allowing a new vaporization would have stimulated some new activity. Another difficulty for the sustainability of the physical volume–weight system as a research program was that it did not find much use in organic chemistry for some time. This was not helpful for its fortunes during those decades in the mid-century during which organic chemistry provided the main excitement in chemical research. Even in Cannizzaro's masterly synthesis and exposition of this system, organic chemistry was given a relatively small role; besides, in the organic realm Cannizzaro was only summarizing others' work to serve as "confirmation" of Avogadro's ideas, rather than using the latter to frame original research (see Cannizzaro [1858] 1910, 5). The physical volume–weight system really took root only because organic chemists came to incorporate it into their work, in a rather round-about fashion (more on this in Sect. 3.2.3). But one can imagine the system being very effective for teaching because of its unified character and uncomplicated realism, and it is no wonder that its most famous promoter, Cannizzaro, was a dedicated teacher. As the editor of the 1910 English translation of his memoir ("J. W.") put it: "The eminence of Cannizzaro as a teacher is plain in every page . . . . The facts are marshalled and their bearing explained with absolute mastery of pedagogic method, and one is impelled to the conclusion that Cannizzaro's students of 1858 must have had clearer conceptions of chemical theory than most of his scientific colleagues of a much later date." (Cannizzaro [1858] 1910, Preface)

3.2.2.4   The Substitution–Type System

Stepping away from the dead-end state of the physical volume–weight system, many leading organic chemists from the 1830s to the 1850s developed the *substitution–type system*. This corresponds roughly to what has usually been called the "type theory", but with my terminology I am, again, trying to emphasize the point that what we have here is not just a theory, but a whole system of practice. The idea of types seems to have originated with Dumas's classic paper of 1840, "On the Law of Substitutions and the Theory of Types". A type was a constitutional template defining a set of related substances. A crucial development was made by Charles Gerhardt (1816–1856), who had studied with both Liebig and Dumas but later found himself an outcast in the French chemical community. In a paper of 1853 Gerhardt identified types on the template of water, hydrogen, and hydrochloric acid. In this work Gerhardt was building not only on Dumas's work, but also on that of his close colleague and fellow outcast Auguste Laurent (1807–1853), who in 1846 had foreshadowed the idea of the water type. Meanwhile Alexander Williamson (1824–1904) at University College London, the remarkable Professor of Practical Chemistry disabled in one eye and one arm,[47] argued that ether and (ethyl) alcohol should be grouped together as belonging to the "water type", seeing both as constituted of two bits held together by oxygen, like water (taken as H_2O) (see Fig. 3.5).

---

[47] On Williamson's background see Brock (1992), 233–234 and Rocke (2010), ch. 1 for a fuller account.

Fig. 3.5 Water, ethyl alcohol and ether as substances belonging to the water type

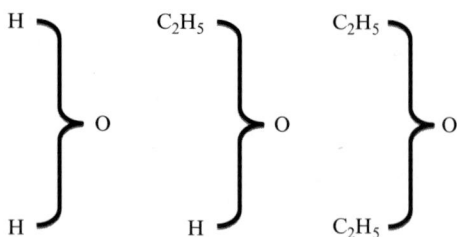

$$
\left.\begin{array}{l} \text{H} \\ \\ \\ \text{H} \end{array}\right\} \text{O} \qquad \left.\begin{array}{l} \text{C}_2\text{H}_5 \\ \\ \\ \text{H} \end{array}\right\} \text{O} \qquad \left.\begin{array}{l} \text{C}_2\text{H}_5 \\ \\ \\ \text{C}_2\text{H}_5 \end{array}\right\} \text{O}
$$

Type-thinking began with a focus on classification, without getting into unnecessary theoretical speculations. During the early period of organic chemistry, classification was both difficult and very much needed. Laurent spoke for many chemists when he said in the preface to his posthumously published treatise (Laurent 1855): "when we reflect upon the absence of all system, all nomenclature, for the classification and denomination of this multitude of bodies, we demand with some anxiety, whether, in a few years' time, it will be possible for us to direct ourselves in the labyrinth of organic chemistry" (quoted in Brock 1992, 211). Therefore, classification was an exciting topic for organic chemists, at the forefront of the research agenda rather like taxonomy in natural history. Classification by types was just the most successful of the many taxonomic ideas that were proposed in nineteenth-century organic chemistry. For example, another innovation issuing from Gerhardt, which has proved its lasting (though limited) power, was that of homologous series of hydrocarbons differing in their compositions by subsequent additions of CH$_2$ (see Brock 1992, 231). Focusing on classification may seem like an overly unambitious attitude for chemists to take. But in my view the type-based classification activity was really the most prudent yet productive enterprise at the time, considering the theoretical difficulty and empirical sterility of the physical volume–weight system, and the challenges that the electrochemical dualist system faced in the organic realm. Types constituted "*paper tools*" *par excellence*, as conceived by Ursula Klein (2003).

Classification into types, however, did not constitute a complete and powerful system of chemical practice until it was combined with the experimental work on substitutions. As mentioned in Sect. 3.2.1.1, it was well known for centuries that one metal could substitute another in salt solutions. Other similar substitution-groups were discovered in the nineteenth century, for instance the newly discovered "halogen" elements: chlorine, bromine, fluorine and iodine. Careful analytical work on the ingredients and products of substitution reactions gave chemists the notion that these substitutions were *atom-by-atom replacements*. This latter notion only became operationally meaningful through some method of atom-counting, and two basic methods were used. Most commonly, previously determined atomic weights were used to enable a deduction of the numbers of atoms from the weight-ratios of elements (determined by gravimetric analysis) in the ingredients and products of reactions. It was also increasingly common, where possible, to use volume-measures to say how many atoms of certain substances came in and out of another substance in a reaction. All of that is to say: in the experimental work on substitutions, both gravimetric and volumetric operationalizations of the atom were employed; however, each was employed as and when convenient, not necessarily both at once as in the physical volume–weight system.

The combination of theoretical type-classifications and the experimental study of substitutions formed the core of what I call the substitution–type system. Classification remained a main activity in this system, and it proceeded in a complex way that combined both theoretical and experimental dimensions. The process might start with the proposal of a classificatory scheme, with an initial attempt to fit a set of known substances into it. If that went well enough, attempts were made to realize reactions that were predicted by the scheme. A major preoccupation in the heyday of this system was to demonstrate the operational reality of one's taxonomy by showing that substances of the same type could be transformed into each other by straightforward substitutions. For example, Hofmann in 1850 identified the "ammonia type", based on the observation by Adolphe Wurtz (1817–1884) that methylamine ($NH_2.CH_3$) and ethylamine ($NH_2.C_2H_5$) were very similar to ammonia, and could in fact be produced by substituting one of the hydrogen atoms in ammonia (NH_3) with a methyl (CH_3) or ethyl (C_2H_5) radical. Hofmann showed that one could replace the hydrogen atoms in ammonia by various other radicals too, thereby producing various other compounds. This whole group of organic compounds constituted the "ammonia type", named after their inorganic template.[48] Laurent had a theory of his own, in a similar but somewhat more realist spirit, based on a scheme of a prism-shaped "nucleus" made up of 8 carbon and 12 hydrogen atoms, from which various molecules could be constructed (at least on paper) by substitutions and attachments (see Ihde 1984, 194–196 for a brief exposition).

Types in their early days, and Laurent's nucleus, were not taken as literal representations of the *shapes* of molecules. Rather, what was assuredly "real" were the facts concerning substitutions, which revealed something about the atomic *constitution* of a substance: the nature and number of units (chemical atoms and groups of chemical atoms) that made up compound substances. But there were various reactions which pointed to various constitutional ideas. Gerhardt in 1856, at the end of his life, gave a trenchant view on the situation: "one and the same body can be represented by two or more rational formulas . . . . when one in a sense freezes a compound into a single formula, one often conceals from oneself chemical relationships that another formula would immediately make evident." (Gerhard quoted in Rocke 2010, 13) Gerhardt might have been in the minority in this instrumentalist and pluralistic view about chemical formulas, and even his partner-in-crime Laurent disagreed with him on this point. But from a certain angle Gerhardt's point is indisputable: with or without uniqueness, operationally certified constitutional formulas were sufficient to give reality to classificatory schemes, and also to stimulate further experimental research. Constitutional formulas suggested numerous new syntheses and substitutions to attempt, opening up many productive lines of practical investigations unencumbered by an overly strong commitment to a particular physical mechanism for atomic combinations. (See Sect. 3.3.2 for a further discussion of realism.)

---

[48] See Lowry (1936), 422–423, on the ammonia type.

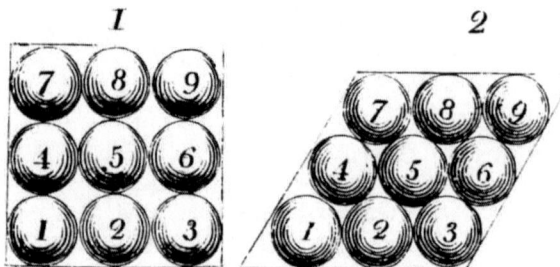

3.2.2.5   The Geometric-Structural System

While chemists working in the physical volume–weight system still struggled
with getting atomic weights and molecular formulas right, and while those in the
substitution–type system were often quite ready to concede the fictional nature of
the structures they postulated, another group of chemists were committed to elucidating
the real geometric structure of molecules. These hyper-realists subscribed to what
I will call the *geometric-structural system*. I will be brief in the description of this
system, since it did not become very active until after the end of the narrative I am
giving in this chapter; however, it must be included since its early steps, which do
fall within my timeframe, were significant enough.

Chemists working in the geometric-structural system in the nineteenth century
did not have much by way of obvious additional operationalization to support their
more ambitious goals. But it would be wrong to think that everything had to wait till
X-ray diffraction and scanning tunneling electron microscopes. The early promise
was given by crystallography, and this proved quite irrepressible despite the lack of
clear chemical success for a long time. Dalton thought in terms of the stacking of
spherical atoms; for example, Fig. 3.6 illustrates how he sought to explain why
water got bulkier when it froze (Dalton 1808, plate 3). Wollaston (1813) in his
physical-atomist mode continued in Haüy's tradition of linking macroscopic
crystalline forms with microscopic molecular shapes. But the geometric-structural
system of atomic chemistry was fruitless in its early days, as it really was premature.
There were good reasons why Wollaston essentially gave up this line of work, and
Dalton never got anywhere very productive with it.

In this system, the main aim was to find the true structure of molecules; concern
about getting the right atomic weights was secondary. Much of its activity consisted
in geometric model-building on the basis of already-known molecular formulas, so
there was a good deal of reliance on the results reached in other systems. However,
sometimes molecular structures did have observable consequences. For example, it
seemed impossible to explain the existence of isomers (substances with different
properties yet same molecular formula) without appealing to geometric hypotheses
about how the same atoms may be arranged differently in order to make different
molecular structures. Also, knowing about certain aspects of molecular structure
could help one determine molecular formulas.

Even in this very realist system of practice, progress was often facilitated by a slight moderation of realism. After the failure of initial attempts by Dalton and others, theoretical molecular-modelers renounced the overly constraining methods of ball- and block-stacking, and allowed shapes made up from lines connecting point-like atoms. This is how ball-and-*stick* models of molecules came to be, and they flourished without any definite ideas about what exactly the sticks represented in terms of physical mechanisms of chemical bonds (see Meinel 2004). Chemical imagination was thereby freed up productively. There is a fascinating story about how the three-dimensional models of molecules came to be invested with reality as stereochemistry developed, but I will not enter into details because this development mostly occurred after the end of my narrative here (see Ramberg 2003 for a comprehensive and up-to-date account).

3.2.3 The H$_2$O Consensus

So the shape of atomic chemistry by the middle of the nineteenth century was very complicated. I have identified five major systems of practice operating in the field at the same time. And although I have tried to delineate the different systems cleanly, the very articulation of these systems has also shown that the boundaries between them were fuzzy and the relations between them complex and dynamic. Various individual chemists constructed numerous idiosyncratic systems of their own, which combined various elements from the idealized systems that I have outlined; the chemists also made modifications in their systems and changed their allegiances as their work progressed. Where was this all going? From a modern point of view it is tempting to say that atomic chemistry before Cannizzaro and Karlsruhe was just a confused field. But that would be to denigrate unjustly the significant achievements of the five systems of atomic chemistry described above—in the discovery of numerous new substances, in the precise analysis of a myriad of new and old substances, in the operational knowledge of atoms, in the classification of organic substances, and in the prediction and explanation of various chemical reactions. All of these are serious achievements in themselves, as well as stepping-stones to the achievements of later chemists.

Yet it is also true that something very significant did happen somewhere around 1860. The "Quiet Revolution" is what Rocke (1992, 1993) has called the 1850s reform of organic chemistry which generated the set of molecular formulas and atomic weights familiar to us today, including H$_2$O. As he points out, this was not at all a quiet event *at the time*; many chemistry textbooks of the 1860s record a palpable excitement, and a clear consciousness that a "revolution" in chemistry had just happened. For instance, take this from Hofmann (1865, v): "No chemist will need to be reminded that, during the last quarter of a century, the science of chemistry has undergone a profound transformation; attended, during its accomplishment, by struggles so convulsive, as to represent what, in political parlance, would be appropriately termed a Revolution." We do need to understand this event.

Again, there are already extensive, detailed and insightful historical accounts that I cannot ever hope to match, but I hope it will not be entirely in vain to attempt to add an instructive philosophical gloss over these events,[49] while not contradicting existing accounts in any significant way.

Two things are certain, at the outset. First, despite all the talk of "revolution", what happened was not the overthrow of one dominant system in favor of another, contrary to what the word implies both in its common usage and in the Kuhnian picture of paradigm-shift. I think what we see here is a much more pluralistic pattern of development, which I will try to characterize as I go along in the rest of this chapter, and then more systematically in Chap. 5. Second, it is undeniable that from the 1860s onward there was a very strong consensus on the set of atomic weights and molecular formulas that we recognize as modern (which I will dub the "H$_2$O consensus" for short). How and why this consensus formed will be the main question driving my inquiry in this section. And then in Sect. 3.2.4 I will follow up with some qualifications showing that this consensus was not a simple happy ending, in fact not a clear ending of any sort at all.

3.2.3.1   Chlorine-Substitution

The last part of the road to H$_2$O began with fuming candles in Paris—and here I am only expressing in an over-dramatic way what seems an agreed view among the best-informed historians, that it was chlorine–hydrogen substitutions that began the series of developments that led to the consensus of the 1860s (Brooke 1973; Ihde 1984, 191ff; Rocke 1984, 191; Brock 1992, 215; Klein 2003, 195ff). This starts with the famous story of a royal ball at the Tuileries, at which the distinguished guests of Charles X were assaulted by mysterious noxious fumes emanating from the chandeliers. Dumas (via his father-in-law Alexandre Brongniart) was called in to investigate the incident, and concluded that the fumes consisted of hydrochloric acid gas. The candles had been whitened with a chlorine-based bleach, and in that process some of the hydrogen in the tallow had been replaced by the chlorine from the bleach. This incident stimulated a whole program of research (see Lowry 1936, 406–407), in which Dumas announced empirical laws concerning hydrogen–chlorine substitution in 1834 and also retrospectively found other experiments that had demonstrated hydrogen–chlorine substitution, going as far back as Gay-Lussac's work in 1815 on the action of chlorine on hydrogen cyanide, which "loses one volume of hydrogen and gains exactly one volume of chlorine", and turns into cyanogen chloride.[50] Faraday in 1821 and Wöhler and Liebig in 1832 had reported other instances.

---

[49] In this I follow Chalmers (2009), ch. 10.

[50] In modern terms, the reaction is HCN + Cl$_2$ → ClCN + HCl.

Chlorine–hydrogen substitution raised serious problems and also opened up some new paths for chemistry. Hydrogen and chlorine having very different properties from each other, it was not clear how one could take the other's place. And to make matters worse, it was also found that bromine and iodine (also highly electronegative) could replace hydrogen, so that the puzzling phenomena could not be attributed to some strange idiosyncracy of chlorine. There were two dimensions to the surprise and discomfort here. First, in some reactions the replacement of hydrogen by chlorine did not seem to change the properties of the substance greatly; for example, there was a very close similarity in chemical properties between acetic acid (the essence of vinegar, modern formula $C_2H_4O_2$) and trichloroacetic acid ($C_2HCl_3O_2$). This led many chemists to entertain more seriously the idea that chemical properties were determined by molecular *structure* as well as (or even in preference to) the nature of the atoms that went into the structures. Second, because hydrogen is highly electropositive and chlorine is highly electronegative, their combination with each other made eminent sense but their substitution for each other raised a serious problem for the electrostatic explanation of chemical combination.

Chlorine–hydrogen substitution had different impacts on different systems of atomic chemistry. The weight-only system was not disturbed; it could just add the newly discovered reactions to its empirical basis, and carry on as before. Advocates of the physical volume–weight system would have found it liberating to learn that chemical combinations were not strictly governed by electrostatic affinities, as it neutralized the strictures against bi-atomic molecules. But chlorine–hydrogen substitution precipitated a crisis for the then-dominant electrochemical dualistic system. Dumas himself vacillated (see Brock 1992, 215–216). Initially, when his former protégé Laurent used his work as a weapon against dualism and invoked Berzelius's wrath, Dumas reacted angrily to point out that he had only noted the input of chlorine and output of hydrogen in such reactions, rather than claiming a direct atomic replacement of hydrogen by chlorine. However, he followed Laurent's atomic interpretation[51] after his own further experimental work on acetic acid in 1838, in which he thought he had managed to substitute all of its hydrogen atoms, one by one, with chlorine atoms. By 1839 "he repudiated, definitely and finally, Berzelius's electrochemical theories"; Dumas claimed that the latter were neither "based on evident facts" nor valuable in "explaining and predicting facts". In the same year Liebig also came out against electrochemical dualism, in notes that he added to Berzelius's own papers. Declaring that Berzelius's views "rest upon a mass of hypothetical assumptions, for the correctness of which proof of every kind is lacking", Liebig expressed his belief in the reality of chlorine–hydrogen substitution (Lowry 1936, 411). The defection of Dumas and Liebig, each perhaps the most influential chemist at the time in France and Germany respectively, must have done a great deal to erode the dominance of the electrochemical dualist system.

Among those who had followed the electrochemical dualist system to any serious degree, this was a crucial moment. It is standard for historians of chemistry to identify

---

[51] Laurent, however, still took pains to distinguish his own view from Dumas's.

chlorine–hydrogen substitution as the locus where "the stream of chemical progress had divided into two branches" (e.g., Brock 1992, 216). The image of two streams is from August Kekulé (1829–1896), famous for his work on the tetravalence of carbon and the ring-structure of benzene, in his retrospective at the 1890 "Benzolfest" marking the 25th anniversary of his publication on the structure of benzene (Kekulé [1890] 1958, 21; see Rocke 2010, ch. 10 for a detailed discussion). The two options were either to stick with electrochemical dualism by means of some uncomfortable adjustments, or to give up on explanations of chemical combination and focus merely on elucidating the constitution of molecules. Even for those who held on to dualism, the facts of chlorine–hydrogen substitution raised a doubt about electrostatic attraction as the real and universal mechanism for chemical bonding. Berzelius's own idea of copula or coordination amounted to an admission that there were some parts of molecules that were not governed by electrostatic forces. For example, after some twists and turns Berzelius concluded that acetic acid ($C_4H_8O_4$) consisted of oxalic acid ($C_2O_3 + H_2O$) with a methyl "copula" (C_2H_6) attached to it (see Brock 1992, 217)[52]:

$$C_2H_6 \underline{\quad} C_2O_3 + H_2O$$

The oxalic acid part of this had the familiar dualistic composition (the two parts of it being positively and negatively charged), but it was left unexplained how the methyl copula was attached to it. The immediate advantage of this strange formulation was that the troublesome chlorine–hydrogen substitution could be banished to the copula, not disturbing the electrochemistry of the oxalic acid. So the formation of trichloroacetic acid involved chlorine replacing the hydrogen atoms in the copula, not those in the oxalic acid:

$$C_2Cl_6 \underline{\quad} C_2O_3 + H_2O$$

Admirable as Berzelius's ingenuity was, the outcome amounted to a twofold admission of defeat for electrochemical dualism: first, chlorine–hydrogen substitution was admitted in the end; second, now there was a well-identified section of many organic molecules expressly not governed by electrostatic forces!

It is important to resist the temptation to write off Berzelius completely at this point. Berzelius's formulation was pleasing in that acetic acid and trichloroacetic acid emerged as completely parallel, as an inspection of the two formulas above shows immediately. Brock (1992, 217) also points out that this parallel is "astonishingly like" what we have in modern structural formulas: $CH_3 \cdot COOH$ for acetic acid and $CCl_3 \cdot COOH$ for trichloroacetic acid, the chlorine replacing the hydrogen atoms in CH_3 but not the one in COOH. But setting our retrospective satisfaction aside, it

---

[52] In the modern formula for acetic acid, we halve the number of all the atoms, to get $C_2H_4O_2$ (or, more structurally, $CH_3 \cdot COOH$). For oxalic acid we have kept Berzelius's $C_2H_2O_4$, but we parse that out as $(COOH)_2$.

has to be admitted that many chemists at the time viewed the Berzelian moves as too complicated and the returns as too meager—as Kekulé ([1890] 1958, 21) put it, this stream "led for the most part through broken boulders, and only later did it again reach fertile country". Why not set aside the Berzelian electrochemical rules altogether, and see what forms of molecular constitution might allow simple and systematic classifications? Those who sailed in this "stream" of work created the substitution–type system, which originated in the late 1830s and flourished throughout the 1840s and the 1850s (Kekulé says that this stream "flowed, chiefly on French soil, through luxuriant flower-decked plains, and those who followed it, with Laurent and Dumas at their head, could reap, during the whole voyage, almost without effort, an abundant harvest.")

As the hold of electrochemical dualism weakened, chemists were freed up to explore the constitution of compounds in ways that were previously unimagined. The substitution–type system could only arise by rejecting the electrochemical dualistic system—or, to be more precise, thanks to the corrosion of the unnecessarily restrictive aspects of electrochemical dualism. No one argued with the operationalized part of dualism (and we still have ionic bonds in modern chemical pedagogy); however, substitution allowed chemists to see that they only had to be bound to electrochemical dualistic reasoning where electrolysis had actually shown the operational reality of dualistic composition.[53] In this connection it should be noted that such relaxation was already happening elsewhere within the electrochemical dualistic system, with the broadening of the notion of "radical" to allow highly electronegative atoms including chlorine and oxygen within radicals. In this sense I think it is wrong to present the "radical theory" and the "type theory" as polar opposites, and also wrong to assume that advocates of the radical theory were all practicing the electrochemical dualistic system. It was merely a short step for Dumas and Liebig from contemplating heterodox radicals to abandoning Berzelian electrochemical dualism altogether. Chlorine–hydrogen substitution was the final push, not the unexpected beginning of a revolution. And the "parting of the stream" was not as clean as Kekulé made it out to be.

The short-term overall effect of chlorine–hydrogen substitution and other similar substitutions on theoretical chemistry may have been only to increase uncertainty about the explanations of chemical bonds and the reality of molecular structures. Through the 1840s and the 1850s there was a slight whiff of "anything goes" in discussions of constitution in organic chemistry. Kekulé (1861, 58) once listed 18 different structural formulas that had been given for acetic acid (see Fig. 3.7). If there was such divergence on that most familiar and simple substance, what hope of unity was there on anything else in organic chemistry? But well-judged uncertainty can be a productive thing, a blessing in disguise. For example, Brock (1992, 217) says that even Berzelian defensive maneuvers resulted in the "astonishingly creative and

---

[53] This is where Hermann Kolbe's struggle on behalf of electrochemical dualism becomes so valuable, because he was attempting to extend the operational basis of dualism by electrolytically isolating organic radicals.

$C_4H_4O_4$ . . . .   empirische Formel.

$C_4H_3O_3$ + HO . . . . . .   dualistische Formel.

$C_4H_3O_4$. H . . . .   Wasserstoffsäure-Theorie.

C_4H_4 + O_4 . . . .   Kerntheorie.

$C_4H_3O_2$ + HO_2 . . . . . .   Longchamp's Ansicht.

C_4H + H_3O_4 . . . . . .   Graham's Ansicht.

$C_4H_3O_2$. O + HO . . . . .   Radicaltheorie

C_4H_3 . O_3 + HO . . . . . .   Radicaltheorie.

$\left.{\begin{matrix}C_4H_3O_2\\ H\end{matrix}}\right\}O_2$ . . . .   Gerhardl. Typentheorie.

$\left.{\begin{matrix}C_4H_3\\ H\end{matrix}}\right\}O_4$ . . . .   Typentheorie (Schischkoff) etc.

C_2O_3 + C_2H_3 + H O . . . .   Berzelius' Paarlingstheorie.

H O . $(C_2H_3)C_2, O_3$ . . . . .   Kolbe 's Ansicht.

H O . $(C_2H_3) C_2 . O . O_3$ . . . .   ditto

$\left.{\begin{matrix}C_2 (C_2H_3)O_2\\ H\end{matrix}}\right\}O_2$ . . . .   Wurtz.

$\left.{\begin{matrix}C_2H_3(C_2O_2)\\ H\end{matrix}}\right\}O_2$ . . . .   Mendius.

$\left.{\begin{matrix}C_2H_2 .HO\\ HO\end{matrix}}\right\}C_2O_2$ . . . .   Geuther.

$C_2\left\{{\begin{matrix}C_2H_3\\ O\\ O\end{matrix}}\right\}O$ + HO . . . .   Rochleder.

$\left(C_2 \dfrac{H_3}{CO} + CO_2\right)$ + HO . Persoz.

$\left.{\begin{matrix}C_2\left\{{\begin{matrix}O_2\\ H\end{matrix}}\right.\\ C_2\left\{{\begin{matrix}\\ H\end{matrix}}\right.\\ \dfrac{H}{H}\end{matrix}}\right\}O_2$ . . . . . .   Buff.

Fig. 3.7 Kekulé's listing of competing formulas for acetic acid

fruitful notion that most organic compounds were copulated, with all substitutions occurring with in the non-electrochemical copulae". And the productiveness of the substitution–type system, which embraced the uncertainty, was obvious. This is just as it happened in an earlier phase of atomic chemistry at the advent of the weight-only system, when it was productive for chemists to step back from Dalton's overly specific certainty about the nature of physical atoms. With the weakening of both Daltonian and Berzelian restrictions on molecular structures, chemists could freely explore various possibilities, including the sort of structures postulated by Avogadro.

This did not mean necessarily going with Avogadro, since chemists still had no idea about why two atoms of the same kind would bond together; it just means that they had less and less dogmatic reason left to prohibit the possibility. So, coming back to water, we might say that chlorine opened the door to H_2O.

3.2.3.2   Atom-Fixing Power

Opening the door is important, but not sufficient. A more positive argument for H_2O and for Avogadro's scheme in general was provided by the establishment of the concept of valency (with values of 1 and 2 for hydrogen and oxygen). This is, again, a story well-known to the expert historians, but there is one aspect of it that does not tend to receive sufficient attention, which I would like to highlight here before moving on to the more general account of valency in the next section. At least from the modern point of view, there is something peculiar about the genesis of the valency concept. According to many latter-day type-theorists, the curly bracket in a type-formula was not just an idle piece of notation; it was a *clasp*, indicating an actual act of binding, the active part being played by the element that sits at the tip of the bracket. For instance, in the type-formula for water (shown in the left-hand side of Fig. 3.5 above), it is the central oxygen atom that holds the two hydrogen atoms together, like a parent holding two children each by the hand. And if the oxygen here has two "hands", nitrogen in the ammonia molecule has three, and carbon in marsh gas (methane) has four. Edward Frankland (1825–1899), German-trained English chemist, was so keen to indicate the special role of such central atoms that he devised a special notation for theoretical ("rational") formulas in which the symbol for a central atom was set in boldface. Thus water was *OH₂*; nitrous acid was *NOHo* (where *Ho* is the hydroxyl radical, which we would now write as *OH*), but nitrous oxide was *ON₂*, and nitric oxide was *NO NO*, with the two *N*'s connected by a curly bracket (Frankland 1866, 17, 61, and *passim*).

The term "atomicity" was used at the time in order to express what I am meta-phorically referring to as the "number of hands" here: the number of atomic units (whether an elementary atom or a radical) that it was able to combine with, and hold together. More literally, Frankland (1866, 18–19) stated that various elementary atoms had different numbers of "bonds", where a "bond" did not mean a link as we might imagine, but instead a "point of attachment … by which it can be united with other element". The number of bonds shaped chemical combinations because "no element, either alone or in combination, can exist with any of its bonds disconnected." He noted that the "combining value of the elementary atoms is usually termed their *atomicity* or *atom-fixing power*."

Seeing one atom in a molecule as more active in any sense than all the other atoms is not the kind of thinking that would be allowed in modern chemistry, or indeed in any fully compositionist system of chemistry (see Chap. 1, Sect. 1.2.3.2 on compositionism). Even setting aside all modern understanding, one might be tempted to ask: wasn't the idea of active atom-fixing power a product of an overactive imagination departing from the formulas OH_2, NH_3, and CH_4, which in themselves say nothing about how the molecules are structured, not to mention which of the atoms play the active role and which others don't? What was there to say that H_2O had to be linked up like H–O–H? Why not O–H–H, with the heaviest atom at the head and the others trailing it, in which case it would not even be the case that the oxygen atom is directly connected with both of the hydrogen atoms? Or why not a closed triangle, in which case each hydrogen atom would also have two connections? The H_2O formula itself (even if all concerned could agree on it, which they hadn't yet) did not decide between these competing structural possibilities. Neither Dumas nor Gerhardt, perhaps the two most important contributors to type theory, meant type-formulas as indications of the actual physical structure of molecules. What warranted a departure from that prudence, especially into a seemingly ill-advised and arcane idea that different parts of a molecule had different ontological statuses?

Once again, the key was operationalization. There were experiments that operationalized the notion of central atoms and their atom-binding powers. And as we shall see, chemists gradually sorted out which notions were really connected with laboratory operations and which ones were merely mixed in with them. A good place to begin is Alexander Williamson's work on "etherification" in 1850, which Alan Rocke identifies as the source of the most decisive experimental evidence in convincing most chemists to adopt the new atomic weights and molecular formulas (Rocke 1992; 2010, ch. 1; 1984, ch. 8 for further detail). Williamson carried out a series of experiments demonstrating that ether ($C_4H_{10}O$) was "a coupled compound containing two ethyl groups, C_2H_5, and not merely the oxide of a single radical, C_4H_{10}" (Williamson 1852, quoted in Lowry 1936, 424). This showed that the oxygen atom in ether was really holding two different atomic bits together, not just one. In the etherification process, two molecules of alcohol (C_2H_6O) become one molecule of ether with the help of sulphuric acid, with one molecule of water as a bi-product. The net reaction is simple:

$$2C_2H_6O \rightarrow C_4H_{10}O + H_2O$$

But from this representation the clumping of the two alcohol molecules and their subsequent division into ether and water is somewhat mysterious, and it is unclear what role the sulphuric acid catalyst could be playing in the process. Without further evidence, the above picture would have had no convincing advantage over the earlier view by Liebig and Dumas that etherification was simply a removal of water from each molecule of alcohol, by the well-known action of sulphuric acid as a dehydrating agent (see Rocke 2010, 19):

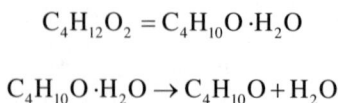

$$C_4H_{12}O_2 = C_4H_{10}O \cdot H_2O$$

$$C_4H_{10}O \cdot H_2O \rightarrow C_4H_{10}O + H_2O$$

Now, what might have seemed like an arbitrary decision in how to write the formula for an organic compound came to have real significance here. The general importance of these organic formulas as "paper tools" has been emphasized by

Fig. 3.8 Williamson's model of "continuous etherification" by two hydrogen–ethyl replacements

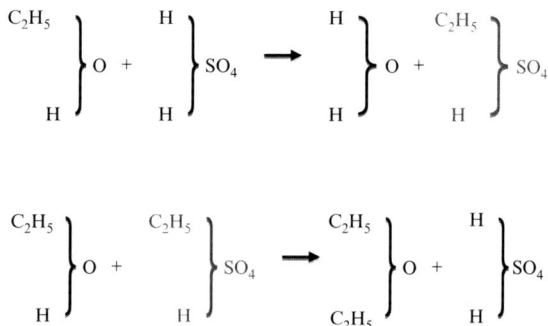

$$\left.\begin{array}{c} C_2H_5 \\ \\ H \end{array}\right\} O \; + \; \left.\begin{array}{c} H \\ \\ H \end{array}\right\} SO_4 \; \longrightarrow \; \left.\begin{array}{c} H \\ \\ H \end{array}\right\} O \; + \; \left.\begin{array}{c} C_2H_5 \\ \\ H \end{array}\right\} SO_4$$

$$\left.\begin{array}{c} C_2H_5 \\ \\ H \end{array}\right\} O \; + \; \left.\begin{array}{c} C_2H_5 \\ \\ H \end{array}\right\} SO_4 \; \longrightarrow \; \left.\begin{array}{c} C_2H_5 \\ \\ C_2H_5 \end{array}\right\} O \; + \; \left.\begin{array}{c} H \\ \\ H \end{array}\right\} SO_4$$

Klein (2003), and this is an excellent example illustrating her case. Williamson wrote the formula for alcohol as C_2H_6O, and Liebig/Dumas wrote it as $C_4H_{12}O_2$ (or, $C_4H_{10}O.H_2O$). The Liebig/Dumas formula was just the Williamson formula with the number of all the atoms doubled. Which was correct? The two views could not be distinguished by empirical analysis, which only gave the ratio of 2:6:1 for the numbers of carbon, hydrogen and oxygen atoms (accepting the same atomic weights). Williamson shed much light on etherification by conceiving of it as a two-step process, as shown in Fig. 3.8 (illustration after Brock 1992, 236). In the first step, the sulphuric acid removes the ethyl radical (C_2H_5) from the alcohol, putting a hydrogen atom in that place; thereby alcohol turns into water (one might call that the inverse-Jesus mechanism). In the second step, the ethyl-loaded sulphuric acid (also known as sulphovinic acid) gives up that ethyl radical to another alcohol molecule, in exchange for hydrogen; thereby the sulphuric acid returns to normal, and the alcohol turns into ether. The beauty of Williamson's account gave credence to his formulas (C_2H_6O for alcohol, $C_4H_{10}O$ for ether, H_2O for water, and H_2SO_4 for sulphuric acid), and operational reality to the water type in which the central oxygen showed its ability to hold together two atomic/radical units in various combinations. But still, hadn't Williamson merely given a pretty story? Was there any direct experimental evidence that all these molecules of the "water type" really had two branches, held together by oxygen? In order to provide such evidence, Williamson made asymmetric ethers (Rocke 2010, 20–21). By contriving to attach different combinations of ethyl, methyl and amyl radicals to the central oxygen of the water type, Williamson produced ethyl–methyl, methyl–amyl and amyl–ethyl ethers at will. No such substances could exist according to Dumas and Liebig's scheme; at best the experiments ought to have only produced mixtures of symmetric ethers: ethyl, methyl or amyl.

Even more convincing evidence of the binding power came when chemists managed to effect substitutions on the central binding atom, not on one of the branches. If oxygen was replaced with sulphur in a water-type molecule, the molecule stayed whole, just transformed into an analogous sulphur-compound. If the oxygen was replaced with chlorine, however, then *two* chlorinated molecules were formed. So it was inferred that oxygen or sulphur could bind together two atoms or radicals, but chlorine could not (it had only one hand, as it were). The judgment that *two* chlorinated molecules had formed, of course, was only possible on the basis of an agreed atom-counting (or molecule-counting) method, the most convincing of which was counting by volumes. The idea of the active central binding atom gradually disappeared in later chemistry as I will explain further in the next section, but it was of crucial importance in the middle of the nineteenth century, as the ladder which chemists kicked away after climbing up to the concept of valency. After that ladder was gone, only the operationally secure notions remained: the atomicity number for each element, and the matter-of-fact about which atoms/radicals were connected with which within a given molecule. All those facts could be discerned and tested through experiment, without assuming any ontological hierarchy or asymmetry in the relationship between different atoms within a molecule.

3.2.3.3   Valency, Realism and Compositionism

There is a good deal of primary and secondary literature on the history of valency (or valence, in modern American parlance), and much dispute about who came up with the idea first (see Partington 1964; Russell 1971; Rocke 1984, etc., for full accounts). Here I will focus on the role of the valency concept on the consolidation of molecular formulas and atomic weights. The crux of my view, already hinted in the last section, is that valency was the compositionist rendition of atom-fixing power. The history of valency is extremely complex, and I just want to highlight two aspects. First, it was the development of type theory itself that began to erode the notion of the central binding atom. When Gerhardt codified the type theory in 1856, he accepted Hofmann's ammonia type and Williamson's water type and also added two others: the hydrochloric acid type, and the hydrogen type, on the templates of HCl and HH (Lowry 1936, 425–426, gives a very quick summary). Now, these two types did not have a central binding atom, their notation simply linking the two elements with a curly bracket but with no atom at the central pointy end of the bracket. With these one-to-one combinations, there was no clear operational sense in which one unit could be seen as doing the binding and the other being bound. So there was a symmetry of bonding in these situations, and I think this must have made the practitioners of the substitution–type system realize that there never was anything operationally significant about the presumed asymmetry between the central binding atom and the other parts of the molecule, even in the other types. The next natural step was the "democratic" valency concept, which had to be satisfied from both sides of a chemical combination. Away with the curly brackets then, just sticks with two ends; out with boldface letters for the active binding atoms, just all atoms equal. Compositionism (Chap. 1, Sect. 1.3.4) had finally come home, after a long period of being mixed in with principlist and other modes of thought in which various chemical substances were seen as having an active agency in chemical combinations.

Secondly, the elimination of the presumed asymmetry in type formulas would have helped clarify chemists' view of what was really operationally significant about what had been recognized as the central binding atom in the water and ammonia types. The significant thing was (chemical) divisibility and indivisibility, going nicely back to the original idea of atoms. The centrality of the oxygen and nitrogen atoms in the water- and ammonia-type molecules lay not in some active power, but simply in being *one* atom bonded with multiple atoms/radicals at once. The operational manifestation of the number of atoms and other atomic units (e.g., radicals) within a molecule, again, came in the form of substitutions. Charles Bloxam, Professor of Practical Chemistry at King's College London and at the Royal Military Academy in Woolwich, put this point very nicely: "the hydrogen in ammonia can be replaced by other bodies *in thirds*, showing that there must be three atoms of hydrogen present, whilst the 14 parts [by weight] of nitrogen cannot be replaced in fractions, so that it must represent a single atom." (Bloxam 1971, 120; emphasis original) It was this inability to replace in fractions that operationally defined an atom, and the essence of the ammonia type boiled down to a single nitrogen atom able to combine with three separate (because separable) atomic units simultaneously. Likewise, the water type came down to a single oxygen atom able to combine with two separate atomic units. The numbers of atoms or atomic units were operationally specifiable, using atom-counting by volume or by the comparison of combining weights.

The identification of the operational basis of valency and type-formulas also led to a higher confidence in their reality. By the 1850s, the practitioners of the substitution–type system were no longer instrumentalists about types. Take Kekulé for example, who declared in the mid-1850s: "It is not merely a difference in formulation but in actual fact that one atom of water contains two atoms of hydrogen and one atom of oxygen, and that the quantity of chlorine equivalent to one indivisible atom of oxygen is itself divisible by two, whereas sulphur, like oxygen, is dibasic so that one atom of sulphur is equivalent to two of chlorine." (quoted in Russell 1971, 56) With this realist confidence, it was possible to build the whole scheme of atomic chemistry as Hofmann presented it in the 1860s (see Sect. 3.1): start from combining volumes of gases; from volumetric atom-counting, deduce the molecular formulas of reaction-products; from the molecular formulas and the observed combining weights, deduce the atomic weights of the elements involved; use those atomic weights to infer other molecular formulas from other combining weights.

As just hinted, the establishment of the substitution–type system on a more realist and fully compositionist footing also meant that a good deal of unification, or at least synthesis, could be achieved in the field of atomic chemistry. Once the types were taken as real representations of molecular constitution, and atom-binding power was taken in the compositionist way, the substitution–type system overlapped very well with the physical volume–weight system. The operational crux of this unification consisted of certain substitution reactions that allowed the simultaneous tracking of volume and weight of well-specified atomic units as they came in and out of larger compounds. To allow this unification, it was important that the substitution–type system had abandoned any essential reliance on electrochemical dualistic restrictions, which prohibited the kind of combinations that the physical volume–weight system required. When it was taken in a realist way, the substitution–type system also became fully compatible with the geometric-structural system, though not identical to it. In fact it breathed new life into the latter, by providing some very useful specific ideas (such as the tetrahedral carbon) to use for the geometric constructions. After all of that, there was truly a "new system of chemical philosophy" built in Dalton's spirit but with a tidied-up phenomenological view on the manner in which atoms combined with each other, though still with no pretensions to pronounce upon the full suite of physical properties of atoms or the deep true cause of chemical bonds.

At the "Benzolfest" of 1890, Kekulé was pleased to look back and celebrate this great unification. The unifying concept was valency, which arose both from the Berzelian radical theory (the electrochemical dualistic system, in my terms) and the type theory (the substitution–type system). In Kekulé's tale of the two streams again: "Suddenly a loud shout of triumph resounded from the host of the adherents of the type theory. The others also had arrived—Frankland at their head. Both sides saw that they had been striving toward the same goal, although by different routes. They exchanged experiences; each side profited by the conquests of the other; and with united force they sailed onward on the reunited stream. One or two held themselves apart and sulked . . . but they too followed the stream." (Kekulé [1890] 1958, 21)[54] But Kekulé's story of happy reunion is exaggerated, and more accurate sociologically than intellectually.[55] Yes, Frankland was a leading proponent of the Berzelian radical theory and he was surely motivated by the noble dualistic dream of isolating organic radicals, but the part of his thinking that led to valency came only when he began to abandon dualistic thinking. Yes, radicals became an integral part of the type theory (as the parts being substituted around), but these "radicals" had lost all real connection to dualistic theory. A more sober and mundane account would recognize that the dualistic system was not compatible with the new valency-centered consensus of types, volumes, weights and structures, as I will explain further in the next section.

3.2.4 Beyond Consensus

The emerging H$_2$O consensus was quite unstoppable by the 1860s, among those who took atoms seriously at all.[56] But my sense is that this consensus did not represent a solution to all the problems that had exercised the atomic chemists. In this section I will attempt to articulate that uneasy feeling. The main point here will be the value of pluralism (indicated at the end of the first section (Sect. 3.1) and also in Sect. 3.2.2). I will argue that the benefits of pluralism were not merely in the realm

---

[54] Crum-Brown had expressed a very similar view, less colorfully, in 1874. See Levere (1971), 195.

[55] A more nuanced view is given by Colin Russell (1971, 42–43) on Frankland and Trevor Levere (1971, 188–189) on Kolbe.

[56] For a discussion of some who remained skeptical about atoms on the whole, see Nye (1976), 253–254 and 262, and Nye (1972), ch. 1.

of heuristics, and that its use was not limited to a period of temporary interaction leading up to a final unified destination, as in Kekulé's streams merging again.

Let us begin by reviewing the demise of the electrochemical dualistic system. Its heart was the electrostatic explanation of chemical bonding. The strong explanatory aim and the simple explanatory scheme that formed the core of the electrochemical dualistic system could not be reconciled with the other three systems that were now merging together. Recall how type-thinking began by denying the central premise of dualism, and how incongruous with dualism Avogadro's bi-atomic elements were. To any remaining true dualists, the emerging consensus of the "Quiet Revolution" must have seemed like a very unholy alliance. The explanation of bonding was not only at the heart of dualism; it was also one of its main merits, and a merit that was unsurpassed by the other systems even as they coalesced with each other into a dominant synthesis. What we have here is not the rejection of an outdated scientific idea, but a story of renunciation, of Kuhn-loss, and of overly hasty monism.

We should not be misled by later triumphalist declarations of unity. Here is a warning from the perspective of a cautious sort of pluralism, which I will elaborate in Chaps. 4 and 5 as "conservationist pluralism": if a system of practice once became well-established for good reasons, it is not easily going to become completely worthless later, and it should only be discarded with great care.[57] (This is as I have argued with regard to the phlogistonist system in Chap. 1.) We would do well to remember Laurent's prophetic assessment in 1837:

> The validity of a theory is judged by the progress in science that it brings about. Now when we consider the immense advantages which the [dualistic] theory possess for nomenclature, for the learning of chemistry, and now its application to organic chemistry, we would still be constrained to use it, even if it should be demonstrated that it is false....[58]

Electrochemical dualism never died; it only emigrated from the mainstream of atomic chemistry, finding a more natural home in the new sub-discipline of physical chemistry. It was a key plank in Arrhenius's achievement, and it lived on in the twentieth-century notion of ionic bonds.

Even aside from the explanation of bonding, there was much that the new consensus did not provide. Recall the structuralist promise of chlorine–hydrogen substitution: it seemed to be structure, rather than the identity of the constituent atoms, that explained the properties of substances. This promise of structural explanations of properties remained largely unfulfilled. The few advances that were made, such as the identification of the benzene ring-structure as essential to aromaticity, only served as a reminder of the unfulfilled promise of the geometric-structural system. It is also not the case that the new valency-centered consensus was completely problem-free. For one thing, it was soon recognized that valency was not constant (see Russell 1971, 171ff).

---

[57] This can be distinguished from pluralism motivated by exploratory or iconoclastic sentiments. All will be brought together in Chap. 5.

[58] Quoted in Brock (1992), 226.

Considering these imperfections of the new consensus, we can understand something that may be puzzling to the modern eye: how the brilliant inventors of valency, right down to Kekulé, remained so cautious for so long about the metaphysical truth of their theories. I believe that they were rightly wary of claiming a final truth about the specific and exciting models of chemical combinations that they were proposing. And after *another* turbulent half-century of developments, the story did turn out quite differently: atoms are divisible, stick-like bonds are fictitious, and valency is a difficult concept to fit exactly into quantum chemistry.

I think the story of atomic chemistry is a splendid illustration of the workings of pluralism in science. There were even some explicit statements of pluralism by the chemists themselves. For instance, Berzelius noted in a letter to Laurent in 1844 that they had diametrically opposite approaches—Laurent attempting to reform inorganic chemistry following organic clues, and Berzelius himself the opposite. Yet, he did not insist that one or the other had to be right: "I am by no means blind to the extension of theoretical knowledge which may follow from the method which you have chosen. It will therefore be best if we each follow our own route amicably, in the hope that science will draw profit from both." (Berzelius, quoted in Levere 1971, 174) And it is also evident that chemists working in one system of practice benefited from engaging with other systems, going beyond peaceful coexistence. Perhaps a prime illustration of this is Berzelius himself. Although usually identified as the fountainhead of the electrochemical dualist system, which he surely was, Berzelius also rose above it and practiced a much broader-minded kind of chemistry. He cast his net widely, and caught a lot more fish, admittedly including some that we now consider, well, fishy (such as his unwillingness to accept the elementary nature of nitrogen and chlorine). Berzelius actually did an enormous amount of work in the weight-only system, making exhaustive analyses of high precision. And in his decisions on atomic weights and molecular formulas, he used a lot of chemical knowledge and analogies. He also made a limited use of atom-counting by volumes (perhaps hard to avoid, when electrolysis yielded gases), though he did not go deeper into the use of volumes and did not accept Avogadro's compound molecules of elementary substances. Berzelius was even willing to make a deep modification into electrochemical dualism itself, as we have seen in his introduction of the copula concept.

But isn't that sort of thing only a symptom of a transitional state? Wasn't monism a more appropriate attitude after the establishment of valency? No, even in such a success story as valency-based structural theory, the limits of scientific achievements were severe enough to warrant keeping multiple systems in operation, with benefits arising from each system in itself and multiple systems in mutual interaction. The story of electrochemical dualism is a good illustration; the hazards of monism in structural theory were only moderated by the *separate* rise of physical chemistry. This is particularly clear to Trevor Levere, perhaps because he worked with a focus on the concept of affinity, unlike most other historians who have told this narrative. While he does stress the unification achieved by the rapprochement of radical and type theories, Levere notes that there was fragmentation going on as well. Electrochemical dualism had provided a definitive and unified account of chemical affinity; in contrast, "the developments of the 1850s contributed to the gradual erosion of precise ideas about chemical affinity", and witnessed "its fragmentation into chemical energetics and chemical structure, the distribution of chemical power and matter respectively." (Levere 1971, 193, 195) In the aftermath, in the 1860s and 1870s, chemists were "much involved with thermochemistry, chemical thermodynamics, structural chemistry, and valence theory—all facets of the anterior blanket concept of chemical affinity." (Levere 1971, 159) John Servos's history of early physical chemistry (1990, ch. 1), identifying Wilhelm Ostwald as its chief pioneer, makes the same point in more illuminating detail. Servos explains (p. 3):

> [Ostwald] sought to redirect chemists' attention from the substances participating in chemical reactions to the reactions themselves. Ostwald thought that chemists had long overemphasized the taxonomic aspects of their science by focusing too narrowly upon the composition, structure, and properties of the species involved in chemical processes. He recognized that this approach had considerable power as amply demonstrated by the rapid growth and achievements of organic chemistry. Yet for all its successes, the taxonomic approach to chemistry left questions regarding the rate, direction, and yield of chemical reactions unanswered.

Still, one might ask: even if we hold firm on the general pluralist line, can we at least admit that there have been *some* permanent achievements, such as the H_2O formula, that are no longer open to question? Here we must consider two issues. First, coherence. As long as we are working within certain systems, not believing H_2O is of course going to create some incoherence in our system of practice. It would not have worked to practice organic structural chemistry after the 1860s while maintaining HO. After the settling of the historical dust, what we can ask is: how far back in the evolutionary tree would we need to go in order to envisage a system of chemistry in which water is not H_2O? There are no absolutely permanent and unalterable achievements in science, but we can trace the boundaries within which a given achievement is going to be secure, and enjoy that security as far as it extends.

The other issue to consider is success. We *could* go back to the weight-only system or the old-style dualistic system and practice a kind of chemistry in which water is HO. There might also be other systems of atomic chemistry in which water is HO, and indeed there have also been non-atomic systems of chemistry. In order to give an exclusive and permanent preference to H_2O, we need to have confidence that none of these alternative systems have been successful, and none of them are likely to be. That may well be the case, although I personally do not have enough knowledge or experience to be sure. Some systems do not work out (i.e., fail to achieve the aims that they set for themselves, despite sustained efforts), and in those cases the gentle voice and staying hand of nature try to keep us from stepping over the edge[59]—her guidance is gentle in the sense that we can so easily ignore or misinterpret it. Occasionally we will be at the mercy of natural selection, but quite often we can go on believing outrageous things, as long as we don't put them into practice in

---

[59] This is a better image than that of surreptitiously or violently extracting her secrets.

deadly ways. And we can always live in hope, as long as we are not eliminated by less gentle forces, that one day a brilliant stroke of luck or genius would come and change the fortunes of our favored system of practice. As in life, so in science, too: we can only do what we sincerely believe will increase our success, and hope for the best. This prospect might seem a bleak one, with nothing we can possibly do in order to guarantee success. Still, the renunciation of guarantees does not amount to a denial of what we have been able to achieve without them.

3.3   From Chemical Complexity to Philosophical Subtlety

3.3.1   Operationalism

In earlier sections of the chapter I have indicated that the operationalist philosophy of Percy Bridgman was a key inspiration for my own take on how nineteenth-century chemists engaged with atoms. More specifically, I claimed that the success of nineteenth-century atomic chemistry was an operationalist triumph, arising from the commitment to take seriously the aspects of atoms that could be operationalized, and only those aspects. My view requires further elaboration and defence, especially because Bridgman has commonly been misunderstood, and I also have quite a particular take on his ideas (see Chang 2009a and also Chang 2004, ch. 3, for further details and my views on other aspects of operationalism). First I will give some relevant background on operationalism, and then address a few specific issues that are especially pertinent to atomic chemistry.

It is important to remember that Bridgman was an experimental physicist, whose pioneering work in the physics of high pressures was rewarded with a Nobel Prize in 1946.[60] His chief scientific contribution was made possible by technical prowess: in his laboratory Bridgman created pressures nearly 100 times higher than anyone else had achieved before him, and investigated the novel behavior of various materials under such high pressures. But Bridgman was placed in a predicament by his own achievements: at such extreme pressures, all previously known pressure gauges broke down; how was he even to know what levels of pressure he had in fact reached? (see Kemble et al. 1970) As he kept breaking his own pressure records, Bridgman had to establish a succession of new measures fit for higher and higher pressures. Therefore it is no surprise that he thought seriously about the groundlessness of concepts for which no available measurement methods existed. Another important stimulus to his philosophical thinking was his encounter with the revolutionary new physics of the early twentieth century. Bridgman's concerns about the definition and meaning of scientific concepts were forged in the general climate of shock suffered

---

[60] On Bridgman's life and work in general, see Walter (1990), Holton (1995), and Moyer (1991).

by physicists at that time from a barrage of phenomena and theoretical ideas that were entirely alien to everyday and classical expectations, including Einstein's theories of relativity, and quantum mechanics and its "Copenhagen" interpretation.

Bridgman's impulse was to provide stability and security to science by grounding it in well-defined and clearly performable operations. To highlight the challenges of the unknown, present even in very prosaic circumstances, he chose to open his discussion of operational analysis with the example of the most mundane of all scientific concepts: length (Bridgman 1927, 5ff). Length is measured with a ruler only when we are dealing with dimensions that are comparable to our human bodies; when we try to extend the concept beyond the familiar domain, we encounter essential physical limitations forcing us to change the method of measurement. To measure, say, the distance to the moon, we need to infer it from the amount of time that light takes to travel that distance and return, for example. For even larger distances we use the unit of "light-year," but we cannot actually use the operation of sending off a light beam to a distant speck of light in the sky and waiting for years on end until hopefully a reflected signal comes back to us (or our descendants). Much more complex reasoning and operations are required for measuring any distances beyond the solar system: "To say that a certain star is 10^5 light years distant is actually and conceptually an entire different *kind* of thing from saying that a certain goal post is 100 meters distant." (pp. 17–18; emphasis original) Thus operational analysis reveals that the length is not one homogeneous concept that applies in the whole range in which we use it: "In *principle* the operations by which length is measured should be *uniquely* specified. If we have more than one set of operations, we have more than one concept, and strictly there should be a separate name to correspond to each different set of operations." (p. 10; emphases original)

With that brief background, let us return to the issue of the relevance of operationalism for nineteenth-century atomic chemistry. First of all, when I say that it was operationalism that made this field successful overall, how is that different from just saying that an empiricist rather than realist approach to atoms bore fruit? About realism I will say more in the next section, but here I want to distinguish operationalism from standard versions of empiricism, as already hinted in Sect. 3.2.1. Empiricism is normally focused on grounding knowledge in what is observable; operationalism has its focus on grounding knowledge in what is *doable*. Ignoring that difference has resulted in much misunderstanding of Bridgman's philosophy. Observability is generally taken as a matter of what can be perceived by human sense-organs, with or without the aid of devices enhancing our senses. In the nineteenth century, there were no ways of making atoms and molecules observable in this sense. But that did not mean they could not be studied scientifically. Chemists learned to operationalize the atom, by learning to measure various properties associated with it; measurement is not passive observation, since it relies on the planned performance of certain well-defined operations.[61] Now, of course, tractable laboratory operations have to be

---

[61] One might indeed argue that there is no such thing as entirely passive observation. That may well be, but that does not pose a difficulty for operationalism, only for standard empiricism.

certified by the senses, but that makes the operations observable, not the presumed objects of measurement. Doing a chlorine–hydrogen substitution reaction and measuring the macroscopic volumes of the chlorine gas absorbed and the hydrogen gas emitted, one *measures* the relative numbers of chlorine and hydrogen atoms involved; this atom-counting is a very solid practice, but it does not make the chlorine and hydrogen atoms observable. The point is that operationalist atomic chemistry can flourish in this manner, even if the atoms themselves remain unobservable. This is how one counts what it unobservable.

The last example is a convenient reminder of a major question regarding operationalism. When we count atoms via macroscopic volumes, what exactly are we doing? The common answer would be that we are making an inference to an unobservable quantity (number of atoms) from an observable one (macroscopic volume), on the basis of a hypothesis (EVEN, or similar). Realists and empiricists would argue about whether the hypothesis involved can be justified. The operationalist position, at first glance, seems to avoid that argument altogether: what is going on is a direct measurement of a quantity (relative number of atoms), through an operation that defines that quantity; there is no reliance on a testable hypothesis here, and no commitment to anything unobservable. As Donald Gillies (1972, 6–7) emphasizes, if we accept the most extreme kind of operationalism, there is no point in asking whether a measurement method is valid; if the measurement method defines the concept and there is nothing more to the meaning of the concept, the measurement method is automatically valid, as a matter of convention or even tautology. And Bridgman's early writings did contain such a line of thought: "we mean by any concept nothing more than a set of operations; the concept is synonymous with the corresponding set of operations." (Bridgman 1927, 5)

But there is a clear difficulty here for the extreme operationalist. Can the meaning of a concept really be reduced entirely to the method of measurement? Even if we eschew deeper questions in the philosophical theory of meaning for the moment, what do we do with the fact that typically one and the same concept will have multiple methods of measurement? As I discussed in Sect. 3.2.1, atom-counting was not only done by volume but by heat as well, and also (and most commonly) by combining weights, once the relevant atomic weights were determined. So now, does the concept "relative number of atoms" have three distinct meanings simultaneously? What Bridgman's critics have not tended to appreciate is that he was very clearly aware of this problem, right from the start. More than that: the lack of one measurement method to cover a given concept in its entire range was at the very heart of Bridgman's scientific experience that drove him to operationalism in the first place, as I've discussed above with the cases of pressure and length. He frankly left it as an unsolved problem. Because he could not solve this problem, he thought it was prudent to presume that different measurement methods specified different concepts.

Come back to the case of length, for a moment: scientists do not recognize multiple concepts of length, and Bridgman was willing to concede that it was allowable to use the same name to represent a series of concepts, if the different measurement operations gave mutually consistent numerical results in the areas of overlap. However, he regarded such numerical convergence between the results of two different operations as merely a "practical justification for retaining the same name" for what the two operations measured (p. 16). Even in such convergent situations, we should be wary of making an unwarranted presumption that the different operations are essentially measuring the same thing, rather than coincidentally returning similar numbers. I think the early atomic chemists, on the whole, had this operationalist caution about them, and I think it was a useful thing in keeping them from rushing into a premature consensus. This helped different systems of atomic chemistry to develop and mature without any of them being terminated carelessly. Later on, chemists did make a synthesis in a way indicated by Bridgman, by crafting clear areas of overlap. Substitution reactions involving gases were very important, since they provided situations in which the number of atoms involved in a given reaction could be determined both by volume and by weight at once. One might call such situations *areas of co-operation*. The success of the volume–weight co-operation was achieved within the substitution–type system, and eagerly endorsed by advocates of the physical volume–weight system and the geometric-structural system.

This is as far as Bridgman takes us. One big difficulty remains, and in order to deal with this one we have to modify and develop Bridgman's ideas a bit, though not drastically. When the mid-century atomic chemists reached the synthesis of systems (described in Sect. 3.2.3) that produced the H_2O consensus, they did not make the synthesis simply as a matter of practical convenience as Bridgman would have it. Yes, the agreement in the results of atom-counting by volume and weight was initially a pleasant and convenient coincidence. But once that coincidence was observed in some key cases, it was elevated to a demand and a requirement that other cases should obey. Unified atom-counting was a practice fully established only through the *corrections* of individual atom-counting methods as and when necessary. The most significant in this was the corrections of atomic weights in order to make the results of counting-by-weight come out "right". Counting-by-volume procedures were also modified in its auxiliary assumptions, such as how many atoms should be contained in molecules of elementary gases (recall S_6, P_4 and Hg, as opposed to H_2, O_2, N_2, etc.). On the side of counting-by-heat, the tidying up was made by acknowledging that the Dulong–Petit law was only approximate at best and had clear exceptions, too.

Such corrections of measurement methods would not be possible under the kind of extreme operationalism that Gillies highlights. The core of the problem here is an overly restrictive notion of meaning, which reduces it to measurement. Although Bridgman was not proposing a general philosophical theory of meaning, he did make remarks that revealed an impulse to do so. One lesson we can take from Bridgman's troubles is that meaning is unruly and promiscuous. The kind of absolute specification of the meaning of scientific concepts that Bridgman wished for is not possible. The most that can be achieved is for the scientific community to agree on an explicit *definition* and to respect it (and that is rarely achieved). But even firm definitions can only *constrain* the uses of a concept. The entire world can agree to define length by the standard meter in Paris (or by the wavelength of a certain atomic radiation), and that still comes nowhere near *exhausting* all that we mean by length. Bridgman himself later specifically admitted that his statement that meanings were synonymous with operations was "obviously going too far when taken out of context" (1938, 117). Especially compared with the notion of "meaning as use," often traced back to the later phase of Ludwig Wittgenstein's work, it is easy to recognize the narrowness of Bridgman's initial ideas. Bridgman's later gloss on his ideas was in fact rather late-Wittgensteinian: "To know the meaning of a term used by me it is evident, I think, that I must know the conditions under which I would use the term" (1938, 116). Since a measurement operation is only one specific way in which a concept is used, it cannot cover all the other ways (including but not restricted to other measurement operations). The correction, even the questioning, of a measurement method becomes legitimate only if the concept possesses a broader meaning than the specification of the method of its measurement. Then the measurement method can be said to be valid if it coheres with the other aspects of the concept's meaning. That way we may also make a judgment about whether an operational definition (or any other kind of definition) is a good one, depending on how well it coheres with other elements of the concept's meaning and how benefi-cially it controls other elements of meaning.

3.3.2   Realism

Modern scientists and philosophers are liable to be puzzled by the philosophical attitudes of many nineteenth-century atomic chemists toward atoms. It often seems difficult to place many of the leading chemists along the realism–antirealism spec-trum. For example Dumas, who was such a major contributor to atomic chemistry, declared in 1836: "If I were master of the situation, I would efface the word atom from Science, persuaded that it goes further than experience and that, in chemistry, we should never go further than experience." (quoted in Nye 1972, 6) For a time Dumas actually *was* very nearly the "master of the situation" in France, becoming Professor of Chemistry at the *Ecole Polytechnique* in 1835 and ever more powerful scientifically and politically after that. Rocke (2010, 12) notes that "nearly coincidentally" with Dumas's disengagement from theoretical chemistry, the French government did elim-inate the word "atom" from its official syllabus for the *lycées* and universities! Perhaps Dumas's sentiment is understandable if it came out of his frustration of first adopting the physical volume–weight system but encountering what seemed to him irresolvable contradictions (see Sect. 3.2.2.3; Nye 1976, 248).

Even harder to understand is the case of Kekulé, who declared in 1867, *after* his successful work on the tetravalence of carbon and the ring structure of the benzene molecule: "The question whether atoms exist or not has but little significance from a chemical point of view; its discussion belongs rather to metaphysics." (quoted in

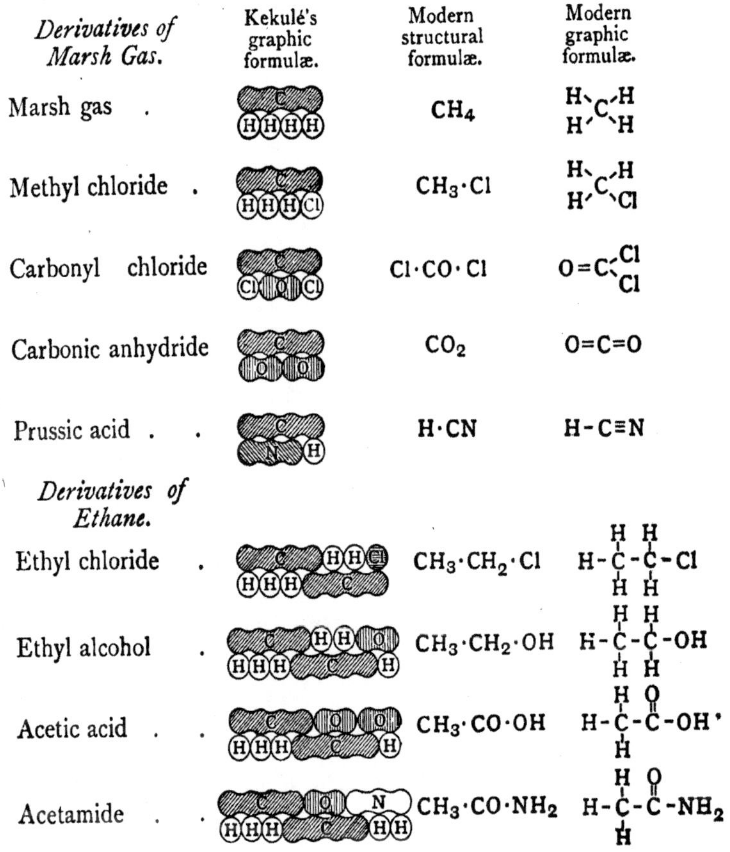

Derivatives of Marsh Gas.	Kekulé's graphic formulæ.	Modern structural formulæ.	Modern graphic formulæ.
Marsh gas		CH_4	$\begin{smallmatrix} H & & H \\ & C & \\ H & & H \end{smallmatrix}$
Methyl chloride		$CH_3 \cdot Cl$	$\begin{smallmatrix} H & & H \\ & C & \\ H & & Cl \end{smallmatrix}$
Carbonyl chloride		$Cl \cdot CO \cdot Cl$	$O = C\begin{smallmatrix} Cl \\ Cl \end{smallmatrix}$
Carbonic anhydride		CO_2	$O=C=O$
Prussic acid		$H \cdot CN$	$H-C\equiv N$
Derivatives of Ethane.			
Ethyl chloride		$CH_3 \cdot CH_2 \cdot Cl$	$H-\overset{H}{\underset{H}{C}}-\overset{H}{\underset{H}{C}}-Cl$
Ethyl alcohol		$CH_3 \cdot CH_2 \cdot OH$	$H-\overset{H}{\underset{H}{C}}-\overset{H}{\underset{H}{C}}-OH$
Acetic acid		$CH_3 \cdot CO \cdot OH$	$H-\overset{H}{\underset{H}{C}}-\overset{O}{C}-OH\,'$
Acetamide		$CH_3 \cdot CO \cdot NH_2$	$H-\overset{H}{\underset{H}{C}}-\overset{O}{C}-NH_2$

Fig. 3.9 Some examples of Kekulé's "sausage" formulas, juxtaposed to modern formulas

Nye 1972, p. 4) How can someone make his scientific fame by elucidating the structure of a complex molecule, and then turn around and say atoms may or may not exist? If atoms do not exist, what sense can there be in speaking of molecular structures that are different configurations of the connections between atoms? And what of Kekulé's earlier declaration it was an "actual fact" that "one atom of water contains two atoms of hydrogen and one atom of oxygen" (quoted in full above, in Sect. 3.2.3.3)?

It is instructive to learn a little bit more about Kekulé's work (see Rocke 2010 for further details). Before he hit upon the hexagonal structure of benzene, he was trying out diagrammatic representations of molecular structures, for example with the "sausage" formulas shown in Fig. 3.9 (from Lowry 1936, 440). It is difficult to believe that Kekulé thought that these sausages really represented the lengths and shapes of real atoms—so why do we so easily think that he would have given very literal significance to the benzene ring? For one thing, all of Kekulé's models were two-dimensional, and he would not have had any good reason to think that real molecular structures were all flat. We now have evidence to believe that the benzene molecule is actually planar, evidence which Kekulé lacked. Kekulé would have thought that there was *something* real to the tetravalence of carbon, but this was nowhere near sufficient to fix the real 3-D geometric structures of organic molecules unequivocally. He explained that he did not "believe in the actual existence of atoms, taking the word in its literal significance of indivisible particles of matter"; rather, he expected that "we shall some day find for what we now call atoms a mathematico-mechanical explanation which will render an account of atomic weight, of atomicity [valence], and of numerous other properties of the so-called atoms." (Kekulé, quoted in Nye 1976, 256)

Even the working out of 3-D structures of molecules did not generate sufficient realist confidence. J. H. Van't Hoff, famous for his pioneering contributions to early stereochemistry (chemistry in three dimensions) who (along with Le Bel) gave us the idea of the tetrahedral carbon atom, stated in a letter to Arrhenius that "the representations themselves, atoms, molecules, their dimensions, and perhaps their shapes, are after all something doubtful, as is the tetrahedron itself."[62] (quoted in Nye 1976, 259) Christoph Meinel (2004) gives a very instructive overview of various material and diagrammatic atomic–molecular models used by nineteenth-century chemists, and how as a group they only very gradually gained confidence that any of their models truly represented geometric structures. The general lesson is that not all structured representations are literal geometric representations of the structures of what one wants to represent. In short, if Kekulé and other chemists puzzle us with their apparently self-contradictory attitudes toward the reality of atoms, that is because we are reading too much modern chemical realism about atoms and molecules into their work. There is no benefit to be gained from that. We can perfectly well celebrate Kekulé and others (whiggishly) as people who made great contributions toward the modern understanding of molecular structure, without attributing an undue degree of realism to their own thinking, or insisting that such degree of realism must have been responsible for their successes.

It seems quite clear that no amount of chemical evidence would have been sufficient to put beyond doubt either the true geometric structures of molecules or the real existence of atoms. At least for the majority of people, this would have required physical evidence, as noted by historians and philosophers such as Michael Gardner (1979). But the way in which this story is sometimes told in less careful sources than Gardner's paper is seriously misleading: it is often imagined that the consolidation of the chemical side of the story in the 1860s (the Karlsruhe Congress and all that) went hand-in-hand with the coming of the kinetic theory of gases and the micro-reduction of thermodynamics to statistical mechanics, generating a formidable basis of conviction in the reality of atoms. Contrary to that picture, I think

---

[62] In his Nobel Lecture, speaking of his other major contribution to science, Van't Hoff expressed the view that molecular collisions only provided "an anyway hypothetical conception of the cause of [osmotic] pressure" (quoted in Nye 1976, 259).

Chalmers (2009, 194) is correct to insist on "the viability of anti-atomism in the nineteenth century", which was compatible with progress in atomic chemistry. In a similar vein, Mary Jo Nye (1976, 252) makes another striking observation that may confuse some realists: fresh skepticism about the reality of atoms arose *after* chemical consensus was reached on atomic weights and molecular formulas in the 1860s: "certainly in the years from roughly 1860 [to] 1895 experimental evidence accumulated which in combination was blatantly inconsistent with any one statement of the atomic hypothesis." These problems included specific heats and spectroscopic data. Paradoxically, it was the more ambitious program of physical atomism, and its relative lack of success in the early days, that brought on these doubts in a more explicit manner. This is the background to the atomic debates that came famously to a head in the clash of titans between Max Planck and Ernst Mach in the early years of the twentieth century.[63]

Returning to the field of atomic chemistry in the earlier parts of the nineteenth century, it probably suffices to say that the field was a very fractured one in relation to the realism question. In terms of the five systems of atomic chemistry that I have identified, some of the dividing lines can be seen quite clearly. In the weight-only system, there was specifically no commitment to the reality of atoms except as some unspecified carriers of weight; even the physical volume-weight system only committed one to weights and volumes[64] as physical properties possessed by atoms. In the electrochemical dualistic system the realist commitment went a bit deeper, picturing well-confined particles possessed of electric charge and the forces that atoms exerted on each other. The substitution–type system was initially nearly as non-committal as the weight-only system, except that it gave reality to the radicals as chemical units. The geometric–structural system was committed to the reality of a topological spatial relationship between the atoms, but as mentioned above this did not completely determine the actual shape of molecules in three dimensions; Peter Ramberg (2000) argues that early stereochemists were not as realist as often presumed about the physical reality of their atomic–molecular models.

Each system of practice, by virtue of the practical and conceptual activities it engaged in, required certain metaphysical presumptions. And then what makes the field even more mixed up is the fact that various chemists had varying additional degrees of commitment to physical atoms going beyond what they were obliged to by virtue of practicing the system of atomic chemistry that they practiced. Dalton and Avogadro stand out as two early pioneers who displayed a nearly reckless degree of commitment to atoms as particles with all physical properties specified. In contrast, many of the early practitioners of the substitution-type system and even of the geometric-structural system did not go any farther than demanded by their practice.

---

[63] See Blackmore (1992), chapter 5, for a convenient reprint of the original papers.

[64] To be precise, I should say "the tendency to occupy certain volumes", as it was undetermined whether atoms themselves took up all the volume occupied by a body, or there was space between atoms contributing to the volume.

3.3.3   Pragmatism

I have identified pragmatism, as well as operationalism, as an important aspect of the productive attitude of nineteenth-century atomic chemists. This requires some elaboration; what I mean is not merely that they were being *pragmatic* in the sense of being practical and sensible, but *pragmatist* in the sense of following the philosophy of pragmatism. It is important for me to take this opportunity to clarify my view of pragmatism, because various pragmatist insights have been bubbling up from my engagement with all three of the historical episodes treated so far, and these insights will also inform the discussion in the next two chapters in crucial ways. So it makes sense to put down a clear statement of what I think pragmatism is and why it is important.

What is usually meant by "pragmatism" is not easy to pin down, and it is useful to begin by reminding ourselves of the broad range of generic definitions out there, before coming to precise philosophical ideas. My *Collins English Dictionary* gives two meanings: "2a. the doctrine that the content of a concept consists only in its practical applicability" (which sounds akin to the extreme version of operationalism); "2b. the doctrine that truth consists not in correspondence with the facts but in successful coherence with experience" (which is referring to the theory of truth usually attributed to William James). Nicholas Rescher defines pragmatism as follows: "The characteristic idea of philosophical pragmatism is that efficacy in practical applications . . . somehow provides a standard for the determination of truth in the case of statements, rightness in the case of actions, and value in the case of appraisals." (Rescher in Honderich 1995, 710) Focusing on the first of Rescher's three strands of pragmatism, Robert Almeder states (2008, 91): "Pragmatists believe that the rational justification of scientific beliefs ultimately depends on whether the method generating the beliefs is the best available for advancing our cognitive goals of explanation and precise prediction."

A presentation of pragmatism that I find most appealing comes in a small work by Hilary Putnam (1995), where he identifies a few different strands in pragmatist philosophy, focusing on James: "holism", which sees "fact, value, and theory . . . as interpenetrating and interdependent" (p. 7); "direct realism, that is, the doctrine that perception is (normally) of objects and events 'out there', and not of private 'sense data'" (p. 7); an active view of knowledge ("the knower is an actor", as James put it) (p. 17); and fallibilism and anti-skepticism (pp. 20–21), the conjunction of which Putnam considers "perhaps *the* basic insight of American Pragmatism." Similar and equally appealing is Richard J. Bernstein's characterization of "the pragmatic *ethos*" by five interrelated themes: anti-foundationalism, fallibilism, the nurturing of critical communities, the awareness of radical contingency, and plurality (Bernstein 1989, 7–10).

I will not attempt a comprehensive view on the meaning and merits of pragmatist philosophy. Rather, I will identify the key features of pragmatism that I consider most important, which will constitute my own definition of pragmatism, and it will be quite apparent that those features are exhibited in the work of the nineteenth-century atomic chemists.

1. Pragmatism as I see it emphasizes that knowledge is rooted in practice, in activity. This is consonant with my analysis of science as consisting in systems of

practice, and harks back to James's view of the knower as an actor. In this sense my thinking is also inspired by a variety of other thinkers, including Michael Polanyi, Marjorie Grene, Percy Bridgman, J. L. Austin, and the later phase of Ludwig Wittgenstein.[65] Many analytic philosophers (e.g. Almeder 2008) give a thoroughly de-activated view of pragmatism, with no mention of practices; this is only a pale reflection of what pragmatism really should mean. It will not be a surprise to hear that most practicing scientists have been pragmatists in my sense, especially in a field like chemistry.

2. Closely related to the above point is an insistence that the concepts employed in our thinking should be operable, that we should be able to *do* something with them, that they should be put to use in some distinct and coherent activities. Here there are clear links with operationalism.[66] The operationalization of the concept of the atom is precisely what I have identified as the key to the atomic chemists' success, as explained especially in Sects. 3.2.1 and 3.3.1. The majority of nineteenth-century chemists did not want to engage in fruitless debates about the existence of atoms, but instead sought various ways to make the concept of the atom usable in concrete experimental and theoretical work.

3. As recognized by most commentators, there is a degree of fallibilism inherent in all versions of pragmatism. This fallibilism is rooted in a basic humility about the capabilities of the human knower, as I will spell out in more detail in Chaps. 4 and 5. I reject the faith generally attributed to Charles Sanders Peirce, that in the long run the paths of inquiry will converge on the truth. I would rather emphasize the point that, as Peirce admits, the "long run" never comes, and inquiry never ends. A truly pragmatist epistemology makes provisions about how we know and live in the here and now (including how we recognize and deal with our own mistakes), instead of straining to prove eventual convergence or focusing on the effort to reduce the meaning of truth to success. Chemists tended not to make explicit philosophical statements concerning fallibilism (or anything else), but the nineteenth-century chemists, as a group, showed a high degree of willingness to revise their views on atoms. This is not to deny the existence of some individuals with very strong views and some mean-spirited disputes (for example, the suppression of Laurent and Gerhardt, and the disputes between Kolbe and others); however, if we take the whole community of chemists in the first half-century of atomic chemistry, there was an evident fallibilist spirit. Even some of the strong-minded individuals were not afraid or unwilling to change their views: for example Dalton kept updating his atomic-weight values; Liebig and Dumas eagerly adopted the dualistic theory and Avogadro's point of view, respectively, and later fell back into the weight-only system. Fallibilism also supported a robust pluralism, which manifested itself in the parallel flourishing of the five systems of atomic chemistry that I have described in Sect. 3.2.2.

---

[65] On Wittgenstein as a pragmatist, see Putnam (1995), ch. 2.

[66] This demand for operability is applied to all concepts, including philosophical ones. In Chap. 4, Sect. 4.3.1, I will give a pragmatist–operationalist analysis of the concept of truth.

4. Lastly, pragmatism concerning science must be rooted in the recognition that science is part of life, and that the aims of science are continuous with concerns of life. This is an attitude that has been deeply ingrained in chemistry in general, and nineteenth-century atomic chemistry was no exception. The converse of this recognition is that life has an inherent epistemic dimension to it; to live as a human being (or as any sufficiently developed life-form) is to know something in the process as well. This is consonant with another dimension of pluralism (see Chap. 5): the recognition and cultivation of multiple aims in science, connected with various desiderata of life. Putnam (1995, 9–10) recalls the various types of "expediency" James saw in factual statements: "usefulness for prediction", "conservation of past doctrine", "simplicity", and "coherence". In contrast to Bas van Fraassen (1980), who separates out the "pragmatic virtues" of a theory from its main aim, namely empirical adequacy, pragmatists would view all aims of science as pragmatic, and recognize that empirical adequacy has various dimensions to it, which are all "pragmatic" in the end.

References

Almeder, Robert. 2008. Pragmatism and science. In *The Routledge companion to the philosophy of science*, ed. Stathis Psillos and Martin Curd, 91–99. Abingdon: Routledge.

Anonymous. 1864. A sad case. *Chemical News,* July 2, 1864, 12.

Anonymous. 1865. Water from a maniacal Point of View. *Chemical News*, October 27, 1865, 206.

Anonymous. 2000. *The Hutchinson dictionary of scientific biography*. Oxford: Helicon.

Avogadro, Amedeo. 1923. Essay on a manner of determining the relative masses of the elementary molecules of bodies and the proportions in which they enter into these compounds. In *Foundations of the molecular theory*, 28–51. Edinburgh: Oliver & Boyd.

Bernstein, Richard J. 1989. Pragmatism, pluralism and the healing of wounds. *Proceedings and Addresses of the American Philosophical Association* 3(63): 5–18.

Berzelius, Jöns Jakob. 1813. Essay on the cause of chemical proportions, and some circumstances relating to them; together with a short and easy method of explaining them [part 1]. *Annals of Philosophy* 2: 443–454.

Berzelius, Jöns Jakob. 1814. Essay on the cause of chemical proportions, and some circumstances relating to them; together with a short and easy method of explaining them [part 2]. *Annals of Philosophy* 3: 51–62.

Blackmore, John T., ed. 1992. *Ernst Mach – A deeper look*. Dordrecht: Kluwer.

Bloxam, Charles Loudon. 1867. *Chemistry inorganic and organic with experiments and a comparison of equivalent and molecular formulae*. London: John Churchill & Sons.

Bradley, John. 1992. *Before and after Cannizzaro*. North Ferriby: J. Bradley.

Bridgman, Percy Williams. 1927. *The logic of modern physics*. New York: Macmillan.

Bridgman, Percy Williams. 1938. Operational analysis. *Philosophy of Science* 5: 114–131.

Brock, William H. 1992. *The Fontana history of chemistry*. London: Fontana Press.

Brock, William H. 1997. *Justus von Liebig: The chemical gatekeeper*. Cambridge: Cambridge University Press.

Brock, William H. 2011. *The case of the poisonous socks: Tales from chemistry*. London: Royal Society of Chemistry.

Brooke, John Hedley. 1973. Chlorine substitution and the future of organic chemistry: Methodological issues in the Laurent–Berzelius correspondence (1843–1844). *Studies in History and Philosophy of Science* 4: 47–94.

Brooke, John Hedley. 1981. Avogadro's hypothesis and its fate: A case-study in the failure of case-studies. *History of Science* 19: 235–273.

Cannizzaro, Stanislao. 1910. *Sketch of a course of chemical philosophy*. Edinburgh: The Alembic Club.

Cardwell, D.S.L., ed. 1968. *John Dalton and the progress of science*. Manchester: Manchester University Press.

Cavendish, Henry. 1784. Experiments on air. *Philosophical Transactions of the Royal Society* 74: 119–153.

Chalmers, Alan. 2009. *The scientist's atom and the philosopher's stone: How science succeeded and philosophy failed to gain knowledge of atoms*. Dordrecht: Springer.

Chang, Hasok. 2004. *Inventing temperature: Measurement and scientific progress*. New York: Oxford University Press.

Chang, Hasok. 2005. A case for old-fashioned observability, and a reconstructed constructive empiricism. *Philosophy of Science* 72: 876–887.

Chang, Hasok. 2007a. Scientific progress: Beyond foundationalism and coherentism. In *Philosophy of science (Royal Institute of Philosophy Supplement 61)*, ed. Anthony O'Hear, 1–20. Cambridge: Cambridge University Press.

Chang, Hasok. 2008. Contingent transcendental arguments for metaphysical principles. In *Kant and the philosophy of science today*, ed. Michela Massimi, 113–133. Cambridge: Cambridge University Press.

Chang, Hasok. 2009a. Operationalism. In *Stanford encyclopedia of philosophy (online)*, Fall 2009 ed., ed. Edward N. Zalta. http://plato.stanford.edu/archives/fall2009/entries/operationalism/

Chang, Hasok. 2009c. Ontological principles and the intelligibility of epistemic activities. In *Scientific understanding: Philosophical perspectives*, ed. Henk De Regt, Sabina Leonelli, and Kai Eigner, 64–82. Pittsburgh: University of Pittsburgh Press.

Dalton, John. 1808. *A new system of chemical philosophy*, vol. 1, part 1. Manchester/London: R. Bickerstaff.

Dalton, John. 1810. *A new system of chemical philosophy*, vol. 1, part 2. Manchester/London: R. Bickerstaff.

Dalton, John. 1827. *A new system of chemical philosophy*, vol. 2, part 1. Manchester/London: George Wilson.

Duhem, Pierre. 2002. *Mixture and chemical combination, and related essays*. Dordrecht: Kluwer.

Dumas, Jean-Baptiste. 1828. *Traité de chimie appliquée aux arts*. Paris: Bechet Jeune.

Dumas, Jean-Baptiste. 1837. *Leçons de philosophie chimique*. Paris: Bechet Jeune.

Dumas, Jean-Baptiste. 1840. Mémoire sur la loi des substitutions et la théorie des types. *Comptes Rendus* 10: 149–178.

Fisher, Nicholas. 1982. Avogadro, the chemists, and historians of chemistry. *History of Science* 20: 77–102, 212–231.

Fox, Robert. 1968. The background to the discovery of Dulong and Petit's Law. *British Journal for the History of Science* 4: 1–22.

Frankland, Edward. 1866. *Lecture notes for chemical students, embracing mineral and organic chemistry*. London: John Van Voorst.

Freund, Ida. 1904. *The study of chemical composition*. Cambridge: Cambridge University Press.

Fruton, Joseph S. 2002. *Methods and styles in the development of chemistry*. Philadelphia: American Philosophical Society.

Gardner, Michael. 1979. Realism and instrumentalism in 19th century atomism. *Philosophy of Science* 46: 1–34.

Gay-Lussac, Joseph-Louis. 1923. Memoir on the combination of gaseous substances with each other. In *Foundations of the molecular theory*, 8–24. Edinburgh: Oliver & Boyd.

Gillies, Donald A. 1972. Operationalism. *Synthese* 25: 1–24.

Gjertsen, Derek. 1984. *The classics of science: A study of twelve enduring scientific works*. New York: Lilian Barber Press, Inc.

Gmelin, L. 1843. *Handbuch der Chemie*, 10 vols. Heidelberg: Karl Winter.

Gregory, Joshua C. 1931. *A short history of atomism from Democrius to Bohr*. London: A. & C. Black.

Hacking, Ian. 1983. *Representing and intervening*. Cambridge: Cambridge University Press.

Hartley, Harold. 1971. *Studies in the history of chemistry*. Oxford: Clarendon.

Hofmann, A.W. 1865. *Introduction to modern chemistry experimental and theoretical, embodying twelve lectures delivered in the Royal College of Chemistry, London*. London: Walton and Maberley.

Holton, Gerald. 1995. Percy W. Bridgman, physicist and philosopher. In *Einstein, history, and other passions*, 221–227. Woodbury: American Institute of Physics Press.

Holton, Gerald, and Stephen G. Brush. 2001. *Physics: The human adventure*. New Brunswick: Rutgers University Press.

Honderich, Ted. 1995. *The Oxford companion to philosophy*. Oxford: Oxford University Press.

Ihde, Aaron J. 1984. *The development of modern chemistry*. New York: Dover.

Jackson, Catherine. 2009. *Analysis and synthesis in nineteenth-century organic chemistry*. Ph.D. dissertation. London: University College London.

Kekulé, August. 1861. *Lehrbuch der organischen Chemie, oder der Chemie der Kohlenstoffverbindungen*, vol. 1. Stuttgart: Erlangen.

Kekulé, August. 1958. August Kekulé and the birth of the structural theory of organic chemistry in 1858 [Kekulé's speech at the "Benzolfest" (trans: O. Theodor Benfey)]. *Journal of Chemical Education* 35: 21–23.

Kemble, Edwin C., Francis Birch, and Gerald Holton. 1970. Bridgman, Percy Williams. *The Dictionary of Scientific Biography* 2: 457–461.

Klein, Ursula. 2001. The creative power of paper tools in early nineteenth-century chemistry. In *Tools and modes of representation in the laboratory sciences*, ed. Ursula Klein, 13–34. Dordrecht/Boston: Kluwer.

Klein, Ursula. 2003. *Experiments, models, paper tools: Cultures of organic chemistry in the nineteenth century*. Stanford: Stanford University Press.

Knight, David. 1967. *Atoms and elements*. London: Hutchinson.

Langford, Cooper H., and Ralph A. Beebe. 1969. *The development of chemical principles*. Reading, MA: Addison-Wesley.

Laurent, Auguste. 1855. *Chemical method* (trans: William Odling). London: The Cavendish Society.

Levere, Trevor. 1971. *Affinity and matter: Elements of chemical philosophy 1800–1865*. Oxford: Clarendon Press.

Liebig, Justus. 1851. *Familiar letters on chemistry, in its relations to physiology, dietetics, agriculture, commerce, and political economy*, 3rd ed. London: Taylor, Walton, & Maberly.

Lowry, T.M. 1936. *Historical introduction to chemistry*, revised ed. London: Macmillan.

Mauskopf, Seymour H. 1969. The atomic structural theories of Ampère and Gaudin: Molecular speculation and Avogadro's hypothesis. *Isis* 60: 61–74.

Mauskopf, Seymour H. 1970. Haüy's model of chemical equivalents: Daltonian doubts exhumed. *Ambix* 21: 208–228.

Meinel, Christoph. 2004. Molecules and croquet balls. In *Models: The third dimension of science*, ed. Soraya de Chadarevian and Nick Hopwood, 247–275. Stanford: Stanford University Press.

Melhado, Evan M. 1980. *Jacob Berzelius: The emergence of his chemical system*. Stockholm: Almqvist & Wiksell International.

Morrell, J.B. 1972. The chemist breeders: The research schools of Liebig and Thomas Thomson. *Ambix* 19: 1–46.

Morselli, Mario. 1984. *Amedeo Avogadro*. Dordrecht: Reidel.

Moyer, Albert E. 1991. P. W. Bridgman's operational perspective on physics. *Studies in History and Philosophy of Science* 22: 237–258, 373–397.

Nye, Mary Jo. 1972. *Molecular reality: A perspective on the scientific work of Jean Perrin*. London/New York: Macdonald/American Elsevier.

Nye, Mary Jo. 1976. The nineteenth-century atomic debates and the dilemma of an 'indifferent hypothesis'. *Studies in History and Philosophy of Science* 7: 245–268.

Odling, William. [1855] 1963. Translator's preface to Laurent's *Chemical Method*. In *Classics in the theory of chemical combination*, ed. O. Theodor Benfey, 40–43. New York: Dover.

Odling, William. 1858a. Remarks on the doctrine of equivalents. *Philosophical Magazine* ser. 4, 16: 37–45.

Odling, William. 1858b. On the atomic weight of oxygen and water. *Journal of the Chemical Society* 11: 107–129.

Partington, J.R. 1964. *A history of chemistry*, vol. 4. London: Macmillan.

Priestley, Joseph. 1969. *Considerations on the doctrine of phlogiston, and the decomposition of water (and two lectures on combustion, etc. By John MacLean)*. New York: Kraus Reprint Co.

Putnam, Hilary. 1995. *Pragmatism: An open question*. Oxford: Blackwell.

Ramberg, Peter J. 2000. Pragmatism, belief, and reduction: Stereoformulas and atomic models in early stereochemistry. *HYLE* 6: 5–61.

Ramberg, Peter J. 2003. *Chemical structure, spatial arrangement: The early history of stereochemistry, 1874–1914*. Aldershot: Ashgate.

Rocke, Alan J. 1984. *Chemical atomism in the nineteenth century: From Dalton to Cannizzaro*. Columbus: Ohio State University Press.

Rocke, Alan J. 1992. The quiet revolution of the 1850s: Social and empirical sources of scientific theory. In *The chemical sciences in the modern world*, ed. Seymour H. Mauskopf, 87–118. Philadelphia: University of Pennsylvania Press.

Rocke, Alan J. 1993. *The quiet revolution: Hermann Kolbe and the science of organic chemistry*. Berkeley/Los Angeles: University of California Press.

Rocke, Alan J. 2001. Chemical atomism and the evolution of chemical theory in the nineteenth century. In *Tools and modes of representation in the laboratory sciences*, ed. Ursula Klein, 1–11. Dordrecht: Kluwer.

Rocke, Alan J. 2010. *Image and reality: Kekulé, Kopp, and the scientific imagination*. Chicago: University of Chicago Press.

Rogers, Eric M. 1960. *Physics for the inquiring mind*. Princeton: Princeton University Press.

Russell, Colin A. 1968. Berzelius and the development of the atomic theory. In *John Dalton and the progress of science*, ed. D.S.L. Cardwell, 259–273. Manchester: Manchester University Press.

Russell, Colin A. 1971. *The history of valency*. Leicester: Leicester University Press.

Servos, John W. 1990. *Physical chemistry from Ostwald to Pauling*. Princeton: Princeton University Press.

Thomson, Thomas. 1807. *A system of chemistry*, 3rd ed., 5 vols. Edinburgh: Bell & Bradfute and E. Balfour.

Thomson, Thomas. 1831. *A system of chemistry of inorganic bodies*, 7th ed., 2 vols. London: Baldwin & Cradock.

Van Fraassen, Bas. 1980. *The scientific image*. Oxford: Clarendon.

Walter, Maila. 1990. *Science and cultural crisis: An intellectual biography of Percy Williams Bridgman (1882–1961)*. Stanford: Stanford University Press.

Williamson, Alexander W. 1852. Theory of etherification. *Journal of the Chemical Society* 4: 106–112, 229–239.

Wollaston, William Hyde. 1813. On the elementary particles of certain crystals: Bakerian Lecture [for 1812]. *Philosophical Transactions of the Royal Society* 103: 51–63.

Wollaston, William Hyde. 1814. A synoptic scale of chemical equivalents. *Philosophical Transactions of the Royal Society* 104: 1–22.

Wollaston, William Hyde. 1822. On the finite extent of the atmosphere. *Philosophical Transactions of the Royal Society* 112: 89–98.

Chapter 4
Active Realism and the Reality of H$_2$O

Abstract Is water really H$_2$O? Did that become a secure piece of scientific knowledge by the 1860s, after the developments that were discussed in the first three chapters of this book? I conclude that water *is* H$_2$O, but also other things, *really*. Inspired by the history of water, I take a new approach to the debate on scientific realism, which argues that realism should be taken as a commitment to maximize our learning from reality, exploring and preserving any promising paths of inquiry. I designate my position as *active scientific realism*, which differs from standard scientific realism but accommodates useful insights from all sides of the realism debate and incorporates key epistemological insights from a wide variety of traditions from falsificationism to pragmatism. I take reality as whatever is not subject to one's will, and knowledge as an ability to act without being frustrated by resistance from reality. This perspective allows an optimistic rendition of the pessimistic induction, which celebrates the fact that we can be successful in science without even knowing the truth. The standard realist argument from success to truth is shown to be ill-defined and flawed. I also reconsider what it means for science to be "mature", and identify humility rather than hubris as the proper basis of maturity. The active realist ideal is not truth or certainty, but a continual and pluralistic pursuit of knowledge.

4.1 Is Water *Really* H$_2$O?

So, is water H$_2$O? Do we have enough grounds for taking that as a simple truth? Is there sufficient reason for us to believe it without qualification? Having followed the last three chapters, I hope you will agree with me that these are reasonable questions to pose. How complex such a simple question as "Is Water H$_2$O?" can turn out to be! Chapter 1 concluded that there was no convincing argument in the Chemical Revolution against the phlogistonist system, which regarded water as an element. Chapter 2 saw that electrolysis also failed to provide a conclusive argument for the

H. Chang, *Is Water H$_2$O?: Evidence, Realism and Pluralism*, Boston Studies
in the Philosophy of Science 293, DOI 10.1007/978-94-007-3932-1_4,
© Springer Science+Business Media B.V. 2012

compound nature of water. Chapter 3 showed that even at the consensus-point reached after half a century of atomic chemistry, there was no absolute proof of that formula H_2O. On the whole, a century of chemistry[1] starting with Lavoisier's work failed to provide sufficient reason to close the debate on the constitution of water.

Yet the debate did close, and "Water is H_2O" became an apparently unshakable scientific fact. Surely, various qualifications have to be added—some of the atoms involved are rare isotopes of hydrogen and oxygen; some of the molecules are dissociated into ions, and then often combined with H_2O molecules to form complex ions; the whole H_2O molecules are also linked up with each other by hydrogen bonds; and so on. But we can perhaps accept these as well-understood and thoroughly agreed "fine print" that does not change the main message, so that "Water is H_2O" appears as an unproblematic approximation to the truth. Even if you have read the preceding chapters of this book with some sympathy, I expect that you still have a clear intuition which makes it nearly impossible to entertain the thought that water might not be H_2O. There seems to be something *undeniably right* about the statement that water is H_2O. But what exactly is the basis of that intuition, if there was no watertight argument that proved the truth of the statement even by the time scientists reached a consensus on it, as we have seen in the first three chapters of this book? That may be an irritating question, but it should not be dismissed out of hand.

For philosophers of science, questions of this kind inevitably lead to the debate concerning scientific realism. What is usually meant by "scientific realism" is the idea that scientists try to discover something about how the universe really is, and that they have been quite successful in that effort.[2] In this chapter I want to place the story of water in the context of the realism debate, and see what it can teach us.[3] Any one case cannot tell us anything general for sure about something as broad-ranging as scientific realism, of course. But H_2O will serve as an important test-case. Anyone who wants to maintain that science gives us the truth about nature ought to be able to convince us that science can at least give us an assurance about very simple and basic things like H_2O, before we start worrying about things like DNA, quarks, black holes, parallel universes, etc. In other words, any doctrine of scientific realism worth considering should be able to handle the case of H_2O; realism must go through a trial by water, as it were. A better metaphor would be to say that the case of water serves as a touchstone for various doctrines concerning scientific realism. It is interesting to recall how a touchstone actually works: to test the purity of gold alloys, one rubs the piece in question on a slab of black siliceous stone and treats the trace left on the stone with nitric acid, to remove impurities; the color of the remaining trace gives a surprisingly accurate indication of the gold content. Any worthwhile doctrine of realism should come out looking properly golden when applied to the case of H_2O.

---

[1] Whether *physics*, later, provided such proof is a whole other issue, and it is difficult to say without having carried out a thorough study.

[2] Bas van Fraassen only uses the first half of this formulation in characterizing realism, which he opposes; most realists themselves include the second half.

[3] For more general considerations on the history–philosophy relationship involved in this kind of attempt, see Chang (2004), ch. 5, and also Chang (2010, 2011e).

After giving an overview of the status of the claim that water is H$_2$O, I will propose a doctrine called *active scientific realism*, which maintains that science should strive to maximize our contact with reality and our learning about it. Active realism is intended as a *normative* doctrine in its dual sense: it captures something good about the norms that have actually governed science, and promotes that good by articulating, defending and developing it for future practice. By means of active realism I also aspire to identify a common core of what is sensible in various opposing sides in the philosophical debates concerning scientific realism. Looking back on the history that I have presented in the preceding three chapters, active realism will help us make better philosophical sense of the seemingly haphazard and untidy development of the sciences of water. As in previous chapters, the first section (Sect. 4.1) makes a brief and accessible presentation of the main points. The second section (Sect. 4.2) will give a fuller, more general and more advanced characterization and defence of active realism, starting with a critique of some common philosophical views and arguments. The third section (Sect. 4.3) contains various in-depth discussions of some contentious issues that will seem absolutely crucial to some and rather pointless to others.

4.1.1 Hypothesis-Testing Within Systems of Practice

Let me now begin with an overview of how we may judge whether claims like "Water is H$_2$O" are really true. Two veneers of conviction protect our belief in many theories of modern science. One layer is generated by testimony and indoctrination: most of us have "H$_2$O" firmly in our heads because that is what we have been told all our lives. In one sense I feel unable to doubt it, in the same way I feel unable to doubt the real existence of, say, Papua New Guinea or the Kalahari Desert even though I have never had any actual contact with those places. In a similar way, I also profess to be certain that I was born on 26 March 1967, even though I have *no recollection at all* of this presumed event. The other layer of conviction is generated by theoretical arguments within our accepted systems of science, which only amount to consistency-checks although they are psychologically comforting and pedagogically useful. This is what happens, for example, in the circular argument in which we derive atomic weights from known molecular formulas, and *vice versa*. Sometimes these theoretical checks are disguised as experimental tests, as in school laboratory work in which students learn to find ways of rationally discarding results that do not confirm the "correct" answer. Those two layers of conviction are important in the perpetuation and protection of an already-established system of knowledge, but they cannot generate a new system, nor justify its initial establishment, so they are distractions from my present concerns. The main question I have been pursuing throughout this book is how scientists came to believe that water is H$_2$O and whether they did so for good reasons, not so much how they have managed to maintain and propagate that belief.

Stripping away these veneers of certainty, let us consider what could have served as the initial source of conviction that water is H$_2$O, drawing from the discussions given in the preceding chapters. Today's scientific and philosophical common sense is that a genuine empirical testing of hypotheses is the best possible source of evidence for what we believe about nature. Lurking just underneath that common sense, threatening to overturn it, is the notorious "Duhem thesis" on the impossibility of testing an isolated hypothesis, attributed to the French physicist–philosopher Pierre Duhem (1861–1916). According to Duhem ([1906] 1962, 182–183), the test of a hypothesis is always the test of a whole theoretical group: "the statement of the result of an experiment implies, in general, an act of faith in a whole group of theories" which underwrite the experiment itself; therefore, an experiment "can never condemn an isolated hypothesis but only a whole theoretical group." And what may seem like a pleasing confirmation of a theory can always be disturbed if the auxiliary hypotheses that had been assumed in the test are rejected.

Put yourself in the position of someone (like Lavoisier) who wanted to make empirical tests to prove the bold new hypothesis that water was a compound. You decompose it using a hot gun barrel, break it down with electricity, synthesize it by sending a spark through oxygen and hydrogen—but no matter what you try, there is some alternative set of auxiliary hypotheses which allow Cavendish, Priestley, Ritter and others to make the experimental outcomes consistent with the hypothesis that water is an element. At the same time, you realize that your own interpretation of events also requires the help of auxiliary hypotheses just as much as the other side. The situation is the same for the testing of hypotheses regarding the specific molecular composition: you cannot test "Water is H$_2$O" without involving unfounded auxiliary assumptions such as EVEN (the equal volume—equal number hypothesis); if you should try to test EVEN itself, there will be other auxiliary assumptions needed for that test. The Duhem problem is one of the most significant sources of despair, frustration and annoyance for those who would like to have scientific or philosophical license to simply believe what science tells us, and get on with life on that basis.[4] How do we move beyond this difficulty?

I think we can transcend the Duhem problem only by accepting and deepening Duhem's insight, not by trying to avoid it. To his insight we should add the recognition that beliefs are inextricably bound up with actions—in science, too, as in life. Not only does the test of a hypothesis involve auxiliary assumptions, but any testing method we employ only makes sense in the context of other accepted *practices*, or epistemic activities, that are coherent with it.[5] Quantitative tests have to be backed up by well-established measurement operations, and also by techniques of statistical analysis for systematic testing. There are also more specific kinds of practices required in each situation. For example, in the middle of the nineteenth century EVEN could only be tested indirectly by the outcomes of its employment, for which the crucial activities were vaporization, density-measurement, and substitution-reactions. And even if test-results are apparently positive, there remains

---

[4] This difficulty has even contributed to the unproductive gulf between realist science and skeptical philosophy; I wish to bridge this gap, *enough* so that the science–philosophy interaction can be productive, while the two fields pursue their own legitimate ends.

[5] Similar insights are applied to give a fresh view on the old "ravens paradox" in Chang and Fisher (2011).

the possibility that the testing-activity itself is misguided due to incorrect assumptions or incoherent practices. To put the point more generally: the activity of hypothesis-testing can only take place within a system of practice,[6] and the validation of a theory only comes as part of the success of a system of practice. Any usable hypothesis or theory comes embedded in a system of practice; it is impossible, and ultimately pointless, to evaluate the correctness of the theory apart from the success of the system in which it operates.

So we need to ask how systems of practice are selected, not just how theories are chosen. I have already hinted at this general view in my discussion of specific episodes in earlier chapters. For example, as explained in Chap. 1 the compound view of water was initially adopted only as part of the general triumph of the oxygenist system of chemical practice over the phlogistonist system. This systemic triumph is also understood better in the context of the even larger story of the increasing dominance of compositionist systems over principlist systems. In Chap. 2 the rejection of the synthesis view of electrolysis was shown to be underwritten by a double systemic commitment: to the oxygenist system inherited from Lavoisier's revolution, and to the new emerging electrochemical dualistic system of Davy and Berzelius. In Chap. 3 the H$_2$O formula for water was seen to emerge as part of the physical volume–weight system of atomic chemistry advocated by Avogadro and his sympathizers, and firmly accepted as a result of the successful synthesis of that system with the substitution–type system. In each case the account of hypothesis-testing and theory-choice would have been crucially incomplete without an account of the relevant systems of practice providing the context and much of the content of the arguments. In each case the theory-choice itself, at least the decisiveness with which it was made, looked mysterious; these decisions made much more sense when we understood which systems of practice were dominant, which were ascendant, and which were waning.

This, however, only pushes the epistemological question one step further: how should systems of practice be chosen? All kinds of factors affect our decision to commit ourselves to a system of practice, and to stay with it as time goes by and we grow up. In a normative epistemic evaluation, we are asking not only how scientists' decisions do come to be made, but whether they contribute to the improvement of knowledge. The ultimate criterion in the choice of systems of practice can only be success: we are making good system-choices if those decisions maximize our success. But success is a rather vacuous concept, just a placeholder for the achievement of all kinds of things that we desire. So the epistemic successfulness of a system of practice can only mean its overall effectiveness in realizing various epistemic values that we have. But scientists hold a multiplicity of epistemic values (ranging from empirical accuracy to theoretical elegance), and they also differ on which values they regard as important and how they translate the values into practice; all this was demonstrated amply in the first three chapters. Now, if the validation of a hypothesis is based on the success of a system of practice, and if success is such a relative matter, it would seem that we could never have a conclusive validation of a hypothesis.

---

[6] For the definition of "system of practice", see Chap. 1, Sect. 1.2.1.1.

But isn't there a rather convincing realist argument that I am ignoring here? If there is a result on which all respectable scientific systems agree, then shouldn't we regard it as true? In Chap. 3, for instance, we saw quite an impressive convergence of three different systems of atomic chemistry, all agreeing on the modern system of atomic weights and molecular formulas including H_2O. There wasn't this sort of convergence in the earlier historical periods on the question of the compound nature of water, but if we know that water is H_2O, then we of course also know that water is not an element, so all the worries about the earlier debate become pointless.[7]

It would seem that a result that is shared by multiple independent systems should receive a higher degree of confidence. This is a familiar argument, with various versions of it associated with an impressive array or scientific and philosophical names ranging from William Whewell to Ian Hacking. But is there really anything in this argument, aside from a psychological assurance? How do we demonstrate that the convergence is not simply a coincidence?[8] Intuitively, the situation can be seen most clearly by taking an extreme scenario first: if *all possible systems* lead to the result in question, then of course the result is necessarily true. The question is whether there are results of science that are *inevitable* in this way.[9] If convergence is taken as an indication that such inevitability holds, then it can be regarded as significant evidence. Of course we can't know about all possible systems, but we can at least make a genuine effort to see if the result in question could be avoided, by attempting to find a viable system in which it does not hold.[10] It would be best for the realist if there were no such systems that can easily be imagined.

4.1.2 Imagine!

Now, it may seem that "Water is H_2O" is a good candidate for such inevitability. Again, even if you have followed all of my unconventional discussion so far, you probably feel that there could not possibly be a good system of science in which water is an element, or a compound whose basic constitution is something other than H_2O. I am now going to engage in some simple conceptual acrobatics in order to de-stabilize that intuition about the inevitability of H_2O. You may ask what the point of this exercise is. It is to loosen up your imagination. I am just trying to show

---

[7] Later in history, there would come an additional convergence-based argument pointing to similar conclusions, regarding the coincidence of the values of Avogadro's number determined in many different ways.

[8] Any calculations designed to demonstrate the raising of probabilities will rest on background assumptions that are themselves subject to non-trivial doubt.

[9] See Hacking (2000) and Soler (2008) for this way of putting the question.

[10] This may include systems in which the statement in question does not occur at all, or systems that do not assign a truth-value to the statement, in addition to ones in which it is flatly contradicted or considered unlikely.

you that there are perfectly rational and sane conceptual universes, fully informed by modern science, in which water is an element, or it is a compound of some other constitution than H_2O. And these "conceptual universes" are simply different ways of thinking about and dealing with the actual universe that we live in.

What is an element? In Chap. 1 I already made a stab at this imaginative exercise, by calling for a better awareness of the subtleties involved in the concepts of "element" and "compound". When we say in today's science that water is a compound and hydrogen and oxygen are elements, there is no pretence or illusion of absoluteness about the element–compound distinction; we all know that hydrogen and oxygen atoms can be broken down and hence they are not "elements" even in Lavoisier's operational sense. We can't even say that in *chemical* reactions the atoms stay intact. It's not just that the line between the physical and the chemical is ultimately fuzzy. The more pertinent point is that the hydrogen atom *routinely* loses an electron in very ordinary chemical situations, while the oxygen atom is prone to gaining an extra one, or two.[11] And proton (hydrogen ion, H^+) is as much a part of any other kind of atom as it is a part of the hydrogen atom. Still, it makes sense to regard hydrogen and oxygen as "elements" in a qualified sense of the word, even in the absence of ultimate immutability in their atoms. In a similar spirit, many chemists in the nineteenth century got used to regarding radicals or groups functionally as "atoms" in the context of organic chemistry, and it was very common for them to speak of "an atom of water", etc.[12] Doesn't water deserve to be regarded as an "element" in this sense in very many situations? What really hinges on calling it an element or a compound in an absolute sense?

"Neutral" is not "natural". Come back to the issue of ionization. In many chemical contexts, shouldn't we say that the natural state of the hydrogen atom and the group 1 alkali-metal atoms is the ionized one, with one degree of net positive electricity? And that the natural state of the chlorine atom (and other halogen atoms) is one degree of negative ionization? What is the reason for regarding the electrically neutral state of these atoms as the natural one? If we consider what the atoms are most prone to doing, there is clearly a point to the old idea that hydrogen is naturally electropositive, and chlorine or oxygen is naturally electronegative.[13] Was Berzelius so wrong then? And was Davy so wrong to think that the neutral hydrogen and chlorine atoms rendered each other electrical when they came into appropriate contact with each other? Neutral is not natural, in these electrochemical contexts. And then, if a proton is the natural state of the hydrogen atom, doesn't it make sense to

---

[11] Two hydrogen atoms are walking down the street.

   Atom A: "I think I've lost an electron!"

   Atom B: "Are you sure?"

   Atom A: "Yes, I'm positive."

It's a pity that quantum mechanics won't let us label identical particles.

[12] This is as noted in Chap. 3, footnote 33.

[13] There is also something to the idea that when they combine, the neutralization of their electricities release energy, which may manifest itself in the form of heat or light.

think of neutral hydrogen gas as a *compound* of "hydrogen" (protons) and negative electricity (in the form of electrons[14])?

Which is the constituent of which? As we have seen in Chap. 2, what is made up of what is not a cut-and-dried issue. Recall from Sect. 2.2.1 in that chapter the representation of the electrolysis of water by Linus and Peter Pauling. The reaction at the negative electrode is:

$$4H_2O + 4e^- \rightarrow 2H_2 + 4OH^-$$

So the production of hydrogen gas can be seen as a result of the combination of water and electrons, with the OH^- ions as a by-product. But just how different is that from saying that hydrogen gas is a compound made up of water and negative electricity, which was Ritter's crazy view?! And then if you allow the identification of phlogiston and negative electricity (or electrons) that I discussed in Chap. 1, then we can also rehabilitate Cavendish's view that hydrogen gas was phlogisticated water. Is the "OH^- as a by-product" part of the story a great big cheat? I think not—nobody blinks an eye when it is said, for example, that the first major step in the Leblanc process for soda manufacture was making sodium sulphate ("salt cake") by combining sulphuric acid and salt, with hydrochloric acid as a by-product (see Brock 1992, 288, for the whole Leblanc process). And there are numerous cases in which water itself participates in crucial ways in chemical reactions without getting much of a mention. What one considers the important ingredients and products in a chemical reaction is a relatively relaxed matter of convention, not of ultimate truth or falsity.

It is also not so absolutely and uniquely true that H_2O is made up of hydrogen and oxygen, if what we mean by "hydrogen" and "oxygen" are the gaseous states of these substances with all their normal properties. In order to make H_2O, at least the oxygen molecules have to be broken up into atomic oxygen, which doe not have the recognizable properties of oxygen. If what we mean by "hydrogen" and "oxygen" are the neutral atoms, then these would not simply combine to make water molecules; they can do so only by transforming themselves first into something ionic (but not simple ions, either, in the case of oxygen). In other words, it is actually difficult to pinpoint how "hydrogen" and "oxygen" exist in any straightforward sense in water, and that is really the same for all chemical combination. Chemical elements are not simple Lego-pieces, after all, even though the ball-and-stick models and the cartoonish graphic representations of molecules may mislead students and consumers of popular science into that way of thinking.

There is something in the HO formula. So far, I have engaged in some changes in viewpoints according to which the H_2O formula can be interpreted variously in relation to the question of what is elementary. Can I also show that the H_2O formula itself is not inevitable? Of course, not *within* atomic chemistry as it was consolidated by Kekulé's generation. Rather, the question is: are there good systems of atomic chemistry imaginable in which we can honestly think of water as having some other

---

[14] And what other form of negative electricity do we ever have in normal life, except for electrons?

constitution than H$_2$O?[15] It is helpful to look back at the history, and recall what it was that the now-rejected systems did well. Imagine—it's easy, if you try—that organic chemistry had not developed in the way it did. For example, imagine that the volumetric and thermal methods of atom- and molecule-counting had not developed as early as they did. Atomic chemistry would not have withered and died; I think it would have continued to flourish in the weight-only system of Thomson, Wollaston and the later Liebig. If that system would have developed further, who knows what kind of interesting organic classifications, even structural theories, it might have led to in the end? And who knows what kind of new syntheses it might have stimulated, by plan or by accident?

Was there any reality to Wollaston's combining weights? Was there something real and objective reflected in that ratio of 1:6:8:14 (roughly) for H:C:O:N that he found, and formulas such as HO for water? There certainly was a functional reality to Wollaston's numbers, within the realm of analytical chemistry he was working in. Even from the modern viewpoint, they do tell us something real about atomic weight *and* valency taken together. Wollaston's 1:8 ratio between hydrogen and oxygen tells us that it is two atoms of hydrogen that like to combine with one atom of oxygen. In that case, what would be so wrong with taking two atoms as the important combining-unit of hydrogen, and regarding water as a 1-to-1 combination of "hydrogen" and oxygen? That sort of system might have ended up with an element having a variable "atomic weight" depending on what it combines with, but is that conceptually so much worse than variable valency?[16]

Which level is "atomic"? Or what if physics had raced ahead of chemistry, so that some direct microscopic weighing would have given us the weight of the particles of gases, before any of the intricate chemical work discussed in Chap. 3? (It isn't impossible to imagine a chemistry-averse civilization, or an easy-to-poison race of beings, that would have invented the mass spectrometer before much development of organic chemistry.) Then the hydrogen and oxygen molecules would have been seen as "atoms", which would split into two halves under certain chemical situations. Sparking a mix of hydrogen and oxygen gases would have been seen as a method of atom-splitting. Assuming EVEN, the reaction would have been seen as $2H + O = 2(H\emptyset)$ (I made up the symbol $\emptyset$ there to indicate half an atom of oxygen). This formulation would have made sense of the fact that in the electrolysis of water the "whole-atom" of hydrogen was much more easily separated as a gas, while the "half-atom" of oxygen, being so much more reactive, tended to form oxides with the metallic anode, much more easily and quickly than whole oxygen "atoms" do

---

[15] When I first drafted this section (on 5 June 2010) I was stumped at this point, and wrote: "I'm having trouble with this, but let's see if I can find a bit more imagination here!" About 15 minutes later: "OK, that wasn't so hard at all!"

[16] It is easily imaginable that the variability of valency might have been foremost in people's minds—even setting nitrogen and phosphorus aside, what if the early chemists had obsessed about cases like CO and CO$_2$? Can't one imagine such cases being so easily taken as a refutation of the whole idea of valency? And without valency, structural chemistry as we know it would not have been possible.

(when metals are exposed to oxygen gas at normal temperatures). Some interesting and useful theory could have been developed, concerning the different chemical behaviors of such whole-atoms and half-atoms. And that would have been a more commodious conceptual framework than the idea of reactive "nascent" hydrogen and oxygen, which people were trying to use in order to explain a whole set of phenomena ranging from the action of chemical bleaches to the formation of secondary products in electrolysis.

A similar system can actually be arrived at without much history-of-science fiction. Simply recognize (in the way nuclear physicists do) that there are various points of stability in the clumping of elementary particles. In the family of substances including what we call hydrogen, we have first of all H^+ (proton), then H_2 (two protons and two electrons), and then countless different hydrogen-compounds. Somehow a single atom of H (one proton and one electron) is chemically not stable and does not normally exist in nature, despite its iconic and foundational status in quantum mechanics. Of the more stable configurations, whether to call H^+ or H_2 the hydrogen "atom" is a conventional decision. We might very reasonably go for H_2, on the grounds that this is the level of organization that displays the common properties that define what we normally mean by "hydrogen". If you're inclined to think that modern knowledge shows that view to be simply absurd, I will ask in return: what is so sacred about these particular groupings of elementary particles in which all the baryons are bound to each other by the strong force, that we should strictly reserve the term "atom" exclusively for them? If we go with H_2 as the hydrogen atom, H^+ would be sub-atomic hydrogen (or hydrogen sub-atom), and it would be an interesting property of the hydrogen atom that it can be broken down into two stable sub-atoms but only by removing electrons from it; similarly the chlorine atom can be broken down into two stable sub-atoms by adding electrons to it. Then something like HCl would be a combination of one hydrogen sub-atom and one chlorine sub-atom. In the same system, the oxygen atom would be what we call O_2, so water would be a compound of one atom of hydrogen and one sub-atom of oxygen.

Why are our imaginations about chemical ontology usually so restricted, and in a particular way? The chemical educationist Keith Taber (2003) argues quite convincingly that many of today's students are misled by "conceptual fossils" present in the standard teaching curriculum, so that they form notions of the atom that are not only constricting but actually incompatible with the up-to-date modern understanding. I think my own musings are quite compatible with Taber's assessment of the situation:

> These conceptual fossils encourage learners to develop an "atomic ontology" (granting atoms "ontological priority" in the molecular model of matter); to make the "assumption of initial atomicity" when considering chemical reactions; and to develop an explanatory framework to rationalise chemical reactions which is based on the desirability of full electron shells. These ideas then act as impediments to the development of a modern chemical perspective on the structure of matter, and an appreciation of the nature of chemical changes at the molecular level. (Taber 2003, 43)

Taber's concern is that modern students should have proper modern understanding; mine is that we should all learn to appreciate all systems that provide good understanding. But he and I would be agreed that people should not be bound by a particular way of thinking that is not even the modern orthodoxy.

4.1.3 H₂O: A Pluralistic Truth

Even if there is a field of phenomena that is reasonably stable for all observers, it is important to recognize that there are many ways of dividing up and ordering that field conceptually and materially. We have to remember that the world itself does not come neatly divided up into taxonomic boxes. We have to invent those boxes ourselves, and there will likely be things that do not fit nicely into any system of boxes we devise.[17] The best general principle of classification is to focus on differences that make a difference, but differences that are very important in some contexts may not have much relevance in other contexts. For example, we do not worry at all in most chemical situations about treating all the isotopes of a given element as the same, though we do know that in other situations the variety of isotope makes all the difference in the world (everything from carbon-dating to the atomic bomb would not work if we didn't have the right isotope, and fish will die in heavy water). In particular, the material world of chemistry does not come pre-classified into neat taxonomic *hierarchies*. No one seems troubled by the fact that the "atom" level familiar to modern science encompasses units containing very different numbers of particles—down to the hydrogen ion, surely one of the most important atomic species in all of chemistry, which does not have any electrons and in fact is just a single elementary particle, unlike all other atoms and ions.

I try to limit my armchair-speculations about how productive systems of science that do not affirm "Water is H₂O" might have been, and could still be now. But I would suggest that such alternative systems would have been able to capture something useful, in the same way the surely fictitious ball-and-stick models of atoms and molecules have done. Who can be so sure that the alternative conceptual possibilities would never have held any significant promise, to stimulate some different theoretical and experimental developments, even to facilitate the discovery of phenomena to which we have not yet paid sufficient attention? I can only go so far at this point in the discussion, because I am actually thinking on the basis of exactly the same set of phenomena that modern chemistry accepts. In the area of atomic–molecular chemistry I have not been able to carry out any experimental work to see if there are any phenomena out there which modern chemistry might have neglected or not found yet. But I have seen enough in other areas to make me generally open-minded at least.[18] Is there really no use in coming back a bit down the evolutionary tree of science for a moment, to consider if there aren't any other branches that hold some nice fruit, or indeed different directions in which branches that we do not have might have grown?

Yes, water is H₂O. It is also an electrostatic combination of electropositive hydrogen and electronegative oxygen, which can be broken up with a battery. It is also a one-to-one combination of hydrogen and oxygen "atoms" (in the weight-only system). It is also an element from which one can produce hydrogen and oxygen gases by the addition or subtraction of phlogiston. And so on. We may choose to interpret these

---

[17] Recall Kuhn (1970, 24): normal science is "an attempt to force nature into the preformed and relatively inflexible box that the paradigm supplies".

[18] Chang (2002, 2007b, 2011c); see also Chap. 2, Sect. 2.3.2 in this book.

statements in such a way as to generate logical contradictions between them, and then force an exclusive choice. Or we may allow them to be *independent* from each other, and appreciate and develop the merits of the systems of practice in which they each occur. Which way we go is our choice. Nature itself will allow a few more conceptual possibilities than we normally allow ourselves.

So my neck is out, by a long way. Historically, I am observing that even three-quarters of the way into the nineteenth century, "Water is H_2O" was not known to be true to the exclusion of all other possibilities. Philosophically and scientifically, I am claiming that even now we should not regard "Water is H_2O" as an inevitable piece of scientific knowledge. Of course, water can't be just *anything* we like to imagine. Once we have defined "H" and "O" in the standard way that we have in modern chemistry, the observed phenomena are such that we have very little choice but to affirm that water is a compound substance consisting of molecules made up of two hydrogen atoms and one oxygen atom (though "consisting" can be quite a complicated matter). In that affirmation we are expressing a truth about nature that we cannot arbitrarily change or deny. However, this truth is internal to various systems of practice in which it is true.[19] This also means that our affirmation of the truth is contingent on our continued employment of these systems, which in turn depends on their continued success. The relationship between truth and success needs to be recognized clearly here: I am advocating neither the standard realist notion that we can infer truth from success ("this theory is so successful, it must be true"), nor the caricature of pragmatism according to which truth merely means success ("whatever is successful *is* true"). Rather, truth as I conceive it means correctness as judged within a specific system of practice, and our decision to adopt a system of practice is determined by its successfulness.

I will say more about the meaning(s) of truth in Sect. 4.3.1, where the notion of truth I employ is listed as "truth$_5$". For now, the most important thing to note is the pluralism inherent in my notion of truth, due to its inherent link with success.[20] Any real-life success is a limited, relative, and provisional thing. Even if the truth of a statement within a system of practice is quite precise and assured, our affirmation of that truth should be only as definitive as our acceptance of the system itself, which is in turn only warranted if the system continues to be successful. Success is a dynamic criterion, and judging relative success is a game of ruling-in, not ruling-out; provisional success is a matter of being "good enough to stay in". What is important is pursuit, not acceptance, as stressed by Larry Laudan (1977, chapter 4). And if I choose a certain system, that does not imply that all other systems should be discontinued and no one else should choose any of them. In most real-life competitions, we do not kill off the runner-up; we give her a silver medal. Water is H_2O in all the systems of practice familiar to modern science. But it has been, and will be, other things in some other systems.

---

[19] In a way, there is nothing really new in what I say in this paragraph. One could easily draw links back to Putnam's "internal realism" or Carnap's "framework relativism".

[20] See Sects. 4.2.1 and 4.2.4 for more considerations on success.

4.1.4   Knowledge, Progress, and Active Realism

At this point I anticipate a set of serious objections from the defenders of Truth, Rationality, Objectivity and Progress, the same kind as Kuhn faced in relation to his concept of incommensurability in scientific revolutions. What is the use of pluralistic truth? What good is "truth", if it is only valid within a particular system of practice? How can scientists be rational, if there are no clear and inviolable rules for choosing between competing systems? If theories, paradigms, or systems of practice come and go, and if the truths known in one system cannot be transferred to another system, in what sense can science retain its achievements and progress in a cumulative way?

For the purpose of understanding and facilitating scientific practice, I would like to suggest a fundamental re-orientation in our conception of knowledge, to think of it in terms of *ability* rather than *belief*.[21] I do not think it is very productive in philosophy of science to pursue lines of thinking arising from the debates concerning the old common sense of epistemologists that knowledge is "justified true belief". It's not that I would presume to dismiss venerable traditions of epistemology focused on truth and belief; rather, I am just modestly expressing a sense of dissatisfaction that those traditions do not help me very much with the kinds of questions that I am attempting to address in this book. At least when considering knowledge as it exists embedded in a system of practice, we can gain new and better insights by thinking of knowledge not as consisting in belief but in ability—an ability to do certain things reliably as intended, without being foiled by resistance from reality. It is unhelpful to postulate a sharp distinction between "knowing that" and "knowing how", especially if it means that we pay exclusive attention to the former. If the practitioners of Lavoisierian chemistry knew that water was a compound of hydrogen and oxygen, that means they knew, for example, how to make water out of hydrogen and oxygen gases, or out of hydrogen gas and a metal oxide; they also knew how to start with water and metal and turn them into hydrogen gas and a metal oxide. If the practitioners of phlogistonist chemistry knew that hydrogen was phlogisticated water, that means they knew, for example, how to make hydrogen gas from a phlogiston-rich substance (such as a metal) and water (such as contained in the solution of an acid, or in the form of steam passing over a hot piece of metal).

Before proceeding further, I should clarify what I mean by nature's *resistance*. Our epistemic activities can be successful only if nature, or reality, does not prevent what we are trying to achieve. If we try to make 17 g of water by combining 1 g of hydrogen and 16 g of oxygen, we will fail. If we try this under the most favorable

---

[21] My view (also cf. Chang 2008) has a good resonance with that of Bernadette Bensaude-Vincent and Jonathan Simon (2008, 201): "Rather than framing the preliminary epistemological debate in terms of the question 'what can one know?' it might be better to pose the question 'what can one do?' and then examine the ontological consequences." With this re-orientation these authors are also able to give subtlety to their historical discussions, avoiding the pigeon-holing of historical figures into facile dichotomies. For example, they give an instructive discussion of Kekulé's puzzling anti-realism about atoms (pp. 188–191; cf. my Chap. 3, Sect. 3.3.2), and remind us that Comte was not an anti-realist (p. 181).

conditions and with all the ingenuity that we can muster and still fail, then it is prudent to conclude that we have failed because nature did not cooperate with our plans. Nature "resisting" or "cooperating" is metaphorical talk, but it may be the best we can do in trying to capture the sense of the existence of a non-arbitrary external reality, and this is probably the most important gut-level basis of the common revulsion against constructivism. From a more positive angle, knowledge can be seen as a state of ability to do things without being foiled by significant resistance from reality. But as Wittgenstein once put it (1969, 66e, §505), "it is always by favour [Gnaden] of Nature that one knows something", and might nature not suddenly turn unsympathetic to us?—there we have the basic fallibilist insight. When there is constancy in phenomena manifested through our observations and actions, that is a gift from nature. When and where we are successful in our activities, to the extent that we seem to have mastered the ways of nature in the relevant ways, we have knowledge, which is more or less synonymous with success, from this point of view. But we must also learn to accept that nature may decide to take away the gift of constancy where we are not expecting that. The ordinary problem of induction is there, before and after the pessimistic meta-induction, to remind us that we are not able to draw any grand conclusions from the ineffable constancy of nature. We go on learning, doing humble inductions, but also expecting that something can always go wrong and eventually will; when it does, that will be the start of another episode of inquiry.[22]

Several important consequences follow from this re-orientation. First of all, the designation of true/false applies to belief, but not to ability (that is to say, ability is not a kind of thing that possesses truth-value). This means that knowledge is not a matter of truth, strange as that may sound. A related point is that ability often comes in continuous degrees, not in black-and-white can/can't. One may be more or less able to achieve something, depending on how perfectly and how reliably one can succeed. The judgment of "how perfectly" would depend on the various criteria of judgment that the task in question calls for, and the judgment of "how reliably" is not only a matter of sheer frequency of success, but also of the range of circumstances and interferences under which one can succeed. These various imprecise and gradu-ated judgments, I believe, are what people who like to talk about truth try to capture when they talk about "approximate truth" or "partial truth", or when they subscribe to fallibilism and allow that our best beliefs can always be mistaken. But I think these attempts are rather self-defeating: if we water down the notion of truth too much in an attempt to preserve it, we may end up with something that is actually not worth preserving. It is more straightforward to come away from the old notion of truth and think in terms of other ideas. The de-coupling of knowledge and true belief also delivers epistemology from the burden of certainty (see Sect. 4.3.2 for more on this).[23] In order to know something, we do not have to know it with the kind of cer-tainty that standard philosophical or everyday notions of truth tend to demand.

---

[22] I am inspired here by John Dewey's view on the nature of inquiry (Dewey 1938, 104ff).

[23] Fallibilism does the same job.

This conception of knowledge gives rise to a new conception of scientific realism, which I will call "active scientific realism" (or, "active realism" for short). In the second section (Sect. 4.2) I will give a full articulation and defence of this doctrine, but here is an intuitive sketch of it. As already indicated, I want to orient the whole discourse on realism away from disputes about truth, and turn it back toward the idea of *reality*, by which I mean whatever exists "out there" that cannot be controlled by one's own will. What better focus for *real*-ism can there be, than exposing ourselves as much as possible to reality and learning as much as possible from that experience? And a proper "ism" should be an ideology (in the broad sense of the term), a doctrine governing one's actions. So "scientific realism" should mean a scientific stance that commits us to expose ourselves to reality, rather than some metaphysical hubris about how we can obtain or have obtained objective truth. Realism in this sense may sound just like empiricism, and that is how it should be—it doesn't make much sense that empiricism and realism have been pitted against each other in debates on scientific realism. Typical scientists, as well as most "normal" people, are both empiricists and realists, and that is not (only) because they are philosophically unsophisticated. There is just one possible difference of emphasis between empiricism and active realism. Empiricism is sometimes taken as a rather passive or defensive doctrine, emphasizing that the only source of knowledge we can have is experience and that we should avoid treating other things as legitimate sources of knowledge; this by itself does not involve much of a recommendation about what kind of experience and how much of it we should try to have. But I think the real spirit of empiricism is an active one, as is that of realism as I mean it; both doctrines recommend that we should seek out contact with reality as much as possible, and in such ways as to maximize our learning.

I think *realistic* people (including most empiricists and pragmatists) should re-claim the label of "realism"! Being realistic means concerning ourselves with what we can plausibly do and know, with conditions of actual inquiry rather than pipe-dreams of ultimate truth and certainty. In this vein, a great slogan for active realism can be taken from Charles Sanders Peirce: "Do not block the way of inquiry." I follow Amy McLaughlin's (2009, 2011) reading of Peirce, which reflects a pluralism that is almost inevitable in active scientific realism.[24] From the active realist point of view, we should want to have multiple scientific systems of practice, because this gives us more angles on reality. That may seem an easy and indisputable thing to say. However, in order to achieve such maximal learning, we must reject what commonly goes by the name of "scientific realism". In particular, we need to abandon the intuition that a successful scientific system in a given domain possesses an exclusive truth that denies the truth of other successful systems if they don't agree with it. In fact, as I will elaborate in Chap. 5, what we really need for the purpose of active realism is the simultaneous cultivation of a set of systems of practice that are as incommensurable

---

[24] Peirce is often interpreted in a way as to suggest that scientific inquiry would eventually yield the one true theory of the universe, but in McLaughlin's interpretation Peirce's ideas were based on the premise that "reality is polymorphic".

as possible from each other! Each system is conducive to revealing particular aspects of reality, and by cultivating multiple incommensurable systems we stand to gain most knowledge. The case of water, as I have presented it in this book, is an extended illustration of the active-realist benefits of pluralism.

And such pluralism is what allows large-scale and long-term progress in science! The *cumulative* growth of scientific knowledge, beyond what is straightforwardly possible within each system, is a consequence of a mostly unspoken *conservationist pluralism* by which we keep various systems of knowledge alive. What we have is an accumulation of system-*dependent* knowledge, which can only happen through an *accumulation of systems*. We can follow much of what Kuhn says, while rejecting his idea that a paradigm in normal science should enjoy a monopoly over an entire field of science. When a useful new paradigm comes into being, we can let it flourish and deliver us new knowledge; contrary to Kuhn's assumption, this can be done, has been done, and should be done, without discarding the old paradigm and the knowledge that is lodged uniquely in it. As discussed in Chaps. 2 and 3, electrochemistry and atomic chemistry made progress through the accumulation of various systems in each field through much of the nineteenth century. And as I have argued in Chap. 1, even though the advent of the oxygenist system of chemistry through the Chemical Revolution resulted in many valuable advances, further progress was made when chemistry later added other systems based on energy and electrons, addressing again some old concerns of the phlogistonists without thereby eliminating the oxygenist advances. This is one of the places where Kuhn's political analogy of revolution breaks down: in science it is not catastrophic to have two or more paradigms side by side in one field, the way it would be to have multiple national governments simultaneously trying to rule the same country. As I will argue in Chap. 5, scientific practice is in fact much more pluralistic than many philosophers imagine and many scientists let on, and it could benefit from becoming even more so.

4.2   Active Scientific Realism

4.2.1   Maximizing Our Learning from Reality

Above I have given an intuitive sketch of a doctrine which I called "active scientific realism", or "active realism" for short. I will now give a more careful and systematic presentation of it. (I would like to make this part of the chapter reasonably self-contained, so references to earlier chapters of the book will be brief and not essential for following the arguments given here.) The core idea of active realism is certainly not my invention, though the terminology is. A most important influence has been the operationalism of Percy Bridgman, as indicated in Chap. 3. In my interpretation of Bridgman (Chang 2009a), his insistence on giving a well-defined method of measurement to every physical concept shows a commitment to make every theoretical statement a site of contact with reality. Karl Popper's injunction for scientists to seek higher falsifiability and more severe tests can also be seen as a demand for more contact with reality.[25] Imre Lakatos's demand for progressiveness in scientific research programs can also be readily interpreted in the active realist vein. For Lakatos "theoretical progress" means the making of novel predictions, which amounts to the creation of contact-points with reality; "empirical progress" consists in some of those predictions being successful, establishing knowledge. A connection that will be more readily recognizable to those who have been following the recent scientific realism debate is with Ian Hacking's "experimental realism", with his famous slogans encouraging active contact with reality: "Don't just peer: interfere"; "you learn to see through a microscope by doing, not just by looking" (Hacking 1983, 189). In Hacking's view, the knowledge of reality comes best of all when we can successfully use parts of it as tools for achieving other ends: "if you can spray them [positrons], they're real" (Hacking 1983, 23). A similar view is expressed by Bernadette Bensaude-Vincent and Jonathan Simon (2008, 206), who present "operational realism" as a common attitude of working chemists that philosophers can perhaps learn from.[26] All of these familiar insights, and more, are synthesized into my doctrine of active scientific realism.

What I am presenting is also an attempt to reframe the whole debate on scientific realism. Specifically, I want to take the debate away from its focus on truth and its attainment by science. Take, for example, Bas van Fraassen's (1980, 8) famous definition of scientific realism: "Science aims to give us, in its theories, a literally true story of what the world is like; and acceptance of a scientific theory involves the belief that it is true. This is the correct statement of scientific realism." I follow this formulation in its framing of realism in terms of the aims of science. However, I do not think that the realist aim of science should be literal truth. While I accept that many realist *philosophers* take ultimate truth ("with a capital T") as the aim of science, such truth does not often guide actual scientific practice, because it is not an operable aim. Truth, in the standard conception of realist philosophers, comes down to a correspondence between what our statements say and how the world is. But what are the methods by which we can judge whether this correspondence obtains in each situation? Recall a piece of old logical positivist wisdom, from Otto Neurath ([1931] 1983, 66): "*Statements are compared with statements*, not with 'experiences', not with a 'world' nor with anything else." The burden of argument is on those who claim or assume that there are methods of judging statement–world correspondence, since there are no obvious ones. Just consider what a useless piece of methodological advice it would be to tell a scientist to "try to make true theories". The standard realist

---

[25] If we can turn the focus away from true/false hypothesis-testing to a broader view of various epistemic activities that can serve as tests of theories, there is actually not so much difference between what Popper recommends and what I advocate.

[26] According to Bensaude-Vincent and Simon, chemists accept "the reality of the tools with which they do their chemical work"; this interestingly contrasts with instrumentalism, which might be seen as dismissive of the reality of the conceptual tools one uses. Bensaude-Vincent and Simon go one step further than Hacking by granting operational reality to abstract concepts as well as concrete entities, as did Bridgman when he spoke of mental and paper-and-pencil operations.

strategy is, of course, to get at truth indirectly; we can pursue truth via other theoretical virtues, if they are truth-conducive. But I think here we are inescapably locked in a vicious circle: if we are not able to judge whether we have truth in each situation, how will we be able to tell which methods have a tendency to lead us to truth? Whether this circularity is really inescapable is the main point of contention in the scientific realism debate. And it is not clear how that question can be sensibly tackled, which I think is the root of the stalemate that we seem to have in the scientific realism debate.[27]

Rather than getting into the discussion of whether we can escape from that truth–method circularity, I would like to find a conception of realism which will allow us to avoid getting into it altogether. For a moment, let's try taking "realism" in a very literal sense, as a commitment to engage with what is real, with *external reality* (or, *reality,* for short). In the context of inquiry (scientific or otherwise), that ought to mean a commitment to maximize our *learning* from reality. (We can debate whether this learning can ever bring us truth, certainty or objectivity, but that is a separate matter.) But what is reality? What do we mean by external reality, and what is involved in learning about it? Instead of entering into serious metaphysics, I want to give you an *operational* definition of reality. I propose to think of external reality as whatever it is that is not subject to one's own will.[28] As pragmatist philosophers have pointed out, nature's *resistance* to our ill-conceived plans is one of the most important sources of our very notion of reality. William James stated, right after his (in)famous definition of truth in terms of success: "Experience, as we know, has ways of *boiling over*, and making us correct our present formulas." He added: "The only objective criterion of reality is coerciveness, in the long run, over thought."[29] There is also an echo here of Michael Polanyi's conception of reality: "To hold a natural law to be true is to believe that its presence will manifest itself in an indefinite range of yet unknown and perhaps yet unthinkable consequences. It is to regard the law as a real feature of nature which, as such, exists beyond our control" (Polanyi 1964, 10); "If anything is believed to be capable of a largely indeterminate range of future manifestations, it is thus believed to be real. . . . It may yet manifest itself inexhaustibly in the future" (Polanyi 1967, 191–192).[30]

And it is not only when our expectations are frustrated that we make contact with reality. Something that we cannot control can also turn out the way we expect, and that is precisely what happens when we make successful predictions. Knowledge is

---

[27] If what I have said about truth so far is so off-putting that it gets in the way of any further productive reading, I would suggest skipping ahead to Sect. 4.3.1 and then coming back to the rest.

[28] This is consonant with the traditional emphasis on the mind-independence of reality, which features prominently, for example, in Anjan Chakravartty's (2011) exposition of the realist commitment.

[29] Quoted in Putnam (1995), 8 and 11. The first quotation is from James, *Pragmatism* and *The Meaning of Truth* (Cambridge, Mass.: Harvard University Press, 1978), 106 (emphasis original), and the second one from "Spencer's Definition of Mind as Correspondence", in *James's Essays in Philosophy* (Cambridge, Mass.: Harvard University Press, 1978), 21.

[30] See also Polanyi (1966), 32.

a state of our being in which we are able to engage in successful epistemic activities, which can only happen if there is insufficient resistance from reality to our plans and expectations.[31] Perhaps sometimes we succeed even if we don't deserve to, but if we do what works, then we have got a hold on reality—deservedly or not. If we want to find an infallible way of succeeding, or find some other-worldly explanation of why it is that we've succeeded, that is another matter. But what reality does, or what it allows us to do, is not governed by how we explain or predict what happens.

What this notion of reality refers to is the fundamental distinction between the self and the world. If everything behaved simply as I willed it, I would not have a sense of external reality. But does that mean that my body, because I can move it around as I wish, is not part of reality? Far from it. When G. E. Moore famously held up his hand and declared it an external object, he would have noted that the color, shape, temperature, and all other properties of the hand are there as they are, out of his own control, except for its motion. My (usual) ability to move the hand about as I wish makes it "my" hand; its other properties (including the rather fixed range of its motion) make it an external object, part of reality. That kind of blend of self and world is the mysterious and wonderful thing that allows me to be in the world and interact with it. And my own will is part of everyone else's external reality, even though it cannot be said to be part of my own external reality.[32] These are the only noises that I am going to be able to make in the realm of metaphysics here. I hasten back to the philosophy of science, in which the important question is what it means to have contact with reality, and to learn from that experience.

The most basic requirement for learning from reality is that we need to place ourselves in situations where things that we cannot control will happen; that is not difficult, indeed quite difficult to avoid. For *learning* to take place, we need to arrange such situations in a way that exposes our senses to the happenings—not only the so-called five senses, but any and all the modalities we have in which information is registered, including muscular tension, so integral to any bodily interactions we have with reality. We must also have relevant operationalized concepts in place, so that experience can *mean* something. And in order to *maximize* our learning, we need to arrange situations in which our expectations are most likely be contradicted; this is the basic intuition underlying Popper's requirement of severe tests and Lakatos's criterion of novel predictions for progressiveness, as well as Peirce's advice that "the optimal way to conduct inquiry is to follow the path of greatest resistance."[33] Kuhn's emphasis on problem-solving ability fits in well here, too;

---

[31] These statements are intended as implicit definitions of terms such as "knowledge", "reality" and "resistance".

[32] And there are certain contexts in which parts of my own mind will fall in the realm of external reality. Emotions and desires will not often be subject to my own control, and introspective psychology would be a meaningless activity if I did not consider my own mental states as part of external reality. In other contexts, more familiar in science, my desires and cognitive states will figure as part of my willing self.

[33] McLaughlin (2011), 353, paraphrasing Peirce.

we have a problem if there has been a falsification; solving the problem is a learning process, in which we learn to produce new and different expectations with the hope of not being contradicted by our further encounters with reality.

In discussing how to maximize our learning from reality, I have begun to touch on the "ism" part of "realism". The realist stance, as I take it, is not only to acknowledge that external reality exists, but to have a commitment to engage with it; the *scientific* realist commitment is to learn as much as possible about reality, deliberately and systematically. Many different things are implied by the suffix "-ism", but I think we should take the strong meaning: an ism is an ideology in the broad sense of that word; this is not the usual spirit in which philosophers take scientific realism.[34] What I will call "standard (scientific) realism" is the belief that accepted scientific theories possess truth, at least approximate truth or partial truth.[35] I will also use the label of "truth realism" where I want to emphasize the focus on truth in standard realism—and just in case it will one day no longer be standard! The basic message of standard realism seems to be just that modern science is doing well. I would rather like to propose a philosophical "ism" that may have non-vacuous advice to offer to science, something with just a slight bit of normative bite. The kind of realism I am attempting to craft here is not merely a description of how scientific theories are, but a guideline that scientists can use in shaping their practices. So scientific realism as I mean it is an *active* doctrine, recommending that we should seek out contact with reality and in such ways as to maximize our learning, rather than an armchair-based description of how we can obtain or have obtained the objective truth about the universe.

The most obvious enemy of active realism is the kind of dogmatism that shuts down lines of inquiry. This is the context in which I can heartily endorse the invectives from Popper and Lakatos against "pseudoscience" with its striving to defend its cherished beliefs at all costs, discarding observations that go against those beliefs and shrinking from making experiments that have a chance of refuting them. I also endorse Philipp Frank's (1949) condemnation of presumed philosophical principles that stand in the way of new scientific theories; in Frank's view these principles are simply "petrified" scientific theories of a bygone era, empirical propositions illegitimately elevated to the status of metaphysical principles. Another type of obstacle to the active-realist process of learning comes from rationalists who seemingly generate science-regulating philosophical principles in their own minds without any contact with external reality.

---

[34] It is interesting to note how certain "ism" terms originating from the West have been translated into my native Korean: "communism" and "capitalism" are rendered as "공산주의" and "자본주의", but "realism" is "실재론". Note that "idealism" is disambiguated as "이상주의" (the pursuit of ideals, as opposed to being practical or selfish) or "관념론" (the belief in the reality of ideas only, as opposed to materialism). The kind of position I advocate might be called "실재주의".

[35] Standard realism, at first glance, is stronger than van Fraassen's (1980, 8) definition of realism, which only says that science *aims* at the truth. However, in practice there isn't such a strong difference, since van Fraassen's definition goes on to state that the "acceptance of a theory implies the belief that it is true", so if there *are* any theories that realists accept, then they do believe that those theories are true.

Where does active realism sit in the field of argument defined by the opposition between standard realism and anti-realism? I will deal with anti-realism first and more briefly, as my main quarrels will actually be with the standard realists. To the extent that anti-realists such as van Fraassen, Duhem or Mach advocate the collection of more and more diverse observations or facts about nature, they fulfill the basic requirement of active realism. It is immaterial to the concerns of active realism that these so-called anti-realists are not interested in pursuing the truth about unobservable entities. If the buzzword for standard realism is truth, it is *progress* for active realism. What is important for active realism, in opposition to some extreme forms of anti-realism, is that we do not discard the heuristic power of theories about unobservables to lead us to more discoveries about reality. In this regard, there is no reason why instrumentalism or constructive empiricism should necessarily get in the way of active realism. Positivism may have a harmful effect, if it means banning the use of statements about unobservables on the basis that they lack independent meaning. On the other hand, positivism or any other form of anti-realism can come to the aid of active realism if it is used to destabilize unnecessary and restrictive standard-realist assumptions about what reality is like and what sort of theories are allowable; recall that Frank was a member of the Vienna Circle. Regarding existence, it would be easiest to take the relaxed "natural ontological attitude" advocated by Arthur Fine (1984). However, positive or negative claims about the existence of certain theoretical entities could actually serve the purpose of active realism, by suggesting certain empirical observations intended for testing such claims.

Explicated as above, it may seem that no empiricist attitude about science and knowledge could ever be contrary to active realism. But, in fact, standard realism can constitute a great hindrance to active realism. This is due to the monism that is inherent in the correspondence notion of truth, which assumes that there is only one truth about any given bit of reality. This monism is apt to slide into the dogmatism that I have identified as the chief enemy of active realism. The Popper–Kuhn debate provides an instructive manifestation of what I have in mind here.[36] Popper was being an active realist when he demanded that each new theory should have higher empirical content than its predecessor, but he added an unreasonable demand for continuity which was quite monist in its flavor: "a new theory, however revolutionary, must always be able to explain fully the success of its predecessor."[37] Popper's demand may have been met in a few evocative moments in the history of physics, such as the transition from Newtonian mechanics to special relativity, in which there was a complete inclusion of the old phenomena in the new paradigm in the numerical sense despite semantic incommensurability. But how many instances of scientific development go like that? The transition from classical mechanics to quantum mechanics is

---

[36] My understanding of the Popper–Kuhn debate has certain resonances with Steve Fuller's (2003) reading which, of course, pre-dates mine.

[37] Popper (1981), 94. In a more detailed statement on what it means for one theory to "correspond better to the facts" than another, he seems to avoid such a strong continuity requirement (Popper 1972, 232).

a good counter-example, because macroscopic rigid bodies in fact cannot be described sensibly in quantum-mechanical terms. Kuhn recognized the limitations of Popper's kind of demand for continuity quite clearly, when he pointed out the incommensurability between different paradigms, and the loss of knowledge that can and does occur in the transition from one paradigm to the next. But even Kuhn's view of science was in the grip of monism, in that he regarded a monopoly enjoyed by the dominant paradigm in a field as a precondition for normal science. In this paradigm-monism, and in his lack of concern about "Kuhn-loss" in paradigm shifts, Kuhn fails the demands of active realism. As indicated at the end of the first section (Sect. 4.1), "conservationist pluralism" is the antidote here: retain previously successful theories and paradigms for what they were (and are) still good at, and *add* new theories and paradigms that will help us make new and fresh contacts with reality.

In order to consolidate my advocacy of active realism further, in the following sections I will examine three of the most important tropes in the discourse of standard truth-realism: the inference from success to truth, the presumed security of that inference in "mature" sciences, and the "pessimistic meta-induction" from the history of science. The first two are taken to be important pillars supporting the edifice of standard scientific realism; I will explode those pillars, and show that a careful re-examination of the notions of success and maturity in science supports active realism instead. As for the pessimistic meta-induction, I will propose a different rendition of it that is fully consonant with active realism, even though it will still work as an argument against the standard realist inference from success to truth. Overall, I aim for a critical moderation of standard scientific realism, which will strengthen the case for active realism and at the same time take the urgency out of the traditional realism debate.

4.2.2   The Optimistic Rendition of the Pessimistic Induction

In re-engineering standard realist intuitions it is crucial to scrutinize the argument for scientific realism from the success of science (or the "argument from success" for short), since it is widely considered the most convincing argument for realism—van Fraassen (1980, 39) even calls it the "Ultimate Argument". It will be helpful to begin with a reminder of the argument itself. The most famous version, due to Hilary Putnam and often dubbed the "no miracle argument", goes as follows:

> *The positive argument* for realism is that it is the only philosophy that doesn't make the success of science a miracle. That terms in *mature* scientific theories typically refer (this formulation is due to Richard Boyd), that the theories accepted in a *mature* science are typically approximately true, that the same term can refer to the same thing even when it occurs in different theories—these statements are viewed by the scientific realist not as necessary truths but as part of the only *scientific explanation* of the success of science, and hence as part of any adequate scientific description of science and its relation to its objects.[38]

---

[38] Putnam (1975a, 73); emphases added, for ease of later reference.

(I think we can understand the *explanans* here more simply as the truth of typical scientific theories, taking the essence of successful reference as the truth of existential statements.) Another oft-cited version of the argument, older than Putnam's, is due to J. J. C. Smart (1963, 39):

> Is it not odd that the phenomena of the world should be such as to make a purely instrumental[39] theory true? On the other hand, if we interpret a theory in the realist way, then we have no need for such a cosmic coincidence: it is not surprising that galvanometers and cloud chambers behave in the sort of way they do, for if there are really electrons, etc., this is just what we should expect.

There is an important distinction that separates these two arguments[40]: Putnam's argument unequivocally concerns the success of science as a whole, as an overall epistemic enterprise. Smart's would apply most naturally to a specific scientific theory, though one can imagine treating all of science as one gigantic theory; one should not be misled here by Smart's talk of a "cosmic" coincidence, as that only reflects the fact that any particular physical theory of sufficient generality would have coordinated implications in many and various parts of the universe. We may think of the difference I'm pointing out as the difference between a global and a local version of the argument from success. Although they are often conflated, they say very different things, as will be clear in my discussion below.[41]

Against the argument from success, what is often regarded as the deadliest objection is the pessimistic (meta-)induction from the history of science, or "the pessimistic induction" for short. I want to argue that the pessimistic induction is nothing for the active realist to worry about (in Sect. 4.2.4 I will have something to say to those who do not accept the argument itself). The pessimistic induction attempts to block the inference from success to truth, by displaying a large number of cases in which successful theories have turned out to be false (that is, false according to the wisdom of current science, which will itself be overturned by later science, in all likelihood). The idea is most often credited to Larry Laudan (1981), but I think its basis was already amply present in Kuhn's work, and Mary Hesse in the mid-1970s already spoke of it as a well-known difficulty.[42] It hardly matters who said it first in our living memory—as I mentioned in Sect. 4.2.3, something similar to the sentiment at the core of the pessimistic induction was already expressed in the late

---

[39] I actually don't think Smart's argument here touches instrumentalism at all. The argument is effective only against the position that our successful theories are actually false. Instrumentalists do not assign truth-value to theories, and they have no need to invoke a cosmic coincidence that allows a false theory to have correct observable consequences.

[40] There are some other differences between these two arguments, too, which are nicely discussed by Stathis Psillos (1999, 72–73).

[41] In fact, Putnam (1978, 19) also gives a clearly local version of the argument.

[42] She says: "there is the possibility, emphasised by revolutionaries, that *all* our theoretical terms will, in the natural course of scientific development, share the demise of phlogiston" (Hesse 1977, 271).

nineteenth century by Arrhenius's mentor Per Cleve, and no doubt by many others. For a pithy and incisive summary of the argument, we can go to Stathis Psillos (1999, 101):

> The history of science is full of theories which at different times and for long periods had been empirically successful, and yet were shown to be false in the deep-structure claims they made about the world. It is similarly full of theoretical terms featuring in successful theories which do not refer. Therefore, by a simple (meta-)induction on scientific theories, our current successful theories are likely to be false . . . and many or most of the theoretical terms featuring in them will turn out to be non-referential. Therefore, the empirical success of a theory provides no warrant for the claim that the theory is approximately true.

I would like to offer an optimistic rendition of the pessimistic induction. Instead of feeling depressed by the fact that success does not give us warrant for assuming that we are in possession of the truth, we should be thinking: how wonderful it is that we can be so successful without even knowing the truth! Recall from Chap. 1 how successful the phlogistonist system of chemistry was, even if phlogiston didn't truly exist; or how Lavoisier did such good chemistry on the basis of his oxygen gas that was the principle of acidity combined with lots of caloric. Recall from Chap. 2 how electrochemists throughout the nineteenth century achieved their successes without even knowing that atoms became ions by losing or gaining electrons. Recall from Chap. 3 all the successes of the first half-century of atomic chemistry achieved on the basis of either very little knowledge of the physical nature or structure of atoms, or such false or fictional conceptions as two-dimensional molecules made up of ball-like atoms connected by stick-like bonds. So, the pessimistic induction can make us happy, if we learn to turn it on its head. Before lamenting the conclusion of the pessimistic induction, savor its *premise* (in Psillos's formulation quoted above): "the history of science is full of theories which at different times and for long periods had been empirically successful". Pause there, and rejoice! Instead of getting worried about the rest of the pessimistic-induction argument, I suggest that we focus more on how to make a proper appreciation of this fact of scientific success (often exaggerated, nonetheless real).

For those who would have an existential crisis if the pessimistic induction were allowed to stand, I would recommend therapy[43]—not just any therapy, but *logotherapy*. And I mean that only partly in jest. Logotherapy, billed as the third Viennese school of psychotherapy (after Freud's and Adler's), was founded by Viktor Frankl, a survivor of a Nazi concentration camp. With his "tragic heroism", Frankl teaches that life's meaning can be found in the appreciation of what we have already achieved, which no one or nothing, neither Hitler nor death, can ever take away, because we have "rescued it into our past". Everything is so fleeting because it is fleeing from the uncertainty of the future to the security of the past. Frankl recommends that we combine an "optimism of the past" with an "activism of the future" (Frankl 1978, 102–113). As in life, so too in science. A successful system of practice

---

[43] I must have picked up this trope from Fine (1984, 102), who recommended the reading of Ernst Mach, *The Analysis of Sensations*, as "effective realism-therapy". And then, of course, there is the later-Wittgensteinian line.

is a secure achievement, and it will be as lasting as anything in life—that is, stable as long as the laws of nature themselves do not change and ordinary induction (of the "sun will rise again tomorrow" type) holds. What is really wrong in the pessimistic induction scenario is the notion that a *successful* theory should be *rejected* if another successful theory comes along and seems to declare the old theory to be false. That is precisely what conservationist pluralism argues against.

In my view, the best way to take the pessimistic induction is as an ironic argument (or it could be rendered as a *reductio ad absurdum*, if one wanted to formulate it in a more strictly logical way). This point is not always appreciated. When we say that modern science shows most of the previously successful theories to have been false, we are assuming that the modern theories are true. But how do we know that? Why, because they are so successful! So the starting premise of the whole argument is the success–truth link—or, to be more precise, the assumption that a successful theory is true. But that premise is contradicted the moment we conclude that the modern theories show that many *successful past theories are false*. The success–truth link destroys itself, given the history of mutually incompatible theories being each sufficiently successful.

The irony of the pessimistic induction has made a great impact. What the pessimistic induction really shows is the implausibility of the attempt to link truth and success too closely, which are very different kinds of things from each other. If two theories contradict each other they cannot both be quite true, but they can both be quite successful. The pessimistic induction helps to break down the presumption of the success–truth link. Severing that link will free the notion of success from the exclusivity inherent in the correspondence notion of truth, and allow us to accept easily that multiple systems can be simultaneously successful. By shifting our attention away from truth we can acquire a more proper and accurate appreciation of the success of science, without unrealistic assumptions about what success must imply.

4.2.3 How the Argument from Success Fails

Going beyond the irony of the pessimistic induction, I would now like to make a more direct critique of the standard-realist argument from the success of science. In the last section I was fairly casual about what is meant by "success", but now I want to raise some serious questions about the notion of success itself, and also about the idea of "explaining" success. The answers I attempt to give to these questions aim for an improved sense of what it is that we can learn philosophically from the fact that (some of) science is successful (to varying degrees). After these considerations, I think active realism will seem all the more sensible and beneficial.

1. Just how successful is science? How do we judge this? How can we be so smug about the success of science, when people keep dying from various horrible diseases that we don't understand well enough, when we can't figure out for sure how the earthly climate works (or how to remove enough CO_2 from the atmosphere), and when the fundamental ontology of the universe remains elusive in spite of

giant accelerators, string theory, and Einstein's genius?[44] Our science is clearly
not as successful as Vulcan science on *Star Trek*. Is it more successful than the
ancient craft of pyramid-building, or modern technologies such as mobile phones,
Google, or GPS, or even the skilled practices of certain animals such as spiders
building their webs? Modern science is only clearly successful in comparison to
older science, and to various dubious human enterprises such as fortune-telling,
witchcraft, investment banking, and politics.

The point is that in all of human history we have not witnessed anything like
complete and utter success, in science or in any other enterprise. All the successes that we have enjoyed or can hope to enjoy are just relative degrees and
different types of success. The inference from success to truth would work well
enough if what we are trying to explain were a complete and utter success of
science, so that at least intuitively we have to say "science is *so* successful that it
must be true". In the actual world such an argument is pointless since there is no
such perfect success there to be explained. What is actually there to be explained
is that science is more successful than many other human activities, and that some
science is more successful than others. On both counts, we are dealing with
comparative degrees of success in different systems of practice.[45] Now, if we say
that theory *A* is successful because it is really true, then what will we say when
we encounter a different theory *B*, which is even more successful than *A*? That
would have to be because *B* is even truer than *A*, but I am not sanguine about
getting a simple metric of truthfulness to support this kind of "truer-than-thou"
one-upmanship. Besides, operating on such a continuous scale of truthfulness
would blur the whole point of realism; if the degree of truthfulness simply maps
on to the degree of success, why shouldn't we just note degrees of success and
leave it at that? What do we gain by sticking the additional label of "degrees of
truth" to degrees of success?

2. Will the success of science be lasting? This question is important in relation to
   the realist argument from success, because fleeting success is not fit for a realist
   explanation. It would be embarrassing if we insisted that the success of a scientific theory (or indeed science as a whole) was an indication that it was getting at
   the truth, only to find out that this success did not last. So any success that we
   want to use as a basis for an inference to truth needs to be lasting.

   This raises a particular problem for the global version of the realist argument
   from success. Although it does seem that the enterprise we call science has been
   increasingly successful on the whole in the last two centuries or so, do we have

---

[44] The excitement in the early summer of 2010 was that there might be as many as five different
Higgs bosons, though none of them had been found. In September 2011 the story all over the scientific news was of neutrinos that apparently travel faster than light. Can anyone claim sufficient
confidence that these latest stories will keep—or that they will not?

[45] I am putting the discussion in terms of "systems of practice", as I have defined in Chap. 1,
Sect. 1.2.1.1, but replacing "system" with "theory" will not affect the present arguments
significantly.

a guarantee that this trend will continue? Really, what grounds do we have for believing such a thing? Can't we imagine, with a mixture of sympathy and ridicule, the ancient Romans being confident that the success of their empire would last forever? Or the medieval Catholic Church unable to imagine any weakening of its power and reach? Or the dinosaurs declaring (if they could) to each other that it was unimaginable that their way of life would ever stop being successful? Why should science be any different from these once-dominant systems that gave way to others?

For the global success of science to be lasting, there must be a capacity inherent in the scientific enterprise to keep generating successful outcomes. In the aftermath of scientists' and philosophers' failure to identify the general scientific method, I have no idea what the miraculous factor guaranteeing the continued success of science would be. It seems reasonable to think that we will probably continue to achieve success because we will demand it of ourselves and do our best to achieve it, by whatever available means. But this is a general fact about human striving, not something uniquely characteristic of science. In fact this also implies that scientists can continue to be successful precisely because they continually adapt the methods of research depending on the objects and circumstances of their work. It is scarcely imaginable, for instance, that the successes of molecular genetics or breakthroughs in high-temperature superconductivity would have been achieved by continuing with scientific methods greatly admired in the early twentieth century, such as Einstein's axiomatization, thought-experiments, and simplicity-seeking. It is simply dinosauric hubris to assume that doing whatever we have been doing will ensure future success (because it somehow gets at the truth).

Local success has a better chance of lasting. Here we should start with an unbiased look at what does tend to be lasting in science: all indications are that lasting success in science has been achieved most credibly in two inter-related realms: various material techniques and technologies, and the empirical adequacy of phenomenological laws extolled by Herbert Feigl (1970), Nancy Cartwright (1983) and others (see Chang 2004, 52, for further discussion and references). I think there are good prospects of retaining operational successes that have already been achieved (*modulo* the problem of induction). The security of achievements already made is a piecemeal thing, a motley collection of successful practices in various parts of science, from which it is going to be very difficult to infer anything about the general character of science.

Such a picture of the success of science will not excite most standard scientific realists. But why shouldn't it? Again, this is because of the almost unconsciously desired link between success and truth. The basic designs of the ruler, the balance, the magnifying glass, etc. have been employed successfully for many, many centuries, and they show no sign of disappearing. (These are things to be marveled at, not taken for granted.) But instruments are not propositions, so no truth-value can be attached to them, and hence little attention is paid to them by truth-obsessed philosophers, and even Feigl and Cartwright chose to present their arguments in terms of the stability of phenomenological laws, which are truth-bearing propositions. Standard realists have been dissatisfied with the latter because a given phenomenological regularity tends to be compatible with, even approximately deducible from, a variety of different "higher-level" theories, some of which may contradict each other. The standard-realist attention then gets focused on deciding which of these alternative theories is really true, and all the successes that we can have by means of the phenomenological laws, and even their lasting truth, get minimized and unappreciated.

3. What exactly do we *mean* by "success"? What is so impressive about science to the proverbial "man in the street" is the production of practical feats (atomic bombs and such). But from that point of view what is really successful is not science but technology or engineering, much of which does not require a true understanding of the underlying scientific principles—consider, for example, the quality of theories of fluid mechanics at the time when flight technology achieved its early triumphs. If we come away a little bit from practical achievements and explicate the success of science as the attainment of empirical adequacy, the realist has to contend with all of van Fraassen's arguments blocking the inference from empirical adequacy to truth. (That will be at least a very long debate.) The main difficulty is the old underdetermination problem: claims to truth being laid by multiple theories, mutually incompatible, all more or less equally successful in their attainment of empirical adequacy. Facing that difficulty, the temptation is, as explained quite clearly by Gerald Doppelt (2005), to demand that what is required is *explanatory* success, so that one can do an inference to the best explanation and block the inference to the truth of all but the best theory. But then, in the absence of a clear agreed notion of what counts as a good explanation (see point (4) below), the whole realist argument from success runs the risk of becoming vacuous, with "success" being *defined* in terms of its apparent ability to support the inference to truth.

Taking a little step back from all this, I think it is futile to attempt to define "success" in any one-dimensional way—we don't try to do that with life in general, and it's not clear to me that we should try it in science. The "success of science" can only really mean the achievement of whatever we value in science—Kuhn (1977, 322), van Fraassen (1980, 87), Lycan (1998, 341) and others give a long and diverse list of epistemic desiderata: accuracy, consistency, simplicity/ elegance, scope/completeness, unifying power, explanatory power, fruitfulness, testability, and even conservativeness. No single one of these is a value that overrides every other. Van Fraassen's statement that empirical adequacy trumps all other values is not accompanied by any explicit argument and, besides, empirical adequacy in itself is a complex virtue incorporating at least accuracy, consistency and scope.

The multi-dimensionality of success has been quite obvious in the historical episodes presented in earlier chapters. For example, Lavoisier's system was very successful in attaining elegance, unifying power and explanatory power, while it was lacking in empirical adequacy as it had many anomalies (Chap. 1). The electrochemical dualistic system of atomic chemistry was successful in providing an explanation of (many cases of) chemical bonding, while the substitution–type system was more successful in the classification of organic compounds (Chap. 3). The multi-dimensionality of success raises a serious difficulty for the realist argument from success, which usually presumes the explanation of success to be univocal; yet if the *explanandum* is not univocal, it is not clear how well a univocal *explanans* can serve.

4. What kind of explanation do we want? Now, the situation regarding this question is really quite bizarre: philosophers of science, who normally demand such a great degree of rigor in debating theories of explanation, suddenly seem to take leave of their senses, or at least their high standards, when they jump into explanations of the success of science relying on vague intuitive notions of what it means to explain something. Putnam is clear, in the passage quoted earlier, that he wants a *scientific* explanation of the success of science; in that case, we should have to apply our best philosophical theories of scientific explanation to this case, too.[46] I see a grave difficulty here. Aside from the sad fact that philosophers have not yet come up with any blindingly successful theories of scientific explanation, I do not see that any of the available theories can be used with good effect for the purpose of the realist argument from success.

For example, it's not plausible to demand a deductive–nomological (D–N) explanation. For one thing, a D–N explanation of success by appeal to truth would require a law enabling the deduction of success from truth; I don't know what field of science would contain such a law. Other types of explanation are equally unpalatable. If we wanted a causal explanation, we would have to show that the truth of a theory is the sort of entity which has a causal power, and the right kind of causal power to cause something like success. I am not a good enough metaphysician to work that one out. Besides, it is clear that truth would not always cause success (see point 6 below). As for a structural explanation of the success of science on the basis of truth, I would not even know how to begin to think about that. Are there other possibilities? I think by this point the burden of argument is squarely on those who want to insist that it makes sense to talk about how truth explains success.

After we consider these difficulties, explanations of the success of science based on something other than truth begin to look much more attractive in comparison. Van Fraassen's (1980, 40) selectionist explanation is unexceptionable as far as it goes, and the explanatory mechanism there is straightforward: science is full of empirically adequate theories because scientists reject empirically inadequate theories.[47] But van Fraassen's selectionist view needs to be extended further. When scientists select theories (and systems of practice) they use all of their

---

[46] This kind of naturalism, which Boyd exhibits even more strongly (see Psillos 1999, 78), saddles us with the burden of reflexivity. For example, Doppelt (2005), 1080, notes that the realist philosophers who take novel predictions as the hallmark of a true scientific theory make no novel predictions in their own work; such irony would not amount to real criticism for a non-naturalistic philosophy.

[47] See K. Brad Wray's recent work (2007, 2010) for more on this issue, and also Stanford (2000) on explanation by predictive similarity.

values, not just empirical adequacy. Some of these values are directly conducive to learning from reality, and others are not; some values, for instance simplicity or elegance, might easily lead to dogmatism. We could begin by distinguishing active-realist values and other types of values. And then it will be possible to see if scientists with active-realist values would tend to achieve higher degrees of success. That way we might get somewhere interesting in explaining success by values, but we would not have an explanation in terms of truth. The best that standard realists could hope for is an explanation of success by reference to scientists' pursuit of truth (as opposed to other values), but that is an empirical question whose answer is not immediately clear to me.

5. Why do we need to explain success? What good is such an explanation? The sense of futility that I have about this reaches its peak with the "surrealist" explanation endorsed by Timothy Lyons (2003, 896). The surrealist explanation goes: "[theory] T is successful because the world is 'as if' T is true." But really, the standard realist explanation is just as pointless, and perhaps to show that is the real point behind advancing the surrealist explanation. If we did somehow manage to explain the success of science, who would benefit from it, and how? Why can't we simply enjoy success and leave it at that? These may seem like impertinent questions that unfairly ignore the inherent value of philosophical inquiry. But I think it is fair enough to demand that philosophical discourse should have some use, and also to try to discern which philosophical questions and answers are more useful than others, and for what purposes exactly. So I regard it as legitimate to ask what we can actually do with an explanation here.

   If the active realist wanted an explanation of the success of science, that would be in the hope that having a good explanation of success might give us the kind of insights that will enable us to have more and better success. Whether there is any such prospect will depend on the type of explanation given. For example, the van Fraassen–type evolutionary explanation can tell us that it is good to proliferate theories that fit with known observations (without worrying about their ultimate truth), in an attempt to hit upon the ones that will remain empirically adequate into the future. Can the standard realist side deliver any useful advice? It is not likely to come from any truth-based explanations of success, since ultimate truth is not an operative category. One could say: "Try to make theories that are genuinely referring and approximately true", but that is only as useful as the ethical injunction to "do the right thing." If we knew how to tell what is right, or true, or genuinely referring, we would not need any advice on how to find it.

6. Finally, can we infer truth from success? After all of the above considerations, I can come profitably to the main line of the realist argument from success. *Is* truth a kind of thing that could be inferred from success? It might seem obvious at least that the truth of a particular theory implies (and therefore explains) its success, and then on that basis we can try running an inference-to-the-best-explanation type argument allowing an inference from success to truth. But I think Timothy Lyons (2003, section 3) has made a convincing argument against even the first step of this scheme. His basic point is that for the truth of a theory to guarantee its success, all the relevant auxiliary assumptions we use in the applying the

theory in question have to be true as well—and that is not likely to be the case. The global version of the argument from success is even more difficult to defend. Putnam's argument is phrased in terms of "typical" truth or reference; how do we get a good enough assurance of success from that?

The success–truth link is fundamentally suspect. Strictly speaking, the question of "success" does not apply to theories, since theories by themselves don't *do* anything; it is what *we do* with theories that may or may not be successful. In other words, successfulness is an attribute that belongs to particular *employments* of a theory; the employments of a theory take place through specific epistemic activities; such activities make up systems of practice. And thinking in terms of activities and systems takes us naturally away from thinking in terms of truth, at least in the sense of a correspondence to reality. An activity, or a system of activities, cannot have the logical relation of correspondence with the world, so we must conceive of some other kind of relation with which to express the relation between our activity and the objective world. For that matter, as Neurath says, it is not clear how a proposition or a theory can "correspond" to anything in the world, either, since the world is not made up of propositions—this, of course, is why the early Wittgenstein opened the *Tractatus* by declaring that the world *was* made up of facts: "The world is everything that is the case." (Wittgenstein 1922, 31, §1) This recognition of the ineptness of the correspondence theory of truth is at the heart of my re-formulation of the realism question.[48] The active-realist take on the situation is to let success be, and to do what we can to have more and better success; our intellectual energies are better spent at the promotion of success, rather than an at arguments aimed at finding a "deep" explanation of it. If that does not seem very "philosophical", so be it.

4.2.4 The Immaturity of Maturity-Talk

Standard realists who subscribe to the argument from success have one powerful line of defence against the pessimistic induction and also against some of the critical points I have raised in the last section. The basic strategy is to declare that any and all inconvenient historical cases are not from "mature sciences". When we only consider the mature sciences, the argument goes, we will see that successful theories have been for the most part preserved; standard realism is safe again, with the success–truth link rescued. Against this position I will argue that a serious examination of the notion of "maturity" reveals developmental patterns that are quite contrary to standard realism and quite consonant with active realism.

Some realists had actually come up with the "mature science" gambit even before Laudan published his pessimistic induction. It is embedded in Putnam's formulation of his argument from success. In the passage quoted at the start of Sect. 4.2.2, he states

---

[48] As Putnam says (1978, 18), "Whatever else realists say, they typically say that they believe in a 'correspondence theory of truth'."

that "terms in mature scientific theories typically refer" and that "the theories accepted in a mature science are typically approximately true".[49] The thinking goes: of course, in its early trial-and-error stages a scientific discipline may produce theories that enjoy initial success rather accidentally on the basis of incorrect conceptions. But in a mature science, these oddities will have been ironed out, and sustained success will only be due to truthfulness. So we can cut out all the annoying counter-examples to the success–truth link that arise from phases of science like pre-Copernican astronomy, pre-Lavoisierian chemistry, pre-Newtonian physics, nineteenth-century ether theories, medicine before the germ theory of disease and modern biochemistry, and so on. This move is intended to reduce Laudan's inductive base significantly (cf. the list given on Laudan 1981, 33).

But has anyone actually given an argument about why mature sciences are likely to contain true and genuinely referring theories? Interestingly, Laudan (1981, 20) already framed his target ("convergent realism") as a doctrine concerning mature sciences. So all of his examples were clearly meant to be from mature sciences, and it is evident that what people consider "mature" can be quite variable. To avoid circularity, it is essential that we have a definition of maturity that is independent from truth and reference. Maturity is invoked much more often than it is explicitly defined, and I turn to Psillos again for a very useful explication (1999, 107–108):

> Realists require that Laudan's list should include only *mature* theories; that is, theories which have passed the 'take-off point' (Boyd) of a specific discipline.[50] This 'take-off point' can be characterised by the presence of a body of well-entrenched background beliefs about the domain of inquiry which, in effect, delineate the boundaries of that domain, inform theoretical research and constrain the proposal of theories and hypotheses. This corpus of beliefs gives a broad identity to the discipline by being, normally, the common ground that rival theories of the phenomena under investigation share.

An example immediately following the passage above in Psillos's text provides a more concrete sense of what sort of thing he has in mind; he says that thermal physics reached its maturity with the establishment of background beliefs such as "the principle of impossibility of perpetual motion, the principle that heat flows only from a warm to a cold body and the laws of Newtonian mechanics".

Both the general statement and the example make it clear that underlying Psillos's notion of maturity is a demand for theoretical unity and stability. The background beliefs that Psillos identifies for thermal physics are actually not the ones that underlay the caloric theory, which he does accept as a mature theory (perhaps the first one in the area of thermal physics). At the crucial background of the caloric theory one finds the conservation of heat rather than the impossibility of perpetual motion, and the laws of Newtonian mechanics had very little to do with anything thermal until

---

[49] He gives the same formulation again in Putnam (1978, 20), attributing it to an unpublished paper by Boyd.

[50] Boyd has various other characterizations, too, but this will do for now.

much later. Why has he made this historical mis-step?[51] First, he is retrospectively identifying principles of fundamental physics that worked toward a grand unification of thermal physics with mechanics and with other sciences in which considerations of energy and entropy could be applied. Second, he is focusing on those principles that remained stable for a long time. I do not think Psillos is unique in these inclinations. It seems to be conventional wisdom among standard realists that science becomes increasingly unified and more stable as it matures—just the kind of attributes that one would wish for in a set of approximately true and genuinely referring theories.

Interestingly, current cutting-edge science gives a very different picture of what happens when science matures. As Kuhn already pointed out (1970, 172), the proliferation of specialized sub-disciplines is a dominant feature of contemporary science. This seems more and more the case, with the great lure of elementary particle physics fading just a bit and various branches of biology in the ascendant, and with diverse technologies of manipulation and simulation in the experimental sciences occupying the center stage of what is considered cutting-edge research. What once seemed to be an inexorable march of unification now appears to have been a peculiar feature of late nineteenth- and early twentieth-century science, particularly physics; as John Dupré puts it (1993, 131), "reductionism is a local condition of scientific research, not an irresistible tide sweeping the whole of science into an increasingly orderly pattern." Even in physics the trend today seems to be toward specialization and fragmentation, and there has always been some degree of that even during the heyday of unification. Really, how much commonality is there in terms of working background assumptions between, say, a general-relativistic calculation of the motion of a gyroscope in space, and a theoretical calculation of the critical temperature in a lanthanide high-temperature superconductor?[52]

There *are* broadly shared elements such as the principle of energy conservation, the periodic table of elements, and certain basic mathematical and computing methods,[53] but these cut across disciplinary boundaries, and by themselves they do not provide enough to support scientific research in a stable way. Consequently, they do not serve to demarcate specific mature scientific disciplines. Rather, these elements are resources that various fields of science use, without thereby being reduced to physics or chemistry. If incorporating such elements were enough to qualify a scientific field as mature, then a lot of very new, uncertain or unstable field of scientific research today would also qualify as mature. These shared elements

---

[51] I have criticized Psillos's handling of the caloric theory elsewhere (Chang 2003), and will not repeat that critique here. More pertinent to my present purposes is to note that he is identifying very general theoretical principles which "inform theoretical research and constrain the proposal of theories and hypotheses", rather than, say, general operational procedures or anything else that could also serve to give the discipline an identity and fix its boundaries.

[52] For an interesting discussion of the diversified state of research in high-temperature superconductivity within itself, see Di Bucchianico (2009).

[53] I thank James Ladyman for prompting me to think about this sort of cases.

also afford unexpected links that criss-cross the landscape of science, which make scientific discovery such an unpredictable and delightful affair (see Holton et al. 1996). But these links, again, go right through disciplinary boundaries, and do not help us very much in defining well-delineated individual fields of mature science.

On the other hand, according to the Boyd–Psillos account of maturity, we actually cannot be confident that even physics, as a whole, has enough of a sufficient shared basis to qualify as a mature science. If the answer is that all the sub-disciplines of physics (elementary particle physics, condensed matter physics, astrophysics, chemical physics, thermodynamics, classical and quantum statistical mechanics, gravitational physics, etc.) should count as separate mature sciences, then so should various bits of now-rejected past science, which in themselves were not lacking in well-entrenched background beliefs. Every little dogmatic school of thought can then lay claim to practicing a "mature science"—which, I take it, is precisely what that concept as articulated by Boyd and Psillos was designed to rule out!

It is also not clear that those sciences that we would intuitively call mature are particularly stable. On the contrary, it is exactly the most mature-looking of theories that have been subject to the most revolutionary upheavals; in Kuhnian terms this makes perfect sense, as paradigms would not typically go into crises until they have matured and reached the limit of their potential. Who could plausibly deny that the following should count as "mature" sciences, in any reasonable sense of the word: Ptolemaic astronomy, Lavoisierian chemistry, Berzelian electro-chemistry, Newtonian mechanics, Newtonian optics, geometric optics, or Maxwellian electro-dynamics? Each of these indisputably mature theories has been overthrown in its fundamental theoretical core, while many of the less glorious empirical laws associated with it have survived. We can also add to the list of unstable fundamental truth and entities more recent items such as superstrings, dark matter and energy, orbitals in atoms, and the Central Dogma of molecular genetics. So, when Boyd (1980, 657) says that successive theories in a mature science are successive approximations, this must be based on some sort of *a priori* conviction, not something derived from an observation of what science actually does.

On the whole, I cannot avoid the feeling that the Boyd–Psillos idea of "mature science" has been rigged up just so that the realist inference from success to truth would appear to work out for mature sciences. And since no one has a direct access to truth, our notion of what constitutes a mature science inevitably gets shaped by what happens to be considered true in the phase of science that we know best about. This would also explain why the image of "mature science" in typical philosophical discourse tends to be rather off the mark in relation to what up-to-date scientific research actually looks like—most philosophers' intuitions have been shaped by exposure to slightly older science, or by classroom science rather than research science. Basing our thinking on the implicit assumption that mid-twentieth-century physics is "really true", we do injustice to other and more current sciences, and of course to the sciences of the more distant past. A final example will make the point vivid. If maturity was not being shaped to fit the argument from success, it is difficult to see what would motivate Psillos (1999, 108) to claim that the maturity criterion rules out the "effluvial theory of static electricity" from Laudan's list. The study of electricity in the eighteenth century was a well-delineated scientific discipline, framed by just the kind of shared fundamental principles that Psillos demands (the material reality of the electrical fluid, its indestructibility, its imponderability, the notion that opposite charges attract and like charges repel each other, and assumptions about how the electrical fluid interacts with ordinary matter). On that basis the "electricians" debated competing theories (starting with the argument between one-fluid and two-fluid theories, and much more), and they attempted to attain theoretical and experimental rigor and precision (as seen in the works by Aepinus and Coulomb).[54] What reason is there that should make us want to declare this branch of science immature, except that we think its deep theory was fundamentally wrong and that granting it the status of a mature theory would mess up the realist argument from the success of science?

Let's think again, about what maturity in science means. Maturity is an everyday concept that has been applied rather metaphorically to science and scientific theories, so any meaning we can discern in the latter context is going to be imprecise and provisional. However, I actually think it is a useful concept that can deliver some very interesting insights. Thinking back to the quotidian origin of the notion, it seems to me that we should distinguish two interrelated aspects to maturity: (1) What does a science look like when it reaches a sufficiently late stage of development? (2) What kind of attitude does a mature scientist or a mature scientific community take toward science?

Regarding the first question, it would be reasonable to say that a mature science is one that has had sufficient time to develop and enough experience to know its proper domain. It would know its own main strengths and weaknesses, having had opportunities to reflect on its successes and failures. It would have a clear and orderly sense (whether right or wrong) of where to seek its own further development, rather than an unsettled excitement that the next great thing could come from anywhere. It would probably also have well-developed areas of specialization within it, having had time to discover that particular areas require particular methods and assumptions for effective handling. All this is not meant to be a precise or thorough definition of maturity. I am just trying to lay out some reasonable sense of what it means for a science to be mature.

In this image of maturity I can see nothing that would be able to guarantee truth or reference. And nothing rules out the possibility of a fortunate or brilliant young science hitting the nail on its head at the start; many realist physicists or philosophers would probably believe this about both the special and the general theories of relativity, for example. The same could be said about the basic ontology of orthodox quantum mechanics, which was fixed within a few years in the initial burst of work by Heisenberg, Schrödinger, Born, Bohr and others starting in 1925. Perhaps the same is true with Newtonian mechanics: although Newton himself took many years

---

[54] For detailed treatments, see the classic works by Heilbron (1979) and Cohen (1956).

germinating his theory in his own head, once the *Principia* was published we could say that all the approximate truth and genuine reference that Newtonian physics was ever going to have was pretty much there. Likewise for the basic truth of the Watson–Crick work on the structure of the DNA, Darwin's theory of natural selection, or any number of other cases. To be sure, there are *some* cases in which a system of practice carefully and slowly evolves its basic ontology and basic theoretical principles, in which truth, if it is ever reached, is only gradually achieved as the discipline matures. The cases of electrochemistry and atomic chemistry that I discussed in Chaps. 2 and 3 are very good examples of such a gradual development. Ironically, these tend not to be the kind of cases typically celebrated by realist philosophers. The lesson from all of this may be disappointingly simple: there are many and varied developmental patterns in science; it will not do to insist on a falsely uniform view of what maturity means.

Even deeper implications follow when we consider what a mature *attitude* toward science would be. So far I have been speaking about a mature science in a personalized way, but let's now consider what maturity in scientists as actual people should be like. I mean both scientists working in a mature scientific field and scientists who are personally mature about their science. Having experienced the ups and downs of scientific development, and having learned to appreciate what can be achieved despite these uncertainties, mature scientists and mature scientific communities would value tolerance, humility and circumspection, combined with a tough questioning attitude. They would display an awareness of human fragility and fallibility and the multifarious complexity of nature, and try to create institutional structures that can handle this awareness. For instance, William Nicholson's work in running his journal, in which he presented various viewpoints to readers and published a wide variety of work, was an independent act of institutional maturity at a very early time during which many of the sciences featured in his journal were quite immature in many senses (see Chap. 2, Sect. 2.1). Thinking back to the main historical episodes covered in earlier chapters of this book, I will say that nineteenth-century atomic chemistry had a mature scientific community, despite some petty personalities within it. In contrast, the Lavoisierian community was not mature; the phlogistonist community was much more mature, though it was very dispersed and not highly cohesive. In the personal realm, Joseph Priestley was a mature scientist, contrary to the common misapprehension of him as a dogmatist; young Humphry Davy was quite immature, but I would say that over the years he did mature; Antoine Lavoisier was youthful in his outlook up to his middle age, and was killed before he had a chance to reach maturity.

Realist philosophers tend to admire the "know-it-all" and "can do" attitude displayed by many theoretical physicists and by other scientists who have emulated them. There is admittedly something noble and admirable in that quest for the ultimate unified true theory of the universe, but that is more like the daring enthusiasm and naïve purity of youth, rather than a sign of maturity. Mature scientists would not only seek a broad and inclusive base of observations and experiments, but also welcome a range of theoretical ideas and assemble a large and versatile methodological toolbox. They would be prepared to be surprised, because they have been surprised before, not only by what nature does but also by human conceptual and technological developments. They would be more concerned about having sufficient knowledge to meet human needs including people's desire for understanding, rather than self-absorbedly chasing the goal of ultimate, all-encompassing knowledge. Maturity needs to be based on the wisdom arising from experience, so a mature attitude would also require a historical sensibility about how science has matured, in addition to the scientists' own personal experiences. So the two aspects of maturity would also merge together in an important way.

All in all, I think a mature science practiced by mature scientists would very nicely fit the vision of science embodied in active realism. If we follow this vision, we will lose the youthful boast about the global and eternal success and truth of science. Rather, we will learn to appreciate particular successes of various kinds within science, and also successes achieved in other traditions. It will seem natural and honorable to try to learn from reality in as many and diverse ways as possible. There is no need to stand around on a crutch of a manufactured notion of "maturity" in a vain attempt to extract truth from success. The conception of reality that I have proposed incorporates some humility on the part of the knower in an essential way. In active realism knowing merges seamlessly with learning, and there can be very little learning without humility. True realism ought to consist in a humble admission of the existence of beings that do not obey us. It is hubris to grant the existence of objective reality and then presume to predict and control it perfectly. The picture of the ideal epistemic agent that I am proposing is not that of the all-seeing and all-controlling system-builder. Rather, the mature epistemic attitude that is sanctioned by the way I have conceived of reality is "un-knowing", as the psychotherapist Ernesto Spinelli puts it (1997, 6): an attempt "to remain as open as possible to whatever presents itself to our relational experience", and "to treat the seemingly familiar, or that of which we are either aware or informed, as novel, unfixed in meaning, accessible to previously unexamined possibility."[55] That is the most fundamental premise adopted by the humble seeker of reality. I want to get away from the epistemic conceit that with the help of science we should always be able to control nature, or at least predict what we cannot control. This is the human mind wishing to play God, or at least to participate in that God's eye view—something that we should have grown out of in reaching our maturity.

4.3   Out of the Standard Realist Fly-Bottle

Having presented my general view of the realism debate, I will now continue with a critical examination of standard realist intuitions from more specific angles. I will focus on several key ideas to which typical scientific realists are quite attached, and attempt to show that they should not be handled in the way they usually are. I aspire

---

[55] I thank Gretchen Siglar for introducing me to Spinelli's work.

to perform the philosophical function articulated by Wittgenstein with the metaphor of "showing the fly out of the bottle", with apologies for the irreverent image. There are glass walls that many a realist philosopher-fly keeps launching into, thinking "there *must* be a way through this—I can just *see* it." My ambition is to show the way to get around the glass and get us out of the bottle, so that we can be free to fly off in more productive directions. I will begin with the great twin mirage of traditional epistemology: truth and certainty. And then I will address a more recent realist preoccupation, namely the preservation of structures across scientific upheavals. I will finish with some brief thoughts on the subject of reference, in the context of which most analytical philosophers think of the statement "Water is H$_2$O", thanks to Putnam's tale of "Twin Earth".

4.3.1   Truth and Its Multiple Meanings

If the realism debate is about truth (whether science does approach it or at least should aim to do so), then realists should worry further about what "truth" means. Following Bridgman's spirit, I would like to offer an operational analysis of the concept of truth. Bridgman didn't say much about such philosophical concepts, but in fact here I can follow for a long way the now-neglected footsteps of J. L. Austin in a similar spirit, which is generally best expressed in Austin (1962). More specifically, in his 1950 paper simply titled "Truth", Austin begins by warning philosophers away from an intractable kind of metaphysical discourse about truth, which would ask whether it is a substance, or a quality, or a relation: "But philosophers should take something more nearly their own size to strain at. What needs discussing rather is the use, or certain uses, of the word 'true'. *In vino*, possibly, '*veritas*', but in a sober symposium '*verum*'." Typical of his "ordinary-language" philosophy, Austin's move is to ask this question: "What is it that we say is true or is false? Or, how does the phrase 'is true' occur in English sentences?" (Austin [1950] 1979, 117) This tradition is continued in the kind of philosophical "anthropology" of language-use advocated by Huw Price (2011). Now, perhaps uncharacteristically, Austin suggests (p. 118) that there are a few "primary forms of expression" of the word "true", all of which get at a "rather boring yet satisfactory relation between words and world" (p. 133), although he admits that the usages of the term "appear at first multifarious" (p. 117). I want to stay a while with Austin's initial impression, to see if there isn't something more to the apparent multifariousness. Perhaps my own inclination is captured better in Richard Rorty's pronouncement that there is no such thing as the "love of truth" (1998, 28–29): "what has been called by that name", Rorty says, "is a mixture of the love of reaching intersubjective agreement, the love of gaining mastery over a recalcitrant set of data, the love of winning arguments, and the love of synthesizing little theories into big theories."

Austin was clear that he was thinking about "English sentences", presumably without presuming that good philosophy only occurs in English. Even though I am writing English-language philosophy here, I would like to make a quick reference back to my native Korean, to introduce an interesting disturbance to our intuitions. Translating the English term "truth" into Korean is not a trivial matter. My trusty *Minjung English–Korean Dictionary* (anonymous 2003) gives the following multitude of translations for "truth": (a)진리; (b1)진실, (b2)사실; (c1)성실, (c2)정직. The Korean–English side of the same dictionary translates (a) back to "truth", and what is meant there is something like "eternal truth", as would pertain to a law of nature. The meaning of (b1) is trickier. It is also translated back to "truth", but it applies more to things that we actually have access to; a criminal investigation or a corruption inquiry might reveal this kind of truth (진실) about what happened. Or this is what a witness in a court of law would be asked to tell; to demand (a)진리 from a witness would in fact make no sense. The Korean word given in (b2)사실 corresponds pretty straightforwardly to the English "fact", and "fact" is the first translation given by the Korean–English part of the dictionary. Meanwhile, (c2)정직 means honesty; (c1)성실 is trickier, rendered as "sincerity; fidelity; faithfulness; honesty" by the Korean–English dictionary; I would describe it as a conscientious kind of reliability in the character of a person.

So our Korean lexicographers have, perhaps inadvertently but most sincerely (성실하게), unravelled at least three different strands of meaning in the English term "truth": eternal Truth, a matter of fact, and an honest report. Let's bring these thoughts back to Austinian ordinary English-language philosophy, keeping in our view the noun "truth" as well as the adjective "true", as I do not think that their uses are so distinct from each other after all, contrary to Austin's suggestion.[56] I think there are clearly distinct uses and meanings, even if we set aside usages that refer purely to people's characters. What I want to do here is consider the linguistic usages in the fuller context of activities, as I hope Austin, Rorty and Price would all approve.

(Truth$_1$) There is the notion of truth that means making a statement that corresponds correctly to what one thinks or feels: "I am telling you the truth (when I say I am hungry, or that I think I saw a snow leopard on the hill)." This meaning is cogent whether or not what I think is true in some ultimate sense ("A snow leopard in London?—surely not"). This is about correspondence, but only between what I say and what I think, which I think is the only operable notion of correspondence we have in relation to truth (this needs to be distinguished sharply from the inoperable notion of correspondence between what I think and how the external world is). This kind of truth may be a matter of personal character and disposition, but it is also an important part of the foundation of knowledge in any empiricist system.

(Truth$_2$) There are truths by definition: "Of course it is true that the standard meter is one meter long"; "One cannot doubt the truth of the statement 'All bachelors are unmarried', as it is a tautology." These are truths that we construct, judge and maintain by making, using and enforcing definitions.[57]

---

[56] However, it is curious to note, the relationship between the Korean equivalents of "truth" and the Korean equivalents of "true" is not so straightforward.

[57] Cf. C. I. Lewis's notion of the *a priori* ([1929] 1956, ch. 8).

(Truth$_3$) Some truths are rendered true by presumption when we *take* them as given and engage in activities for which they are preconditions. When the presumption is made consciously and explicitly, we call these truths "axioms" or "postulates". For example, when Einstein declared that the speed of light was the same regardless of the motion of either the observer or the source, this by no means followed from the definitions of "light" or "speed" or "observer" that were in place before Einstein's work.

(Truth$_4$) In the context of logic, propositions are true if they can be deduced from other true propositions according to the axioms of the logical system in which one operates. "If proposition P is true, so is its contrapositive." This sense of truth most clearly has the exclusive property: if P is true, and if Q implies *Not P*, then Q is not true.

(Truth$_5$) We acknowledge a statement as true within a system of practice, if it passes, *contingently*, the tests of correctness operative within that system. Is it true that the atomic weight of chlorine is roughly 35.5? Within a specific system of atomic chemistry that we operate in, we have specific procedures for assessing atomic weights, and we can say whether the statement is true or not. These procedures of judgment can give us definite true/false verdicts, but we should not pretend that these judgments are absolute or universal. These verdicts are definite only within a given system, and truth in this sense rests first and foremost on coherence with the rest of the system; however, in addition, good empirical judgment procedures would also establish ways in which the resistance from reality can be shown.

How do these different meanings of truth bear on the realism debate? Truth$_5$ is the core notion of truth that supports my doctrine of active realism, which is a continual and humble search for this truth. But truth$_1$ to truth$_4$ are also each linked to various epistemic activities that are indispensable to inquiry: reports of experience, definitions of concepts, adoption of enabling assumptions, and logical deductions. So while the five meanings of truth are distinct, they arise from activities that are harmoniously linked up with each other in effective inquiry. Within each system, the search for truth$_5$ will hopefully yield knowledge about reality. Truths$_5$ confirmed in different systems can hold various relations with each other; they may be consistent or incommensurable with each other, or have little to do with each other. Active realism advocates the search for truth$_5$ in each system, and also the cultivation of various systems in each of which truth$_5$ can effectively be sought.

How about standard scientific realism? The universal and timeless variety of "truth" by correspondence with reality, which standard realism requires, is conspicuous in its absence in the list above. What epistemic activities does that notion of truth arise from? How is it that we might discover and judge such truths? I say it can only be the combined work of imagination, metaphor and authority: by imagination we come up with the idea of the external world which has some real properties unobservable to us; by metaphor, we conceive of the correspondence between that world and our statements, parallel to the correspondence between thought and utterance (truth$_1$); and to find out what is actually true, we read the Bible, or listen to Einstein. I take comfort in Hilary Putnam's assessment of this matter (Putnam 1995, 10): "To say that truth is 'correspondence to reality' is not false but *empty*, as long as nothing is said about what the 'correspondence' is. If the 'correspondence' is supposed to be utterly independent of the ways in which we confirm the assertions we make . . . then the 'correspondence' is an occult one, and our supposed grasp of it is also occult." (As for Tarski, "his work does *nothing* to explicate the notion of truth".) Then Putnam ends with a paraphrase of William James, which captures the sense of my operational view of truth: "Truth . . . must be such that we can say how it is possible for us to grasp what it is." Otherwise, the notion of truth cannot even serve as a regulative ideal in practice.

4.3.2   The Certainty Trap

For some people, the ideal of certainty seems as difficult to abandon as the ideal of truth. And as with truth, there is no reason to abandon it if we apply the notion only as appropriate and refrain from imposing it where it does not belong. Many philosophers, realists and anti-realists alike, have put certainty in our claims about reality at the center of the realism debate, and I believe this is a significant mistake. The focus on certainty makes us conflate the question of realism with the question of skepticism. When we demand absolute certainty, the defeat of radical skepticism is what we require; if realism has to defeat radical skepticism, then the enterprise is doomed. If we are looking for absolute certainty, realism cannot deliver it any more than any other stance toward science can. A healthy degree of humility should make us wary of any claims of certainty, or any demands that we should be seeking certainty.

There are some things that we can be certain about, but these are things that we *make* or *take as* true, namely truth$_2$ and truth$_3$ as articulated in Sect. 4.3.1. We can also turn candidates for truth$_5$ into truth$_2$ or truth$_3$, but that is not recommended according to active realism. About empirical matters (that is, those propositions that remain candidates for truth$_5$), the question is not how we become certain, but how we get on while accepting uncertainty. It may be a healthy trend that philosophers and scientists are now more likely to speak about probability rather than certainty. However, particularly in certain strands of Bayesianism, there is an impulse to treat probability as a stand-in for both truth and certainty, with a focus on demonstrating the increase of probability approaching 1 as we continue inquiry. In this enterprise, the concept of probability loses its *use*, and only ends up playing the same empty role as notions of "approximate truth". No, the real point of probability is precisely that it can guide our actions when its value is far from 1 or 0; the non-extreme values of probability (such as 1/6 for getting 2 on a roll of a die) serve as useful tools of planning. When we do not have certainty, what we should try to achieve with Bayesian analysis is to arrive at *stable* probabilities for planning, not to prop up the vain hope that the probability values will approach 0 or 1.

Many philosophers that I greatly admire have unfortunately fallen into the certainty trap. Bridgman is a good example; skeptical scrutiny is useful when it explodes overblown claims of certainty, but it cannot be a positive program of work. In the end, operations do not give us any more certainty than passive observation; both are aspects of direct experience, and ultimately cannot be de-coupled so cleanly from each other. Similarly, Hacking (1983) fell into the certainty trap when he tried to argue that the knowledge of experiments or the knowledge about entities was more certain than the knowledge of theories. As David Resnik's (1994) critique shows, Hacking's claims for higher certainty cannot be sustained (see also van Fraassen in Churchland and Hooker (1985), 297–300). Hacking's focus is not quite right when he drives his arguments toward conclusions of the form "We know for certain that X is real". Rather, the main appeal of Hacking's experimental realism, as well as Bridgman's operationalism, should be seen as the exhortation *to go find more ways of engaging with reality.* Popper in his youth seems to have been seduced by a false asymmetry between the certainty of falsification and uncertainty of verification. As long as Popper continued to claim that he had solved the problem of induction, he tied himself to the unproductive search for certainty. (The Popper who used falsifiability as a demarcation criterion did not have this problem.) Popper denounced perfectly legitimate inductive methods for their lack of certainty, while in the end admitting that falsification lacked certainty, too, and bringing in "corroboration" as an uncomfortable proxy for inductive confirmation.

Certainty, if we could reach it, would in fact be the end of inquiry, which is the very antithesis of active realism. Recall Kuhn's observation that if ever there was a paradigm that could eliminate all anomalies, it would cease to support research (Kuhn 1970, 79):

> what we previously called the puzzles that constitute normal science exist only because no paradigm that provides a basis for scientific research ever completely resolves all its problems. The very few that have ever seemed to do so (e.g., geometric optics) have shortly ceased to yield research problems at all and have instead become tools for engineering. Excepting those that are exclusively instrumental, every problem that normal science sees as a puzzle can be seen, from another viewpoint, as a counterinstance and thus as a source of crisis. The only plausible way to reach even reasonable certainty in empirical science seems to be to put an excessive and pernicious restriction of the scope of enquiry. In the kind of sciences that I have been discussing in this book, which are both broad-ranging and full of avenues for new discoveries, certainty would only seem like a pipe-dream, not even a productive regulative ideal.

4.3.3   Structure

With the kind of timid view of knowledge that I am proposing in this chapter, it may seem that we may never be able to express the sense that successful systems of practice in science do embody *some* correct knowledge about nature. And in order to capture that sense, wouldn't we have to try to save some version of the realist argument from success? The persistent feeling of there being "something right" in successful science has been expressed by many people including Pierre Duhem and Henri Poincaré. Duhem ([1906] 1962, 28) thought that theories that are successful in making novel predictions should be regarded as employing a "natural classification", in which "the relations established by our reason among abstract notions truly correspond to relations among things". Poincaré, nearly at the same time and even more famously, opined that the relations postulated in successful scientific theories were robust and real, despite the "ruins accumulated on ruins" created by the routine abandonment of "theories . . . which pretend to teach us what things are" (see Psillos 149–151).

John Worrall (1989) and others have identified Poincaré as the chief originator of their current doctrine of structural realism, which they regard as the most robust form of realism that can be defended.[58] There has been so much attention to structural realism lately that it would be egregious to finish my discussion without some comment on it. James Ladyman (2009) defines structural realism, as first introduced by Worrall (1989), as the position that "we should . . . epistemically commit ourselves only to the mathematical or structural content of our theories." As I understand it, the impulse behind structural realism is along the lines of what I have called "preservative realism" (Chang 2003): structuralists from Poincaré onward have been impressed by the continuity of structures in science, by the fact that some formal structures seem to survive through even extreme instances of scientific change. So it would seem that the knowledge of structures is the secure aspect of scientific knowledge that scientists can discover and keep forever. This is not the place to enter into a detailed appraisal of various structural realist doctrines, nor am I the right person to do so. Rather, I only wish to express some basic frustration I have with the whole debate, in order to explain why I have not discussed structural realism seriously in this work (or in any other works of mine so far).

I will frame my dissatisfaction as a dilemma, which might be called "the structural realist's dilemma": either the structure identified is observable (in which case trust in the structure only amounts to empiricism), or the preservation of the structure is willful (in which case there is no warrant for taking it as an element of external reality). To illustrate the first horn of the dilemma, consider Worrall's favorite example, namely Fresnel's optical equations that are saved in the subsequent Maxwellian theory. Worrall (1989, 117) states that "this was much more than a simple question of carrying over the successful empirical content into the new theory", but I am not convinced about this. Fresnel's equations discussed by Worrall are phenomenological laws, as they are mathematical relations between observable variables, namely the intensities of the incident, reflected and refracted light beams and the angles that these beams make with the reflecting surface. One would expect that sort of thing to be preserved going from one theory to the next, or even from one paradigm to the next as long as the incommensurability is partial as Kuhn allowed. Here we are ultimately only talking about the structure of data-sets, which antirealist empiricists would be very happy to accept. As Ladyman (2009, section 2) puts it: "The most minimal form of structuralism focuses on empirical structure, and

---

[58] See Psillos (1999), ch. 7, for a helpful exposition.

as such is best thought of as a defence of the cumulative nature of science in the face of Kuhnian worries about revolutions". Ladyman rightly includes van Fraassen among the advocates of such structuralism, and Otávio Bueno (1999, also 2011) had made a clear articulation of "structural empiricism" along the same lines. All this goes to show that structuralism does not necessarily fall on the realist side of the standard realism–antirealism divide.

The other horn of the dilemma is conveniently illustrated by the case of Copernicus and Ptolemy (for historical details see Kuhn 1957). Even though Ptolemaic and Copernican theories were very different from each other, uniform circular motion was an essential structural part of both theories (and they even used similar devices, such as epicycles, in order to accommodate observations that seem to deviate from uniform circular motion). Is this structural continuity impressive? Yes, but only in terms of how the obsession with uniform circular motion could have lasted from Ptolemy through to Copernicus. The rigidity, obstinacy or uniformity of scientists' way of thinking, by itself, reveals nothing about the nature of external reality. Instead, the constancy of structures may only be an indication of what we *hold* fixed because of our mathematical or esthetic preferences. This consideration points to the kind of structuralism that Claude Levi-Strauss articulated, which is about structures in our perceptions and conceptions of reality—structures imposed on reality by the human mind, not structures inherent in reality itself. At best, the continuity of structures could be a manifestation of the Kantian synthetic *a priori*, not any support for scientific realism.

My reservations about structural realism should not be mistaken as a denial of the structural continuity that we often do observe in the course of scientific development, nor as a negative assessment of the value of such continuity. It is only that I do not think we should imagine that structuralism will save the realist argument from the success of science. There is no general warrant for regarding the structural aspect of a successful scientific system to be solely or even mainly responsible for its success. An inference from success to structure is going to be just as unsafe as the troubled inference from success to truth. For one thing, it is always going to be very uncertain business to try to pick out the success-generating elements of a successful system; this only returns us to the Duhem problem. And if the attention to structural continuity is combined with the pursuit of truth that is typical of standard realism, then we can easily arrive at a dogmatic position which declares that modern physics has found the correct structures that exist in the world, and nothing else can be real. Such structural realism would be as inimical to active realism as any other version of standard realism can be.

4.3.4 Reference (Farewell to Twin Earth)

Finally, I give up: I have to say something about Hilary Putnam's philosophical tale of "Twin Earth", since I have spoken to so many philosophers who thought that was what my book would be about, upon hearing its title. Twin Earth is just like real earth, except that its oceans, rivers, etc. are filled with a complex chemical "XYZ", which behaves in all observable ways like H_2O. On Twin Earth, does the word "water" refer to H_2O, or XYZ?[59]

Putnam's thought experiment was intended to bolster the intuition that meanings "just ain't in the head" (1975b, 227). That was in favor of semantic externalism. One can follow the causal theory of reference here if one likes, and say that the extension of "water" is the set of all the bodies that bears a particular "sameness relation" to the initial samples that people christened "water". For natural kind terms, Kyle Stanford and Philip Kitcher (2000, 108, 114) cash out this "sameness" relation in terms of having the same "inner constitution" that is causally relevant to producing the characteristic observed properties of the substance. Putnam identifies the molecular formula H_2O as this inner constitution for "water". It is not that people who first started using the term "water" should have had any conception of its inner constitution. Stanford and Kitcher (2000, 114) give a refined version of the causal theory of reference which allows "people who are ignorant of underlying structures to partition the total cause"; according to this conception, "term introducers make stabs in the dark", and "*conjecture* that there's some underlying property (or 'inner structure') that figures as a common constituent of the total causes of each of the properties" that are typically exhibited by the substance in question. Putnam (1975b, 225) also made it clear that the exact nature of the sameness relation in question was a matter to be decided by scientific investigation.

I follow Paul Needham (2000, 2002), Jaap van Brakel (2000, ch. 4), and Erik Curiel (forthcoming) in doubting the workability of rigid designation for theoretical terms. In my view, the biggest problem is stability. If the relevant sameness relation is subject to the verdict of scientific investigation, then it cannot be guaranteed to be stable enough to fix reference with sufficient certainty and permanence. Setting aside the intricate philosophical details of the arguments about rigid designation and the causal theory of reference, to which I cannot possibly do justice in this brief discussion, I want to make two broader critical points about the whole enterprise, which will constitute a justification for not getting sucked into the debate.

First, I have objections to philosophers' fixation on reference-fixation. Reference often seems to serve as a proxy for truth, in order to rescue standard realism: if we can't have the correspondence between statements and facts, the hope is that at least we might have a correspondence between words and things. We might speak of a "correspondence theory of reference", which aspires to render reference as a purely extensional correspondence between a word and a set of objects with no necessary or essential mediation between them by intensional semantics. But if our goal is to understand scientific or quotidian linguistic practice, the correspondence theory of reference is futile, because reference to bits of unobservable reality is just as inoperable as "Truth with a capital T". The causal theory of reference fails in making the

---

[59] Putnam's own presentation can be found in Putnam (1973, 1975b). Hendry (2008, 522–524), gives a convenient entry-point to the debate.

correspondence operable, except possibly for proper names. When it comes to the reference of substance terms, it ultimately leans on theoretical knowledge. What this comes down to is anchoring the theory of reference on standard truth-realism, which I think is ill-advised. The real problem comes when people try to use this correspondence theory of reference in order to bolster standard realism. Then what we have is a tightly circular mutual justification of the correspondence theories of truth and reference; the burden of argument is on those who want to claim that this circle somehow latches on to reality.

Second, I have a deep discomfort about "naturalistic" philosophy, if what naturalism means is an unthinking deference to science. At the very least, if we are going to be slavish naturalists, we should follow the latest and best science as judged by scientists themselves, not some tired old approximation like "Water is H_2O", which scientists who research on water have moved beyond a long time ago. So it will not do to follow Putnam's notion (1975b, 224), in itself 40 years old and already outdated even then, that the extension of "water" is "the set of all wholes consisting of H_2O molecules". Rather, we should at least start by taking note of the notion of water current in chemistry, which Hendry sums up succinctly (2008, 523): "macroscopic bodies of water are complex and dynamic congeries of different molecular species, in which there is a constant dissociation of individual molecules, re-association of ions, and formation, growth and dissociation of oligomers."[60] Without such complex and dynamic interactions within and between H_2O molecules, water would not have the properties that make us grant that it is water. As Curiel (forthcoming, 4) puts provocatively: "no portion of water, no matter how pure or small and no matter in what state or environment, consists of water molecules." We can't call just any bunch of H_2O molecules thrown together "water," without doing violence to up-to-date science.

References

Anonymous. 2003. *Minjung's Handy English–Korean Korean–English dictionary*, 6th ed. Seoul: Minjungseorim, Co.

Austin, J.L. 1962. *How to do things with words: The William James lectures delivered at Harvard University in 1955*. Oxford: Clarendon Press.

Austin, J.L. 1979. Truth. In *Philosophical papers*, 3rd ed, ed. J.O. Urmson and G.J. Warnock, 117–133. Oxford: Oxford University Press.

Bensaude-Vincent, Bernadette, and Jonathan Simon. 2008. *Chemistry: The impure science*. London: Imperial College Press.

Boyd, Richard. 1980. Scientific realism and naturalistic epistemology. In *PSA 1980: Proceedings of the Biennial Meeting of the Philosophy of Science Association 1980*, vol. 2 (Symposia), 613–662.

Brock, William H. 1992. *The Fontana history of chemistry*. London: Fontana Press.

Bueno, Otávio. 1999. What is structural empiricism? Scientific change in an empiricist setting. *Erkenntnis* 50: 59–85.

---

[60] See also Weisberg (2006) and VandeWall (2007).

Bueno, Otávio. 2011. Structural empiricism, again. In *Scientific structuralism*, ed. Alisa Bokulich and Peter Bokulich, 81–103. Dordrecht: Springer.

Cartwright, Nancy. 1983. *How the laws of physics lie*. Oxford: Clarendon Press.

Chakravartty, Anjan. 2004. Review of van Fraassen, *The empirical stance*. *Studies in History and Philosophy of Science* 35A: 173–184.

Chakravartty, Anjan. 2011. Scientific realism. In *Stanford encyclopedia of philosophy (online)*, Summer 2011 ed., ed. Edward N. Zalta. http://plato.stanford.edu/archives/sum2011/entries/scientific-realism/

Chang, Hasok. 2002. Rumford and the reflection of radiant cold: Historical reflections and metaphysical reflexes. *Physics in Perspective* 4: 127–169.

Chang, Hasok. 2003. Preservative realism and its discontents: Revisiting caloric. *Philosophy of Science* 70: 902–912.

Chang, Hasok. 2004. *Inventing temperature: Measurement and scientific progress*. New York: Oxford University Press.

Chang, Hasok. 2007b. *The myth of the boiling point*. http://www.cam.ac.uk/hps/chang/boiling. First posted on 18 Oct 2007.

Chang, Hasok. 2008. Contingent transcendental arguments for metaphysical principles. In *Kant and the philosophy of science today*, ed. Michela Massimi, 113–133. Cambridge: Cambridge University Press.

Chang, Hasok. 2009a. Operationalism. In *Stanford encyclopedia of philosophy (online)*, Fall 2009 ed., ed. Edward N. Zalta. http://plato.stanford.edu/archives/fall2009/entries/operationalism/

Chang, Hasok. 2010. The hidden history of phlogiston: How philosophical failure can generate historiographical refinement. *HYLE* 16(2): 47–79.

Chang, Hasok. 2011c. How historical experiments can improve scientific knowledge and science education: The cases of boiling water and electrochemistry. *Science and Education* 20: 317–341.

Chang, Hasok. 2011e. Beyond case-studies: History as philosophy. In *Integrating history and philosophy of science*, ed. Seymour H. Mauskopf and Tad Schmaltz, 109–124. Dordrecht: Springer.

Chang, Hasok, and Grant Fisher. 2011. What the ravens really teach us: The inherent contextuality of evidence. In *Evidence, inference and enquiry*, ed. William Twining, Philip Dawid, and Mimi Vasilaki, 341–366. Oxford: Oxford University Press and the British Academy.

Churchland, Paul M., and Clifford A. Hooker, eds. 1985. *Images of science: Essays on realism and empiricism, with a reply from Bas C. van Fraassen*. Chicago: University of Chicago Press.

Cohen, I. Bernard. 1956. *Franklin and Newton: An inquiry into speculative Newtonian experimental science and Franklin's work in electricity as an example thereof*. Philadelphia: American Philosophical Society.

Curiel, Erik. Forthcoming. Why rigid designation and the causal theory of reference cannot stand.

Dewey, John. 1938. *Logic: The theory of inquiry*. New York: Holt, Reinhardt & Winston.

Di Bucchianico, Maria Elena. 2009. *Modelling high temperature superconductivity: A philosophical inquiry in theory, experiment and dissent*. Ph.D. dissertation. London: London School of Economics.

Doppelt, Gerald. 2005. Empirical success or explanatory success: What does current scientific realism need to explain? *Philosophy of Science* 72: 1076–1087.

Duhem, Pierre. 1962. *The aim and structure of physical theory*. New York: Atheneum.

Dupré, John. 1993. *The disorder of things: Metaphysical foundations of the disunity of science*. Cambridge, MA: Harvard University Press.

Feigl, Herbert. 1970. The 'orthodox' view of theories: Remarks in defense as well as critique. In *Analyses of theories and methods of physics and psychology*, ed. Michael Radner and Stephen Winokur, 3–16. Minneapolis: University of Minnesota Press.

Fine, Arthur. 1984. The natural ontological attitude. In *Scientific realism*, ed. Jarrett Leplin, 83–107. Berkeley/Los Angeles: University of California Press.

Frank, Philipp. 1949. Why do scientists and philosophers so often disagree about the merits of a new theory? In *Modern science and its philosophy*, 207–215. Cambridge, MA: Harvard University Press.

Frankl, Viktor. 1978. *The unheard cry for meaning: Psychotherapy and humanism*. New York: Simon and Schuster.

Fuller, Steve. 2003. *Kuhn vs. Popper: The struggle for the soul of science*. Cambridge: Icon Books.

Hacking, Ian. 1983. *Representing and intervening*. Cambridge: Cambridge University Press.

Hacking, Ian. 2000. How inevitable are the results of successful science? *Philosophy of Science* 67: 58–71.

Heilbron, John L. 1979. *Electricity in the 17th and the 18th centuries: A study of early modern physics*. Berkeley/Los Angeles: University of California Press.

Hendry, Robin Findlay. 2008. Chemistry. In *The Routledge companion to the philosophy of science*, ed. Stathis Psillos and Martin Curd, 520–530. London: Routledge.

Hesse, Mary. 1977. Truth and the growth of scientific knowledge. In *PSA 1976: Proceedings of the 1976 biennial meeting of the philosophy of science association, vol. 2 (Symposia)*, ed. Frederick Suppe and Peter D. Asquith, 261–281. East Lansing: Philosophy of Science Association.

Holton, Gerald, Hasok Chang, and Edward Jurkowitz. 1996. How a scientific discovery is made: A case history. *American Scientist* 84: 364–375.

Kuhn, Thomas S. 1957. *The Copernican Revolution: Planetary astronomy in the development of Western thought*. Cambridge, MA: Harvard University Press.

Kuhn, Thomas S. 1970. *The structure of scientific revolutions*, 2nd ed. Chicago: University of Chicago Press.

Kuhn, Thomas S. 1977. Objectivity, value judgment, and theory choice. In *The essential tension: Selected studies in scientific tradition and theory change*, 320–339. Chicago: University of Chicago Press.

Ladyman, James. 2009. Structural realism. In *Stanford encyclopedia of philosophy (online)*, Summer 2009 ed., ed. Edward N. Zalta. http://plato.stanford.edu/archives/sum2009/entries/structural-realism/

Laudan, Larry. 1977. *Progress and its problems: Towards a theory of scientific growth*. London/Henley: Routledge & Kegan Paul.

Laudan, Larry. 1981. A confutation of convergent realism. *Philosophy of Science* 48: 19–49.

Lewis, Clarence Irving. 1956. *Mind and the world order: Outline of a theory of knowledge*. New York: Dover.

Lycan, William G. 1998. Theoretical (epistemic) virtues. In *Routledge encyclopedia of philosophy*, ed. Edward Craig, vol. 9, 340–343. London: Routledge.

Lyons, Timothy D. 2003. Explaining the success of a scientific theory. *Philosophy of Science* 70: 891–901.

McLaughlin, Amy L. 2009. Peircean polymorphism: Between realism and anti-realism. *Transactions of the Charles S. Peirce Society* 45: 402–421.

McLaughlin, Amy L. 2011. In pursuit of resistance: Pragmatic recommendations for doing science within one's means. *European Journal for Philosophy of Science* 1: 353–371.

Needham, Paul. 2000. What is water? *Analysis* 60: 13–21.

Needham, Paul. 2002. The discovery that water is H$_2$O. *International Studies in the Philosophy of Science* 16: 205–226.

Neurath, Otto. [1931] 1983. Sociology in the framework of physicalism. In *Philosophical papers 1913–1946*, ed. Robert S. Cohen and Marie Neurath, 58–90. Dordrecht: Reidel.

Polanyi, Michael. 1964. *Science, faith and society*. Chicago: University of Chicago Press.

Polanyi, Michael. 1966. *The tacit dimension*. London: Routledge.

Polanyi, Michael. 1967. Science and reality. *The British Journal for the Philosophy of Science* 18: 177–196.

Popper, Karl. 1972. *Conjectures and refutations: The growth of scientific knowledge*, 4th ed. London/Henley: Routledge & Kegan Paul.

Popper, Karl. 1981. The rationality of scientific revolutions. In *Scientific revolutions*, ed. Ian Hacking, 80–106. Oxford: Oxford University Press.

Price, Huw. 2011. *Naturalism without mirrors*. New York/Oxford: Oxford University Press.

Psillos, Stathis. 1999. *Scientific realism: How science tracks truth*. London/New York: Routledge.

Putnam, Hilary. 1973. Meaning and reference. *Journal of Philosophy* 70: 699–711.

Putnam, Hilary. 1975a. What is mathematical truth? In *Mathematics, matter and method*, Philosophical papers, vol. 1, 60–78. Cambridge: Cambridge University Press.

Putnam, Hilary. 1975b. The meaning of 'meaning'. In *Mind, language and reality*, Philosophical papers, vol. 2, 215–271. Cambridge: Cambridge University Press.

Putnam, Hilary. 1978. *Meaning and the moral sciences*. London/Henley/Boston: Routledge & Kegan Paul.

Putnam, Hilary. 1995. *Pragmatism: An open question*. Oxford: Blackwell.

Resnik, David B. 1994. Hacking's experimental realism. *Canadian Journal of Philosophy* 24: 395–412.

Rorty, Richard. 1998. Pragmatism as romantic polytheism. In *The revival of pragmatism*, ed. Morris Dickstein, 21–36. Durham, NC: Duke University Press.

Smart, J.J.C. 1963. *Philosophy and scientific realism*. London: RKP.

Soler, Léna. 2008. Revealing the analytical structure and some intrinsic major difficulties of the contingentist/inevitabilist issue. *Studies in History and Philosophy of Science* 39: 230–241.

Spinelli, Ernesto. 1997. *Tales of un-knowing: Therapeutic encounters from an existential perspective*. London: Duckworth.

Stanford, P. Kyle. 2000. An antirealist explanation of the success of science. *Philosophy of Science* 67: 266–284.

Stanford, P. Kyle, and Philip Kitcher. 2000. Refining the causal theory of reference for natural kind terms. *Philosophical Studies* 97: 99–129.

Taber, Keith S. 2003. The atom in the chemistry curriculum: Fundamental concept, teaching model or epistemological obstacle? *Foundations of Chemistry* 5: 43–84.

Van Brakel, Jaap. 2000. *Philosophy of chemistry*. Leuven: Leuven University Press.

Van Fraassen, Bas. 1980. *The scientific image*. Oxford: Clarendon Press.

VandeWall, Holly. 2007. Why water is not H_2O, and other critiques of essentialist ontology from the philosophy of chemistry. *Philosophy of Science* 74: 906–919.

Weisberg, Michael. 2006. Water is not H_2O. In *Philosophy of chemistry: Synthesis of a new discipline*, ed. Davis Baird, Eric R. Scerri, and Lee C. McIntyre, 337–345. Dordrecht: Springer.

Wittgenstein, Ludwig. 1922. *Tractatus Logico–Philosophicus* (trans: C.K. Ogden, intro: Bertrand Russell). London/Boston/Henley: Routledge & Kegan Paul Ltd.

Wittgenstein, Ludwig. 1969. *On certainty*. New York: Harper & Row.

Worrall, John. 1989. Structural realism: The best of both worlds? *Dialectica* 43: 99–124.

Wray, K. Brad. 2007. A selectionist explanation for the success and failures of science. *Erkenntnis* 67: 81–89.

Wray, K. Brad. 2010. Selection and predictive success. *Erkenntnis* 72: 365–377.

Chapter 5
Pluralism in Science: A Call to Action

Abstract In this chapter I present a sustained and systematic defence of pluralism in science, building on various hints from earlier chapters. I define my position as "active normative epistemic pluralism". Based on the recognition of the benefits of having multiple systems of practice in each field of study, pluralism as I intend it is an active stance committed to the cultivation of plurality. There are two types of benefits of plurality. Benefits of toleration arise from simply allowing multiple systems simultaneously, which provides insurance against unpredictability, compensation for the limitations of each system, and multiple satisfaction of any given aim. Benefits of interaction arise from the integration of different systems for specific purposes, the co-optation of beneficial elements across systems, and the productive competition between systems. Pluralism should not be confused with an abdication of judgment: each pluralist has the freedom and responsibility to evaluate the quality and value of scientific work. Pluralism can deliver its benefits without a paralyzing relativism or an uncontrolled dissipation of resources. In practice, the kind of pluralism I advocate comes down to a directive to proliferate valuable systems of knowledge: this has concrete implications for scientific practice, and also gives new purpose and approach to the history and philosophy of science, in line with my vision of history and philosophy of science as "complementary science".

5.1 Can Science Be Pluralistic?

5.1.1 Plurality: From Acceptance to Celebration

I became a pluralist about science because I could not honestly convince myself that the phlogiston theory was simply wrong—or even genuinely inferior to Lavoisier's oxygen-based chemical theory. OK, the story is not quite so simple as that, but I really was pulled into a pluralist way of thinking about science by a set of historical

H. Chang, *Is Water H₂O?: Evidence, Realism and Pluralism*, Boston Studies
in the Philosophy of Science 293, DOI 10.1007/978-94-007-3932-1_5,
© Springer Science+Business Media B.V. 2012

episodes in which discarded past theories turned out not to be obviously absurd on a closer look. More positively, in the course of doing the research for this book, I became convinced that there was something worth preserving in Priestley's phlogiston, in Ritter's elementary water, in Dalton's HO formula for water, and so on, without denying the merits of the newer ideas that came to replace them. My previous work had already prepared me in this direction, for example when I realized that the caloric theory of heat had much to recommend it, and even some merits that made it superior to the early kinetic theories of heat for many decades until the middle of the nineteenth century. Of course it would be unwise to make generalizations from a few particular studies, but they were too suggestive to ignore. Like an itch demanding a scratch, they made a persistent call for a re-examination of some fundamental assumptions about the nature of science that were deeply ingrained into my own thinking. They made me seriously call into question the common intuition that there could only be one right answer to a scientific question, and that once science has answered a question definitively its verdict is final.

In each of the earlier chapters I have given strong hints at a pluralism concerning science, indicating various ways in which it seemed beneficial for science to maintain multiple approaches in the same area of study.[1] Now it is time to consolidate those hints into a coherent and systematic statement of a philosophical position. In this chapter I advance a case for pluralism based on general and abstract arguments, supported by reference to as many cases as I can invoke and also buttressed by the arguments concerning scientific realism made in Chap. 4. I will refer to examples drawn from Chaps. 1, 2, and 3, but I will also introduce various other brief examples, partly because I want to make this chapter reasonably self-contained and understandable to those who have not studied the previous chapters closely. Most of my examples will still be confined to the physical sciences, and that is simply a reflection of my own limitations; whether my arguments hold up with respect to other sciences is something I leave to the judgment of better-informed scholars. As before, there will be three parts to the chapter. The first section (Sect. 5.1) motivates and states the general themes in a way that is accessible to non-specialists; the second section (Sect. 5.2) contains a systematic and thorough argument for my position; the third section (Sect. 5.3) addresses some important specialist or in-depth questions that would have gotten in the way of the flow of thought in the earlier sections.

As indicated above, I began with a grudging fascination with plurality in science. But the longer I examined this troublesome plurality, the more I became positively excited about it. All of the historical episodes that I have presented in earlier chapters exhibit the benefits of having multiple systems operating simultaneously in a respectable and exciting area of science. In Chap. 1 I ended up presenting an unorthodox view of the Chemical Revolution, according to which the realization of the full potential of late eighteenth-century chemistry was delayed because of the unjustified termination of the phlogistonist system. Even though I also showed that there was

---

[1] See especially Chap. 1, Sect. 1.2.4.2; Chap. 2, Sect. 2.2.3.3; Chap. 3, Sects. 3.1, 3.2.2 and 3.2.4; Chap. 4, Sect. 4.1.

more plurality than meets the eye in chemistry during and after Lavoisier's time, I argued that it still would have been beneficial to let the phlogistonist system survive longer and more robustly than it did. In Chaps. 2 and 3, I described some truly long-lasting periods of real plurality in the fields of electrochemistry and atomic chemistry that continued for much of the nineteenth century. These pluralistic periods were in fact very productive. And as I argued in the first section of Chap. 4 (Sect. 4.1), various reflections arising from Chaps. 2 to 3 suggest that there is nothing sacrosanct or inevitable about the specific formulation that water is H_2O. These thoughts are not crazy, and I have found them to provide a refreshing and provocative view on how scientific work is, could be, and should be conducted.

In the process of thinking through these cases, I began to see general reasons for which plurality would be necessary and beneficial in science. I will give a systematic presentation of these reasons in the second section of this chapter (Sect. 5.2), but here are some intuitive highlights with some suggestive metaphors. The most fundamental motivation for pluralism is *humility*: we are limited beings trying to understand and engage with an external reality that seems vastly complex, apparently inexhaustible, and ultimately unpredictable.[2] If we are not likely to find *the* perfect system of science, it makes sense to foster multiple ones, each of which will have its own unique strengths. If we are like the proverbial blind people feeling the elephant, not only should we learn not to generalize too much from our own particular experience, but we should also recruit more collaborators in an attempt to get at all the different parts of the elephant.[3]

It is possible that science initially needed to be launched on the strength of hubris, which made the whole enterprise seem both doable and worth doing: that we *could* grasp *the* truth about nature! Perhaps early scientists needed to believe that nature was fundamentally simple enough for them to be able to understand it. Newton had God on his side, and proceeded with the faith that there was one truth about God's creation and with the grace of God he could find it—how else would anyone muster the conviction that one simple equation could cover all of the universe? But after centuries of success modern science has reached its maturity, and no longer needs the crutches of faith and hubris. We can now afford to be more humble, yet confident that we will be able to continue learning about reality.

Joseph Priestley had a particularly instructive notion of epistemic humility, which was dynamic: "every discovery brings to our view many things of which we had no intimation before". He had a wonderful image for this: "The greater is the circle of light, the greater is the boundary of the darkness by which it is confined." (See Fig. 5.1 for my rendition of it.) As knowledge grows, so does ignorance—or rather, the range of ignorance that we are aware of. "But," Priestley continued, "notwithstanding this, the more light we get, the more thankful we ought to be. For by this means we have the greater range for satisfactory contemplation. In time the bounds of light will be

---

[2] Here I seem to be following in the footsteps of William Wimsatt (2007). I regret that I have not been able to build on his work, as yet.

[3] And what makes us confident that *seeing* gets at all the different aspects of the elephant?

Fig. 5.1 A graphic
representation of Priestley's
metaphor concerning the
growth of knowledge and
ignorance

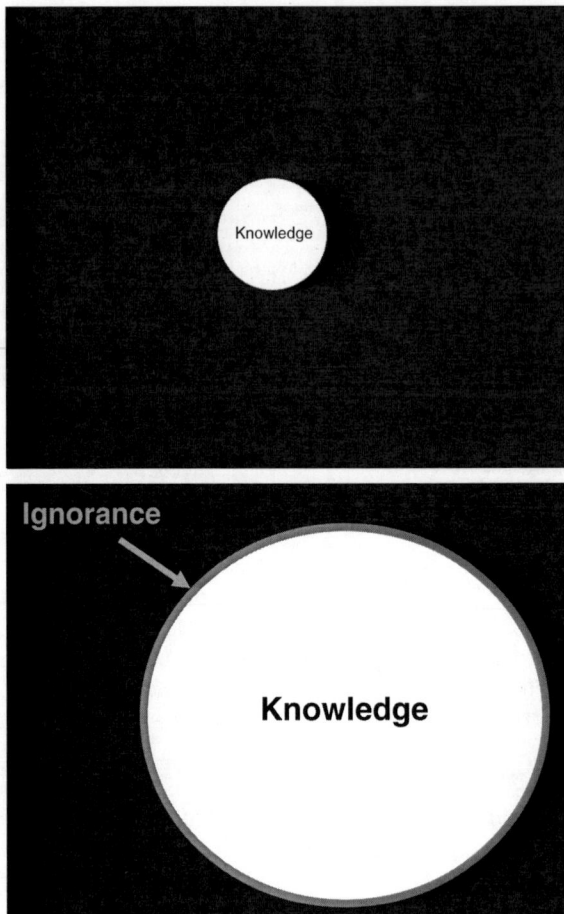

still farther extended; and from the infinity of the divine nature and the divine works, we may promise ourselves an endless progress in our investigation of them: a prospect truly sublime and glorious." (Priestley 1790, 1: xviii–xix) Michael Faraday had a similar vision, which he expressed in one of his papers on electrolysis: "Indeed, it is the great beauty of our science, CHEMISTRY, that advancement in it, whether in a degree great or small, instead of exhausting the subjects of research, opens the doors to further and more abundant knowledge, overflowing with beauty and utility, to those who will be at the easy personal pains of undertaking its experimental investigation." (1834, 122, §871; quoted in Hartley 1971, 184)

Priestley and Faraday were guided by their religious convictions, but for non-believers this picture of unending abundance of natural inquiry may simply be taken as a fact of life in science. There is no need to invoke highfalutin metaphysical doctrines, either. It just does seem that nature holds an indefinitely large number and diverse types of facts there to be revealed, and this makes it likely that each different system of practice could tap into a different part of that inexhaustible reservoir, and continue to tap into more of it. When science loses sight of this abundant potential, I think it ends up restricting itself unnecessarily. Much of successful scientific practice seems to be based on a perfectly functional kind of messiness that have been persuasively described by authors such as Nancy Cartwright (1999), Mike Fortun and Herbert Bernstein (1998), and Andrew Pickering (1984).

In coming to terms with the plurality in science, we may also think in terms of the complexity of nature, rather than its plenitude. It seems that any domain of nature we choose to study reveals an indefinite degree of complexity, while human minds can only handle relatively simple schemes, no matter how much help we have from increasing computing power. So what we need is a set of various simple schemes to get at specific aspects of the phenomena in question. Against this vision of complexity many scientists and philosophers have put up a notion of reduction: all complex structures can be broken down into simpler ones, and all things in nature ultimately consist of a small number of simple physical units, so knowing the truth about those simple units can tell us everything there is to know about nature. This is the basis of the notion that elementary particle physics is the one and only science we should ever need; Ernest Rutherford reportedly said "all science is either physics or stamp collecting" (quoted in Birks 1962, 108)—perhaps it was fitting punishment that he was given the Nobel Prize in *Chemistry* in 1908. Reductionists would grant that it may be expedient to use a plurality of approximate theories when we are trying to deal directly with complex levels of phenomena, but they would insist that in principle we only need one good theory of the simple level. I cannot enter a full discussion of the reductionism question here, but I will make three observations which tend to suggest that it is unrealistic to expect the reductionist strategy to work in general. First, there does not seem to be an end to the process of going to more and more basic units; it would have been very nice to be able to stop at the trio of protons, neutrons and electrons, but it did not turn out that way—will we be able to stop at superstrings? Second, as we go further and further down to more basic levels, the physics involved does not seem to get any simpler. Finally, wholes can be simpler than their parts, depending on our conceptual interests; for example, consider the geometric simplicity of a clean triangular piece of plastic, and the horribly complicated molecular structure of that piece of plastic.

Or the point can be put in terms of prudence, which is based on a realistic pessimism about life in general: not everything will go to plan, and some things we attempt will fail—that is how reality surprises us (the way I conceive of "reality" was explained in Chap. 4, Sect. 4.2.1). There will be contingencies, so we need to have a kind of science that has some capacity for handling surprises, so that if one thing fails not everything else fails as a consequence. This is reminiscent of the argument against monoculture in agriculture. Peter Galison has argued that the resilient strength of modern physics has owed much to the "intercalation" of theory, experiment and instrumentation. Charles Sanders Peirce (1839–1914) provided part of Galison's inspiration, and Peirce's point was a more general one: in philosophy, as well as science, we ought to "trust rather to the multitude and variety of its arguments than to the conclusiveness of any one. [Our] reasoning should not form a chain which is no stronger than its weakest link, but a cable whose fibers may be ever so slender, provided they are sufficiently numerous and intimately connected."[4]

It is instructive to view this need for prudence in terms of the dynamics of scientific progress. Science is an inherently *progressivist* enterprise, which always strives to improve things, even if it should fail.[5] Basic humility should lead us to expect that any successful system of practice will hit upon its limitations sooner or later. In Kuhn's view of science (1970), this expectation takes on a tinge of inevitability. Scientists' desire for increasing precision and scope will force almost every paradigm into failure by revealing fresh anomalies. Success encourages ambition, and as our ambition grows, so does the scope for inadequacy. The growing imperfections in the ruling paradigm will eventually produce a crisis, which precipitates the appearance of a new paradigm. Kuhn's view serves as a very welcome antidote to the hubris of counting on the uninterrupted success of one's own scientific system. However, when it comes to what should happen after the onset of crisis, two limitations to Kuhn's view become clear. First, as Popper (1970) and Watkins (1970) argued, without at least an underlying plurality in the normal state of science, it is difficult to see how a new paradigm could suddenly arise when it is needed. Second, when we have managed to create a successful system for doing something, it is unreasonable to demand that it should be discarded completely just because it fails on the edges when pushed too far; workable systems are not easy to come by, and they should be preserved as much as possible (see Chap. 4, Sect. 4.1.4 on this point). On both counts, the Kuhnian view of scientific dynamics would be improved by an injection of pluralism.

5.1.2 Monism and Pluralism

In politics, hardly anyone would deny that some degree of pluralism lies at the foundation of liberal democracy. As Hilary Putnam (1995, 1) stresses, this is a relatively new modern insight:

> Today we tend to take the ideas of tolerance and pluralism for granted. If we are aware that there was diversity of views and the clash of different opinions in ancient Athens, for example, or in the late Roman empire, we are likely to regard that activity as a sign of vitality in those societies. Few people realize that that is not how those societies themselves saw the mater. Classical thinkers saw diversity of opinions as a sign of decay and heresy; only since the Enlightenment have we been able to see it as a positive good.

Might there not be a similar pattern of development regarding science, so that we can now appreciate pluralism in a way that was not possible before? The modernist project of scientism, which tried to shape society on the model of science, has not really worked out. Why not try the converse, shaping science on the model of what

---

[4] Quoted in Bernstein (1989), 9, from Charles S. Peirce, *Collected Papers*, 5.265.
[5] That is to say, even when science does not manage to be progressive, it is still progressivist.

we consider a good social and political system?[6] There are a few initial reasons for taking that suggestion seriously. Most obviously, the scientific community is a society after all, to which any general principles of good governance ought to apply. Scholarship, including science and philosophy, is founded on *dialogue*, which is a fundamental principle of social communication (cf. Bernstein 1989). And there is an epistemic dimension to political pluralism, too—if different religions and cultures are allowed to co-exist, so will the various beliefs that they carry within them.

But you may still object: "surely pluralism can't be applied to the *products* of scientific work?" There is a prevalent assumption of *monism* regarding science, which I now want to face down explicitly in order to create a viable philosophical space for pluralism. Monism about scientific knowledge springs from the notion that science is the search for the truth about nature; since there is only one world, there is only one truth about it, and only one science that should seek it. Say, *either* the universe began with a Big Bang *or* not—there is one right answer to each well-formed question, and science tries to find out that right answer, employing the one best scientific method known and employed by the relevant mainstream scientific community. Monism is widespread among scientists, especially physicists. For example, take Steven Weinberg (1992, 3), master elementary-particle physicist: "Our present theories are of only limited validity, still tentative and incomplete. But behind them now and then we catch glimpses of a final theory, one that would be of unlimited validity and entirely satisfying in its completeness and consistency."

Even Philip Anderson, renowned condensed-matter physicist well-known for his objections to the assumption that the only really important science is the study of fundamental laws to which all else is reduced, makes his basic monist commitment clear: "In order to maintain our daily lives we have to accept the objective reality of the world and that it is the same world for everyone." (Anderson 2001, 492) This statement occurs in his violent objection to Nancy Cartwright's pluralistic view on science. So as to leave no doubt about where he stands, he also quotes from the opening of his own classic paper, which Cartwright had cited in an anti-reductionist vein: "The reductionist hypothesis may still be a topic for controversy among philosophers, but among the great majority of scientists it is accepted without question. The workings of our minds and bodies, and of all matter . . . are assumed to be controlled by the same set of fundamental laws, which . . . we know pretty well" (Anderson 1972, quoted in Anderson 2001, 489) Anderson, of course, still recognizes that it is often very complicated business working out what the fundamental laws imply about concrete situations. But he maintains that all science is connected into one "seamless web". I agree wholeheartedly (as would Cartwright) that the structure of scientific knowledge is reticular, but I do not think that the web is so seamless as Anderson believes. The web is, and should be, also multi-layered in places, if I may stretch the metaphor a bit further.

---

[6] For much more considered views than I can offer here see, for example, Kitcher (2011) and the articles collected in Wylie (2006).

To give more precision to the discussion, I will take a specific definition of monism, given by Stephen Kellert, Helen Longino and Ken Waters in their recent edited collection on scientific pluralism (2006, x). In their five-point definition of monism, two points are especially pertinent to my purposes: "(1) the ultimate aim of a science is to establish a single, complete, and comprehensive account of the natural world (or the part of the world investigated by the science) based on a single set of fundamental principles" and "(4) methods of inquiry are to be accepted on the basis of whether they can yield such an account".[7] I would add that monists also typically suppose that there is one best method of inquiry at least in each domain (though that is not strictly required by the definition by Kellert, Longino and Waters).

I would counter this monist position in two steps. First of all, what we want science to do is to give us an account of the natural world that serves whatever ultimate aims we may have; the monistic character of the account should not in itself be our ultimate aim. This is obvious once you've said it, but not many monists seem to think of it. The second step is to show that the aims of science can be served better in general by cultivating multiple interacting accounts. This step requires careful argument, and that is what I will attempt to provide in the remainder of this chapter.

In place of monism, I offer pluralism as an ideal of science. I would define pluralism in science as the doctrine advocating the cultivation of multiple systems of practice in any given field of science. By a "system of practice" I mean a coherent and interacting set of epistemic activities performed with a view to achieve certain aims (see Chap. 1, Sect. 1.2.1.1 for a more detailed account). Each system of scientific practice embodies an account of the aspect of reality that is its subject area, and methods for creating and using such accounts. In Chaps. 1, 2, and 3 I have shown different systems of practice in each given field of science developing in productive interaction with each other. And it is important to note that pluralism (or monism) as I intend it is not merely a descriptive statement about how science is,[8] or not even an armchair-normative statement about how science should be. As I said in relation to realism in Chap. 4, a proper "ism" should be an ideology, which implies a commitment to action. So, pluralism about science is a commitment to promote the presence of multiple systems of scientific knowledge. It is not an idle pronouncement to "let a hundred flowers bloom", but the effort of actively cultivating the other 99 flowers. (In advocating pluralism in science I am building on a considerable body of literature. In the third section (Sect. 5.3), I will say a few words about what it is that I want to add to what has already been said by my eminent predecessors, and how my views differ from theirs.)

---

[7] The rest of Kellert, Longino and Waters's definition of scientific monism continues as follows: "(2) the nature of the world is such that it can, at least in principle, be completely described or explained by such an account; (3) there exist, at least in principle, methods of inquiry that if correctly pursued will yield such an account"; and "(5) individual theories and models in science are to be evaluated in large part on the basis of whether they can provide (or come close to providing) a comprehensive and complete account based on fundamental principles."

[8] It is especially not to be taken as a scientific theory in itself; this is one reason I avoid the phrase "scientific pluralism", preferring to spell it out as "pluralism in science" or "pluralism regarding science".

5.1.3 Why Pluralism Is Not Relativism

Having made an initial statement of pluralism, I now hasten to anticipate and defuse some obvious worries and objections. When I present my pluralist ideas, especially to philosophers, I often meet a vociferous objection: "But isn't it just relativism?" Exactly why relativism should be such a dirty word is a whole other issue, which I can't fully go into here.[9] More urgently, I want to distinguish pluralism from relativism. The most fundamental difference is that relativism involves a renunciation of judgment and commitment at least to a degree, which pluralism most definitely does not. The mature pluralist attitude is to engage productively with what one disagrees with, which is very far from the feared caricature of relativism in which one says "Whatever". Curiously, although it may seem that relativism is a stronger and more radical doctrine than pluralism, relativism does not necessarily imply pluralism. If relativism only insists on the equal treatment of any alternatives that *do* exist, there is no requirement that there should be multiple alternatives. If everyone actually agrees on something and no one seeks any alternatives, relativism has no strong way to oppose that state of affairs. The following may sound like a stupid point, but it needs to be stated clearly: the demand for *plurality* is the most crucial feature of *pluralism*. Pluralism is about the benefits of actually having multiple systems in co-existence. So, my slogan for pluralism is not "Anything goes", but "Many things go." Pluralism takes a clear stance against absolutism, in a way that relativism actually can't do easily. A system of practice that denies the rights of other systems to exist would have to be banned in a pluralist scientific regime. This is just as a truly free society needs to impose constraints on individuals and groups to restrain them from restricting the freedom of others.

The fear of relativism, and its conflation with pluralism, will not go away easily. The objection comes back, in a different guise: "If you go with pluralism, how do you choose what to believe?" Well, how *do* you choose, in any case? If you want to be a pluralist, do whatever you would do in choosing the winner in a monist scheme, and just pick *two* winners at the end, or put in a second prize—that would be a fine start. Pick three winners, if you can afford to have a third prize, too. Of course you would want to be less crude and facetious than that, but I am trying to make the point that neither monism nor pluralism delivers us from the responsibility of judgment. If anything, monists have a heavier burden of choice than pluralists, since they can't stop the process until everything except one option is eliminated. You may think that monism does not require any real choice because you can rely on The Scientific Method, which automatically delivers the verdict. If you have been involved in peer review, either at the giving or the receiving end, you know that scientific choices are not made by algorithm-following automata. And besides, who chose The Scientific Method, and how? It's choices all the way down, unless you ask God or a dictator to come in and just tell us what to do.

---

[9] For a recent and considered view of the matter, see Bloor (2007).

"But, but... how do you keep the crazies out?" The objection keeps coming, from many scientists and others who are concerned about the erosion of scientific authority. They worry that pluralism would result in schools teaching Biblical creationism (or intelligent design) alongside evolution, climate-change skeptics having equal voice with majority scientists in determining environmental policy, alternative medicine gaining a foothold in the medical establishment, and so on. As John Norton once put it to me: it may sound fine to cultivate a hundred flowers, but how do you keep the weeds out? The metaphors keep coming—this, from Sandy Mitchell and Peter Machamer: how do you decide who gets to come to the table?[10] This is surely an unavoidable question that we must take seriously in any discussion of scientific methodology, and my critics are unhappy because pluralism is not able to answer it. But this unhappiness is based on a category mistake: pluralism is a doctrine about how many places we should have at the table; it cannot be expected to answer a wholly different question, which is about the guest list. And monism doesn't answer the latter question, either! Deciding that there will only be one place at the table does not determine who gets to sit there. "Me, of course", is the unspoken presumption. And how do we keep from ending up with a room full of tables-for-one—sad, and no more productive than having one big table with an uncontrolled guest list? It should be plain that either pluralism or monism, in itself, cannot determine what we actually believe in science, or how we decide what to believe. We do need to have ideas about how we make such choices, and if relativism is a doctrine that says we should make them randomly or not make them at all, then pluralism and monism are equally remote from it.

There is another version of the same objection which is subtler: in a monist regime, *if* we figure out the right way to get answers, we can just have the right answer and eliminate everything else; in a pluralist regime, we would still have to allow other, inferior answers to exist and confuse things. I think this is a very big "if", again a manifestation of hubris. But many people do find it a realistic prospect, about *some* scientific questions. For example, they feel that it is impossible that Darwinian evolutionary theory is not fundamentally correct, and that to allow any platform to something as patently absurd as creationism or intelligent design would be nothing short of criminal. All right, let's talk about creationism seriously, though briefly, since this issue arouses such passions and cannot be ignored. In short, I think that a fundamentalist insistence on the exclusive truth of neo-Darwinian evolutionary theory is an attitude that is just as immature as fundamentalism on the religious side.[11]

---

[10] These comments were part of the discussion after I presented an earlier version of these ideas at the Center for Philosophy of Science at the University of Pittsburgh on 13 November 2009.

[11] I certainly would not try to claim that the points I make here are really new. However, as I haven't studied these debates in depth and detail, I will not try to say who made what point first. As a pluralist I try to study many things, but this is not one that I have managed to fit in. Those who can devote the time and energy to follow this topic might start with Steve Fuller's defence of intelligent design (2008b) and all the furore that followed; see, for example, A. C. Grayling's debate with Fuller in *The New Humanist* (http://newhumanist.org.uk/1856/origin-of-the-specious)

The first thing to remind the evolutionary fundamentalists is that creationism *in itself* is not such an irrational and absurd thing to believe. Hundreds and thousands of great scientists and other respectable thinkers—including practically everyone in Europe before Darwin's theory took root—believed it, for lack of a credible alternative. Let's name some names: Boyle, Newton, Dalton, Faraday, Kant, Jefferson, and on and on. Of course, the rationality of a belief is largely a function of what other conceivable alternatives there are, and in the presence of the evolutionary alternative it is quite irrational to insist that only intelligent design can explain the origin and functioning of life.[12] But why not also accept that there is a lasting mystery about how the marvelous harmony and coordination in living bodies and ecosystems could really have arisen spontaneously, and insist that Darwinian evolutionary biology is making great progress and that it is probably the best available way to solve that mystery, while not insisting that no other method could possibly work and should not even be entertained? Why would it be so terrible to mention in biology classes a whole set of alternatives ranging from the neo-Darwinian orthodoxy through neo-Lamarckism to Biblical creationism, and give a frank assessment of how credible we think each alternative is?[13] What do we really gain by restricting the focus of science teaching? Are we thereby producing a citizenry that is so wonderfully educated in science and excited about it? Are we making a big dent in the spread of religious fundamentalism in the areas where it has been strong? What is most objectionable about creationism is dogmatism, the commitment to taking things on faith and refusing to consider alternatives. By shutting down debate within science we only manage to pervert science into the same objectionable dogmatism.

It must be possible to engage the creationists in productive ways. To be honest, I have not worked on that issue in my own life because I have not had occasions to care enough about it in comparison to many other issues that I feel more strongly about. But if you do care enough, why not join the others who have at least dignified the creationist position with an argument against it? To say that it is not even worth debating is, again, hubris. There must be *some* reason why millions of people find creationism compelling. If you think they shouldn't, then try to talk them out of it; if you think it is a mass delusion arising from insanity, go into psychiatry! More plausibly and creatively, why not encourage the creationists to come up with concrete methods of testing their ideas, from which we might all learn something, rather than insisting that their ideas are not testable? Why not point out the uninformativeness of saying "God designed it that way" as an explanation, and encourage the devising of more specific explanations? Why not encourage pluralistic debate among creationists, by highlighting the serious contrast between those Christians who take the account in the Book of Genesis very literally and those who take the Biblical account as metaphorical? And if you are feeling particularly belligerent, why not take the pluralist fight to the churches and demand that their teachings should include evolution as well as divine

---

[12] This is similar to the line that Paul Thagard (1978) takes on astrology.

[13] An excellent example of such an attempt is by Del Ratzsch (1996); see also the historical perspective given by Michael Ruse (2005).

creation? There are plenty of enlightened churches and even creationists who would, and actually do, allow this. And the others can be reminded that even the Catholic Church has a record of changing its mind on important scientific matters, for example about heliocentrism and the condemnation of Galileo.

With these considerations, we come back again to the inevitable political dimension of knowledge, and the ineliminable link between knowledge and politics, between science and policy. And again, scientists and others who extol the virtue of science might take a humble lesson from the messy world of politics, in which people have learned some valuable lessons over the centuries through the unspeakable suffering of millions caused by failed political systems. Without pretending that the current forms of pluralist liberal democracy are anywhere near perfect, we should also acknowledge that they are protecting us from far worse excesses. There is a simple and crude pluralist lesson: at least have a two-party system, not a one-party system; yes, pluralism is less efficient than totalitarianism in many ways, but we have to remember that efficiency creates a nightmare if it serves a nefarious aim. Science has also learned some basic lessons about its governance, including the principle of peer review. But science has not yet figured out how to prevent the system of peer review from turning into oligarchy or mob-rule, except by relying on the good will and the good judgment of the individual scientists who have made it into the establishment. We need pluralist science policies, and I do not pretend to have the answers there. But I think the act of doing the kind of concrete work presented in this book is a valuable preparatory step.

5.1.4 Is Pluralism Paralyzing?

Even if we can rest assured that pluralism will not drag science into a chaos dominated by crackpots and madmen, there is another kind of chaos that monists worry about. In one sense, this is a point about human psychology: scientists can only focus down on esoteric questions if they are not unduly distracted; monism is the best mind-set for this activity. This is a valid point, at least about *some* people's psychology. But such necessity for a narrow mental focus is quite compatible at least with a minimal sort of pluralism: it is OK for pluralism if individuals or groups pursuing their own systems of knowledge are monists at heart, as long as no one prevents anyone else from pursuing their own schemes (see Sect. 5.3.3 for more comments on this issue). That way, all the benefits of Kuhnian normal science can be had within each paradigm, while we retain multiple paradigms. Again, a political parallel may be helpful: a democratic society can let various individuals and groups pursue all sorts of outrageous views and activities, as long as they do not actively prevent others from pursuing their own. In this kind of situation, what we require is an overall social authority that enforces toleration; that is just the sort of thing for which we need a government.

What I have described so far is a minimal solution, and we can also go beyond it. Surely it is possible to train our minds to be more capable of switching between different systems of knowledge as needed for solving different types of problems? We have all learned to see the duck and then the rabbit, and then the duck again, in looking at the duck–rabbit; it is harder to learn to flip the Necker cube back and forth, but most of us can do it. And we can even learn to think simultaneously in terms of different systems. I have no hesitation in saying that I enjoy thinking about physical situations according to a whole set of different viewpoints. The work involved in the research leading to each of Chaps. 1, 2, and 3 of this book has been a journey of learning such pluralistic thinking. So I am by now quite adept at thinking about basic chemical reactions in both phlogistonist and oxygenist terms; I can easily switch between thinking of electrochemical reactions in terms of the shuffling of ions and in terms of electricity making compounds with other substances, or in both of those terms if I equate electrons with the negative electrical fluid. I can also think in all five systems of atomic chemistry described in Chap. 3, and follow various nineteenth-century chemists who employed different systems successively or in combination.

To take a more everyday sort of situation: I sit down on the cold floor and feel the flow of cold seeping into my body; but I also know that what's going on is a flow of caloric out of my body; at the same time I can calculate the rate of energy transfer that makes the molecules of the floor underneath me vibrate so much harder. Is there a problem here? Whence the fear of co-existence? There was a time when many immigrants to the United States did not teach their native languages to their children, for fear that this would confuse the children and retard their learning of English. This fear has largely disappeared, and bilingual upbringing is now widely considered a good and useful thing. If there were such a difficulty in learning and using two languages simultaneously, life would be impossible in places like Montreal. Language is only an imperfect metaphor for scientific thought, but it is very suggestive. In fact most of us know how to do conceptual frame-switching and frame-blending effortlessly, in scientific thinking as well as everyday communication: we view a photograph of the round earth floating around in space, full of admiration about the truth shown by modern science and technology—all the while standing on our feet firmly secured on what we think and feel is the immovable, flat ground.

Even so, the monists will object: pluralism may be fine in the realm of thinking, but at the point of action it has to be reined in, because we cannot act effectively if we get mixed up by taking our cues from different systems all at once. But this objection is based on a misconception. Yes, effective action does need to be coherent within itself, but this does not mean that all of the beliefs we ever hold have to be of a piece, or that we have to do all the things we do according to one and the same method. Rather, what we need to do is identify reasonably self-contained chunks of activity, to each of which we give as much coherence as we can, and within which we can act effectively. We will sometimes have to move from one such coherent domain to another, and there will be disruptions, but that is just part of life—this is what happens when we move house, change jobs, marry or divorce, lose a loved one, grow up and leave home, and so on through major life-events. There are distinct action-domains in scientific work, too: that is precisely what I mean by scientific "systems of practice"; after all, action and knowledge are not so separate from each other.

If there are different scientific systems offering different advice in the same domain of life, those who need to use the knowledge can and will make the choice. Plurality in science provides opportunities rather than hindrance, as long as those who apply science are willing to make their own judgments. For instance, ordinary people in various parts of Asia do not have a crisis about whether to use traditional or Western medicine; the choice is available to them, and the decision is up to them, for each particular ailment. One may go to the hospital or the acupuncturist for a sprained ankle; it is not uncommon for terminally ill patients to turn to traditional remedies; there is nothing incoherent about these decisions. Sometimes what we want or need is a case-by-case integration of different systems, as Mitchell recommends (see Sect. 5.2.3.1). A nice cutting-edge high-tech example of such integration is the global positioning system (GPS): by means of satellites kept in place by Newtonian physics, and atomic clocks ruled by quantum mechanics and corrected by special and general relativity, this system maps the spherical surface of the round earth on a geocentric grid (or rather, a geostatic grid), and gives advice to people on the ground from a flat-earth point of view. All of these choices and integrations are judgments to be made, enabled by a pluralist science. There is no reason to fear that pluralism will paralyze science or its application to problems of life.

5.1.5   Can We Afford It All?

Apart from the question of how much plurality scientists will be able to handle effectively, there is a more material kind of objection to pluralism: scientific research requires a great deal of time, money and talent, and it is not possible for society to support all lines of inquiry; so resources need to be pooled into one line of inquiry in each field. I have four layers of responses to this objection.

1. First of all, it is a simple logical fallacy to jump from admitting that we cannot afford *all* lines of inquiry to claiming that we can only afford *one* line of inquiry in each field. In fact I don't think that anyone who carefully considers this issue would argue in that way. In the modern era science is not so under-resourced, despite protests from many scientists. Surely there are enough resources to go beyond a strict monopoly; the pertinent question is just how pluralistic we can afford to go.
2. It may take a great deal of resources to pursue a line of inquiry in an intensive and focused way, but it usually does not take very much to keep them alive. Exploratory research is often very cheap, and all that is required may be salaries and academic freedom given to some unorthodox thinkers, or simply a little bit of kind encouragement to amateurs and enthusiasts who would devote their own resources to keep up their work.[14] Many great discoveries and inventions in

---

[14] On this point, and many other points about how research should be supported and financed, see Gillies (2008).

science, technology and medicine, at least up to the late nineteenth century, were
made with just these kinds of low-key support. Again, let's name some names:
Priestley, Dalton, Jenner, Fleming, Tesla, the young Edison, the young Einstein
working at the Swiss patent office, and so on. There is no reason why that kind
of work should stop completely even in this day and age, especially in theoretical
work. I remember once reading a plea from James Lovelock, the author of the
Gaia hypothesis, asking just 1% of the national science budget to be given to all
the unorthodox schemes. This is not a crackpot suggestion, whatever one might
think of Lovelock's own particular scientific ideas. In fact official funding agencies
have been entertaining similar thoughts. Consider the U.S. National Institutes of
Health (NIH) Director's Pioneer Award (NDPA) Program, whose intention is to
support highly innovative and paradigm-changing research proposals; it takes up
a small portion of the NIH budget, but it could make a real difference in outlook.[15]
The U.K. government's Engineering and Physical Sciences Research Council
(EPSRC) is another example: its "Transformative Research" funding line is
designed for "stimulating creativity and adventure" in research.[16]

3. In concentrating resources on a particular line of inquiry, we may well hit points
   of diminishing returns. I think that modern scientists have actually tended to put
   too much investment into monopolistic lines of work. Have we, for example, got
   a good return for our investment by pushing so much of the best talent in theoretical physics in recent decades into string theory? Is the current concentration
   on synthetic pharmaceutical research the most productive investment for health
   and well-being? Maybe, maybe not. What seems clear is that having too many
   people trying the same fashionable approach can be actually wasteful. Popper
   (1981, 96) already lamented that "too many dollars may chase too few ideas." So
   we always need to be asking whether we are making the maximally efficient
   distribution of resources into different lines of work. We can and should question
   whether we can afford pluralism, but we should also be asking whether we can
   afford monism.

4. It is a pessimistic fallacy to assume that the amount of resources that society
   devotes to science will remain the same and the only thing we can decide is
   how we distribute the fixed amount of resources. Science is not a zero-sum game.
   If we inspire people, we will increase both the number of people going into
   science and the amount of funding that public and private institutions are willing
   to give to science. Currently it does not seem that scientists in any nation are
   doing very well in inspiring young people to come into science or in getting the
   public to develop a true appreciation of science, despite some strenuous efforts.

---

[15] See the scheme notes (http://commonfund.nih.gov/pioneer/) and also the NIH's own assessment
of early results, *Outcome Evaluation of the National Institutes of Health (NIH) Director's Pioneer
Award (NDPA), FY 2004–2005* (https://commonfund.nih.gov/pdf/Pioneer_Award_Outcome%20
Evaluation_FY2004-2005.pdf), both documents last accessed on 4 October 2011.

[16] See the scheme notes (http://www.epsrc.ac.uk/funding/grants/network/ideas/Pages/default.
aspx), last accessed on 4 October 2011.

Is it absurd to think that this is because monism has made science narrow-minded and stultifying? Is it impossible that we might interest a wider variety of people in science if we provided a plurality of ways of learning and practicing science, in a more open-minded fashion? It is true that a more pluralistic science will be mired in more debates, some of which may turn out to be fruitless, and this can be considered a waste of resources. But won't the presence of some lively arguments actually get more people excited about science? You say we don't have time for such wrangles? We do have time—on the whole, people in the developed countries have plenty of time. Yes, some of us are very busy, and some others of us can't afford any time to do anything but make a living and look after each other, but think about all the time people spend in playing computer games, gossiping, and watching reality shows on TV. In terms of resources, consider what a staggering amount we devote to war and other forms of destruction. It cannot be denied that a great deal of time, effort and money would be required in establishing and maintaining a pluralistic system of science. But I can hardly think of a more worthwhile and necessary long-term investment for the future of human culture.

5.2 Benefits of Plurality, and How to Attain Them

5.2.1 What Is Pluralism?

It has been said that there are as many pluralisms as there are pluralists. While that may be quite appropriate in a reflexive way since pluralists celebrate diversity, it is certainly confusing. I have given a brief definition of pluralism in the first section (Sect. 5.1), but for the systematic discussion that I wish to give in this section of the chapter, I must begin with a more precise statement of the kind of pluralism that I am advocating. Pluralism as I mean it is an ideology of science aimed at promoting plurality in order to reap its benefits. To give a more informative label, I will designate my position as *active normative epistemic pluralism*.

My position is "epistemic", as opposed to "metaphysical", in the sense that it is aimed at improving the ways in which we go about acquiring knowledge, rather than at elucidating the fundamental ontology of nature. Without denying the inevitable linkage between epistemology and metaphysics, I want to show that there are strong arguments for epistemic pluralism, almost no matter what the world is really like. So, for example, while accepting Sandra Mitchell's (2003, 2009) view that the complexity of the biological, ecological and social domains demands a pluralist methodology, I don't want to tie the general arguments for pluralism too strongly to the special complexity of those domains, which would make them inapplicable to much of the physical sciences. Similarly, while I do not share Nancy Cartwright's (1999) positive conviction about the ontological "dappledness" of the universe, I think the kind of pluralist epistemology she advocates still has sufficient justification. (See Sect. 5.3.2 for more on this issue.)

The pluralism I advocate is unapologetically *normative*. I do have great sympathy with descriptive pluralism, which argues that in fact scientific practice has been more pluralistic than often imagined. But the descriptive thesis is not my main focus. Besides, the normative argument is largely independent of the descriptive: my position is that if we should find a field of science which is quite monistic, then that is quite likely not healthy, and we should consider reforming it. A normative argument must begin with the clarification of the relevant aims and values; putting a clear emphasis on this point is the first contribution I would like to make to the development of pluralism. And I want to cast my axiological net widely, to argue that pluralism is more beneficial to science than monism, given any reasonable position regarding the aims of science and the fundamental values operating in science. In this method of argument, I am inspired by Feyerabend's declaration that "anarchism helps to achieve progress in any of the senses [of progress] one cares to choose." (1975, 27) But unlike Feyerabend, I want to do this in a systematic fashion, by surveying all the various things that one might think science should desire to achieve. (Here I am using "aims" and "values" almost synonymously; this is not ideal, but not so problematic if we take the achievement of something valued as an aim.)

Also, going beyond the value judgments of a spectator, I advocate pluralism as an *active* stance. A passive version of normative pluralism would simply point out the benefits of having multiple systems of practice in a given area of science. Active pluralism engages in actually cultivating multiple systems. The pursuit of active pluralism has clear implications about how we should practice history and philosophy of science, about which I will say more in Sect. 5.2.4.

There are two broad categories of potential benefits of plurality: *benefits of toleration*, and *benefits of interaction*. Accordingly, there are also two different categories of pluralism, focused on achieving each type of benefit. Sections 5.2.2 and 5.2.3 will discuss these benefits in more detail, thereby providing arguments in favor of pluralism. The benefits of toleration arise from allowing the flourishing of multiple systems of practice, each making its own distinct set of contributions. Such plurality can be achieved either by spontaneous mutual toleration, or by a more centralized structure sanctioning diversity; realistically, probably a bit of both is needed. The main feature of *tolerant pluralism* is quite simply to allow different systems to co-exist, with respect and toleration for each other. It is not required that the different systems should have any interaction with each other, and the practitioners of each system may even be strong monists (see Sect. 5.3.4 for more on this point); what is important is that each system is allowed to exist and pursue its potential. Going beyond toleration, *interactive pluralism* also seeks benefits from having different systems interact with each other, rather than standing separately and delivering separate sets of contributions. For the benefits of interaction to be realized, additional conditions are necessary: at least some of the parties involved need to be pluralist in their outlook, so as to be willing to engage with others or at least use others' works productively; some level of common language is also required, to allow sufficiently insightful communication. If the slogan for tolerant pluralism is "Let a hundred flowers bloom", interactive pluralism says: "Yes, let them all bloom, and also cross-fertilize."

5.2.2   Benefits of Toleration

There are various types of benefits of toleration, appropriate for various different purposes. I will describe four types of benefits.

5.2.2.1   Hedging the Bet

It is a common conception that science has only one ultimate aim, and that this aim is Truth ("with a capital T"), which is objective and univocal. In Chap. 4 I made an argument against that vision of science (and see Sect. 4.3.1 in that chapter for a detailed discussion of the meanings of "truth"). Setting that discussion aside for the moment, I want to show that pluralism is a more productive strategy than monism even for those who regard Truth as the aim of science. The basic point is simple: since the course of scientific development is unpredictable, it will be helpful to keep multiple lines of inquiry open in the hope that one of them will lead us to the right answer.

The most obvious difficulty with the search for Truth is that we can never be sure whether we have got it, or if we are even approaching it. The history of scientific progress shows that today's favored path may not remain the most promising one tomorrow. As I've discussed in Chap. 4, Sect. 4.2.2, Laudan's (1981) pessimistic meta-induction from the history of science does point up the basic insecurity of our theoretical positions in science. Kyle Stanford's (2006) "problem of unconceived alternatives" has the same disturbing effect on any alleged security about the the-ory-choices that scientists make. From recent science, the case of prions as the cause of "mad cow disease" and other brain disorders provides a very vivid example of an unconceived alternative; prior to prions all known infectious agents were organisms containing genetic material.[17] The unpredictability about the direction of scientific development was already emphasized by Kuhn (1970, 206–207):

> I do not doubt, for example, that Newton's mechanics improves on Aristotle's and that Einstein's improves on Newton's as instruments for puzzle-solving. But I can see in their succession no coherent direction of ontological development. On the contrary, in some important respects, though by no means all, Einstein's general theory of relativity is closer to Aristotle's than either of them is to Newton's.

In Chap. 4 I have argued against the standard scientific realist moves attempting to show that the most important elements of scientific theories have been securely preserved through developmental upheavals. But even if there were such security, it would do nothing to prevent the unpredictability of future developments in which entirely new facets of reality are revealed. For example, the survival of the equations of Newtonian mechanics as a limiting case of special relativity in a low-velocity regime did nothing at all to preserve the cogency of the concepts of absolute space and time in the new theory. The overall success of Newtonian mechanics was no

---

[17] See the instructive historical, sociological and philosophical analysis by Kiheung Kim (2006).

guarantee that the subsequent best approach to Truth would not lie in an entirely different line of inquiry.

Faced with an insurmountable unpredictability, what rational agents have to do is clear: hedge our bets. Given that we do not know which line of inquiry will ultimately lead to our destination, we should keep multiple lines open, instead of pursuing one line faithfully to its dead end, only then to try a different one. To put it in Bayesian terms, all theories with non-negligible prior probabilities should be monitored for signs of life (that is, increases in posterior probabilities) as further evidence comes in. Why Bayesians don't usually think like that is not clear to me. It is most irrational to insist that only the theory with the highest probability *at the moment* should be preserved and all others killed off. Once eliminated and forgotten, avenues of inquiry will be very difficult and costly to re-invent (consider, as I discussed in Chap. 1, Sect. 1.2.4, how long it took chemists to get back on to some productive lines of inquiry that the preservation of phlogiston would have facilitated much more easily). To use a home-spun analogy: if we are looking for someone lost in the wilderness and we don't know which direction he/she has gone, would we round up all the available people into one search party and send them in the direction we conjecture to be most likely? Or would we spread people out a bit? (One might object that maintaining multiple lines of inquiry is not plausible in the world of limited resources, but I have already dealt with this objection in Sect. 5.1. Most of all, it is important not to close our scientific minds so much as to think that certain ideas have been forever discredited and cannot come back in any way whatsoever. This reasoning applies to all possible aims concerning future attainments, not just to ultimate Truth. For example, if empirical adequacy is our aim, we should pursue various lines of inquiry because we do not know which line will deliver the most empirically adequate theory in the end (recall that van Fraassen's definition of empirical adequacy includes being correct about all possible observations, including future ones.)

5.2.2.2 Division of Domain

Hedging the bet is a strategy aimed at an ultimate outcome. Not everyone thinks that it is meaningful to think in terms of such clear end-points in science, when definite and good-for-ever verdicts will be given. Whatever we may think about the "end" of science, I think everyone would agree that science does need to deal with the here-and-now, whether that is just while we await the scientific Judgment Day or because the here-and-now is all we are ever going to have. In the here-and-now of science, ultimate truth is not an operable aim at all. Instead of Truth, we pursue aims whose achievement we can actually assess as we go on, and ideals toward which we can strive knowing whether or not we are getting closer to them, even if we cannot actually reach them. When we consider the ongoing satisfaction of scientific aims, there are further arguments for pluralism that apply in addition to that arising from the unpredictability of future development.

One obvious argument for pluralism is that different systems may fulfill a given aim partially, so that jointly they serve that aim more completely than each one can. Another way of thinking about this is in terms of a beneficial division of labor among co-existing systems serving the same aim in different domains.[18] This is very easy to see, for example, in relation to empirical adequacy. At the large scale, we have a division of labor between different sciences; even within a given science, different systems (even mutually incommensurable ones) are used to cover different phenomena, or different aspects of the same phenomena. If our best theory covers only *some* of the known observable phenomena, then we need to have other theories that cover the rest. Scientific development proceeds in this pluralistic pattern more often than people often realize.

I have given extensive discussions of two such pluralist developments, in electro-chemistry and atomic chemistry, in Chaps. 2 and 3 of this book. To take another example, much more briefly: it is a myth that classical mechanics is no longer necessary because it has been reduced to quantum mechanics. We still need classical mechanics to handle what was always its domain, and this is not just a matter of practical convenience, arising from the lack of adequate mathematical techniques or computing power for solving enormously complicated equations. We actually cannot set up the Schrödinger equation in any sensible forms for macroscopic rigid bodies or fluids (not to mention not knowing how to combine it with general relativity properly). Even if we pretended that a small object was an extension-less and structure-less point-mass, quantum mechanics would not be able to handle the precise and simultaneous specification of its position and momentum that classical mechanics requires for writing and solving any equations (this is one place where semantic incommensurability really comes in).[19] It makes sense that modern physicists have retained classical mechanics (in all its various branches), special relativity, general relativity, ordinary quantum mechanics, quantum field theory, and beyond.

More generally speaking, it has always been the case that even our best scientific theory fails to cover all phenomena. If we are realistic, we should be prepared for the possibility that there will always be need for a plurality of theories to deliver the best possible empirical adequacy. If we concern ourselves with the here-and-now, that need for plurality is present and clear. It will not do to reject all theories while we wait for the single theory with complete coverage; in van Fraassen's terms, that would be to privilege simplicity (a mere pragmatic virtue) over empirical adequacy. And there is no absolute reason why even logical consistency should be placed above empirical adequacy. As Putnam puts it, for the sake of predictive ability,

---

[18] For a sophisticated discussion of the cognitive division of labor, see Kitcher (1993), chapter 8. My points are much more simple-minded.

[19] Ehrenfest's theorem comes in to tell us that the average values of dynamic variables, as evolving according to quantum mechanics, obey classical laws, but that leaves the setup of the classical problem entirely unspecified as to how to distribute the overall uncertainty between position and momentum. See Chang (1995, 1997) for a discussion of some pertinent cases.

plurality of theory would be best "even if their conjunction were not consistent" (1995, 14). What Putnam does not point out is that it is often only the hubris of presumed indefinite applicability that makes us think that there is any meaningful conjunction at all between these theories.

All of the above is an in-principle matter; if we actually consider pragmatics, we will also see that even if one system can in principle cover a whole domain, different systems are used to deliver empirical adequacy conveniently and effectively throughout the domain. In classical mechanics we often rely on the Lagrangian or Hamiltonian formulations for a simpler solution of a problem that would be very tedious to handle in the original Newtonian formulation. In quantum mechanics, there are some problems that are most easily solved using the Schrödinger wave equation, and others that are most easily solved using the Heisenberg matrix formulation. In my own work (1997) I have once encountered a curious case of a very simple problem (an electron going up a linear electrostatic potential slope) that is practically insoluble except by going to the Feynman path-integral formulation. In that respect, pluralism in science is just as natural as wanting to have various types of tools in our toolbox, or having different types of shoes in our cabinet to suit different occasions.

The applicability of this type of pluralist arguments, based on the division of domain, is not restricted to concerns of empirical adequacy. A domain-based division of labor can be beneficial in the pursuit of any aim whose satisfaction can come piecemeal, or at least to different degrees in different places. For example, different idealized models may provide good intuitive understanding in different domains, where no single model can provide it everywhere. Different theories may provide unification in relation to different aspects of phenomena. Different measurement methods may deliver high precision at different points on the scale of a quantity; this is why, for instance, the official International Scale of Temperature is a patch-work relying on different standards in different parts of the scale (see Preston-Thomas 1990). And different approximations may deliver greater simplicity in different domains. In all of these cases, our aims can best be satisfied overall by maintaining multiple systems, each of which works well within a limited domain and not so well outside of it.

5.2.2.3 Satisfaction of Different Aims

So far, in Sects. 5.2.2.1 and 5.2.2.2, I have considered what I call "uni-axial" regimes of science, with one value overriding all others in each regime. I have argued that a given aim of science may be served best by a simultaneously co-existing multitude of systems of practice. Now I want to open up my thinking further by considering "pluri-axial regimes", in which there are multiple legitimate values/aims that drive scientific work. There is no convincing reason to think that science has only one overriding value or aim. While Kuhn privileged problem-solving ability as the key value in some places (e.g., Kuhn 1970, 205), in other places he gave us the oft-cited list of accuracy, simplicity, consistency, fruitfulness, and scope (e.g., Kuhn 1977, 322). Van Fraassen (1980, 87) identified a whole list of pragmatic virtues including elegance, simplicity, completeness, unifying power and explanatory power, and I don't think he gave a convincing reason for regarding these as secondary in importance to empirical adequacy.[20] Each and every scientist is driven by a set of epistemic values simultaneously. To pretend that they are or should be devoted to the pursuit of only one epistemic value would be foolish. And when we consider a whole scientific community, it is clear that there will be a whole range of values that shape and guide the collective actions and decisions of the community, not to mention the values of external agents who affect the community.

In a pluri-axial regime, all of the arguments for pluralism given so far still apply, in relation to each value. In addition, there are arguments for pluralism that arise from the multiplicity of values. Once we grant that there are multiple human needs that science is called upon to satisfy, it is easy to recognize that we will most likely not be able to come up with *the* perfect scientific system that satisfies all needs. Call it pessimism, but I do not think it is unwarranted pessimism. As discussed in the first section (Sect. 5.1), I would rather think of it as reasonable humility concerning human ingenuity, or a recognition of the complexity of life and nature. So there is a type of division-of-labor argument for covering the spectrum of values, in addition to the earlier division-of-labor argument for covering the whole domain of phenomena in relation to a particular value.

In each of the episodes investigated in Chaps. 1, 2, and 3 of this book, there was a divergence of values and aims that contributed to disagreements between groups of scientists. Generally philosophers get worried about such lack of consensus, but I am suggesting that we look at such situations in a happier light, as those in which various values and aims can all be satisfied *thanks to* the existence of multiple systems, and different people can choose different systems to practice depending on their personal priorities. In Chap. 3 we saw different groups of atomic chemists prizing different aims such as literal truth, explanatory power, empirical adequacy, and convenience of classification. In Chap. 1 we saw a contrast between Lavoisier's pursuit of theoretical elegance and Priestley's commitment to empirical comprehensiveness.

Even when people profess to share a common value or aim, they might actually mean quite different things by the same name, so in effect there are different values at play; this was another point stressed by Kuhn (1977, 331). It is perhaps the easiest to see this point with respect to the aim of understanding. Although unfashionable among analytic philosophers, understanding is still often cited by scientists themselves as the ultimate aim of science, so it makes sense to consider it here.[21] Different people will derive understanding from different types of systems of knowledge. As Duhem infamously put it, the "ample and weak" mind of the English physicist could only understand something if a mechanical model could be made of it; the "strong and narrow" mind of the French physicist derived all the necessary understanding

---

[20] See Chap. 4, Sect. 4.2.3, for more on this.

[21] What I mean by "understanding" incorporates any kind of explanation that goes beyond mere subsumption. De Regt et al. (2009) is an excellent collection of recent work on various dimensions of scientific understanding.

from formal mathematical systems, with no need for childish models (Duhem [1906] 1962, 64ff). But if the overall aim of science is the greatest understanding by the greatest number of people (including even English people), then pluralism is needed since it is not likely that any single system will be able to provide this sort of intuitive understanding to everyone. With other aims, too, there would be different renditions given by different people. More generally, whenever there is a subjective element to an aim, we will have divergence in its pursuit. For example, simplicity was rendered as the economy of thought by Ernst Mach, and as formal elegance by the likes of Albert Einstein and Paul Dirac.

Again, in earlier chapters we have seen various instances illustrating the divergent manners of satisfying a given aim. For example, the electrostatic explanations of chemical combination, so popular with most who were interested in physical mechanisms behind chemical phenomena, did not appeal to Faraday at all. Various parties wanted theoretical unity in electrochemistry in various ways: Clausius wanted reduction to physics, Davy desired fewer elements, and Berzelius aimed to place all substances on the spectrum of electropositivity and electronegativity (Chap. 2, Sect. 2.2.3.2). In the Chemical Revolution, both oxygenists and phlogistonists argued that their own system had a superior kind of unity, systematicity, and empiricism (Chap. 1, Sect. 1.2.1.4).

The case of quantum mechanics also makes a very instructive episode in illustrating the divergence of scientific aims and the divergent manners of their satisfaction.[22] First of all, consider the contrast between those physicists who worry about the interpretation of quantum mechanics, and those who don't. The former group is deeply concerned with both empirical adequacy and understanding, and the latter group with empirical adequacy and not so much with understanding (or at least they are content with a structural or mathematical kind of understanding). Among those who concern themselves with interpretations, there are widely divergent notions of understanding at play, scattering people over the whole spectrum of interpretations ranging from the fully deterministic picture given by David Bohm's reformulation of the theory to the entirely non-causal mode of Hugh Everett's relative-state interpretation (or the many-worlds version of it due to John Wheeler). The interpretive debates and disagreements were already very present in the jostling between Heisenberg and Schrödinger with their alternative formulations of the original quantum theory, and there is little sign that these disputes are going to cease.

5.2.2.4 Multiple Satisfaction

So far, my arguments for plurality have all been based on some sort of limitation on the part of human inquirers.[23] In Sect. 5.2.2.1 plurality was recommended for prudence in the face of our inability to predict the future course of scientific development.

---

[22] Many of these issues are discussed in a convenient and accessible recent collection by Evans and Thorndike (2007).

[23] In this vein, I regret that I have not been able to build on William Wimsatt's (2007) vision of "re-engineering philosophy for limited beings", as yet.

In Sect. 5.2.2.2 plurality was prescribed as a remedy to the limited applicability of our systems of knowledge. In Sect. 5.2.2.3 plurality was presented as a way of satisfying diverse needs, in the face of our inability to come up with a single solution satisfying all of them.

Now I want to give a more exuberant and less defensive argument for plurality, whose main point is that plurality enriches knowledge.[24] Even when one system can serve our aims quite adequately, other systems may also be able to serve the same aims, in new ways. Such epistemic abundance should delight us. Even if our aim is truth and even if we have attained it, we can still demand more pluralist abundance. Science does not have to end even if we have the true theory of the universe—we can try to make another one! Who says that two truths about the same subject matter have to be exactly equivalent to each other?[25] All that logic requires is that two true theories do not directly contradict each other.

In general, even if we have a system of practice that satisfies a certain aim quite well, we can always benefit from adding another system that satisfies the same aim in a different manner. For example, we can enjoy pluralist abundance in understanding: this is not just a matter of different people having different "taste in world-making" (Thomas Beddoes, quoted in Knight 1967, 28). In a mature state of science, one and the same person can appreciate and enjoy different types of understanding of the same phenomena drawn from different systems of practice. So, I like knowing how to frame and solve the same problem in Newtonian, Lagrangian, and Hamiltonian formulations of mechanics (and would add the Hertzian formulation to my arsenal if I had the time).[26] I want both teleological and mechanistic accounts of the same phenomenon, whether it be the propagation of light or the development of embryos. It delights me to learn that the phlogiston theory could explain exactly all the chemical phenomena that Lavoisier explained according to his own theory, and that Kepler's laws of planetary motion could be derived from a version of Descartes's vortex theory.[27] Such delight does not come from abandoning Lavoisier's and Newton's theories, but from learning additional ways of making sense of things. Give me the Heisenberg, Schrödinger, Feynman, *and* Bohm versions of quantum mechanics: so many

---

[24] See the end of the first section in Chap. 4 (Sect. 4.1.4) and Sect. 4.2.5, for a detailed account of how I conceive the meaning of knowledge.

[25] The point is illustrated perfectly by a joke, for which I thank Elva Siglar: "*Teacher:* Clyde, your composition on 'My Dog' is exactly the same as your brother's. Did you copy his?" "*Clyde:* No, Sir. It's the same dog." We laugh about this, but when it comes to science we tend to get very ceremonious in dismissing any suggestion that there might be two different and equally valid and good stories about the same object.

[26] Heinrich Hertz (1899) himself famously noted that these formulations were equivalent yet gave different "images" of mechanics and had distinct uses; see also the discussion of this point by Ian Hacking (1983, 143).

[27] On the latter case see Shea (1987), 166ff, where he discusses the work of Johann Bernoulli and Joseph Privat de Molière in this context, around 1730 shortly after Newton's death.

different ways of appreciating the physical world, more windows on nature, more enriched understanding of it. Why is this any worse (or better) than having galleries full of the crucifixion of Jesus Christ depicted in so many different ways by so many wonderful artists?

Contrast this pluralist attitude to Albert Einstein's famous dissatisfaction over having two parallel explanations of electromagnetic induction, which he voiced at the very beginning of his 1905 paper on the special theory of relativity: "The observable phenomena here depend only on the relative motion of the conductor and the magnet, whereas the customary conception draws a sharp distinction between the two cases in which either the one or the other of these bodies is in motion." (Einstein, in Miller 1998, 370). It is difficult to deny that this sort of consideration must have played a positive heuristic role in spurring Einstein on to the principle of relativity, and on to the rest of the special theory of relativity. However, I do not see why having two parallel explanations was so intolerable as Einstein made it out to be. It rather seems to me a wonderful luxury to have two different ways of understanding the same phenomenon, each of them cogent in itself. Any cheap philosophical lesson one might draw from the special theory of relativity itself would, in the end, be a pluralist one: descriptions and explanations given in any inertial frame of reference fully make sense; there is no fact of the matter about which frame of reference is correct, so we can say that the coil is moving or that the magnet is moving, and either way will be fine and coherent. The actual philosophical lesson I would like to draw from the history of relativity theory is somewhat different, but still a pluralist one: the coming of special relativity did not eliminate explanations of electromagnetic phenomena situated in absolute space and time. Classical electromagnetism still serves very useful descriptive and explanatory roles in physics and engineering, and that exists side-by-side with relativistic descriptions and explanations. Each provides useful understanding. Einstein was actually careful enough to say that the postulation of the ether will be "superfluous" (Einstein, in Miller 1998, 371), not that he had proven its non-existence. What monists denigrate as superfluousness, pluralists may celebrate as abundance.

Even if we consider aims that are not so apparently subjective as intuitive understanding, there are benefits of having a plurality of systems all delivering the goods in the same area. There is an optimistic side to the underdetermination of theory by evidence: the same body of observations may be accounted for by multiple theories—how nice that we can have so many good theories! Even factual learning itself can be achieved in multiple ways, if there is observational incommensurability creating distinct bodies of observations about the same domain of nature. In that case we would want to have different theories that can cover the different bodies of observations, so we can have empirical adequacy in all those ways. Some may deny that there is such incommensurability in observations, and I do not wish to rehearse all of Kuhn's arguments to the effect that there is, but I should highlight what seems incontrovertible among Kuhn's arguments. First, it is practically true (even if avoidable in principle) that observations that do not fit prevailing assumptions tend to be obscured, ignored, or explained away (e.g. how pre-Copernican European astronomers, assuming the immutability of the heavens, recorded no novae or supernovae, while

Chinese astronomers recorded them freely). Second, facts collected in one context will not be straightforwardly translatable for use in another context (e.g. statistical data gathered in a previous study, without the particular controls one now wants). Making one observable properly accessible may even make another one inaccessible. Third, observations have different degrees of significance depending on the theoretical frameworks in which they are fitted. For example, consider the precise measurements of atomic weights, which the radiochemist Frederick Soddy (1877–1956) described as "the life work of that distinguished galaxy of nineteenth-century chemists, rightly revered by their contemporaries as representing the crown and perfection of accurate scientific measurement". But after the discovery of isotopes, it was recognized that the numbers so carefully determined were just average values reflecting the rather random proportions of the various isotopes of an element that we happen to have on earth. Soddy, writing in 1932, lamented this as a tragedy: "Their hard won results" now appeared to be "of as little interest and significance as the determination of the average weight of a collection of bottles, some of them full and some of them more or less empty."[28]

The overall consequence of these aspects of observational incommensurability is that each paradigm (or system of practice, more generally) will tend to elicit and highlight its own distinct set of observations, to reveal and retain different facts about nature. This multiple-satisfaction pluralism about factual learning should be distinguished from the division-of-domain pluralism discussed in Sect. 5.2.2.2 above, in which case there is a commonly accepted body of observations which none of the available theories can cover completely. The present argument is that even when we have an adequate body of observations about a given domain of phenomena which are adequately covered by a theory, we should want to go further and produce an *abundance* of observational knowledge. Recall that the production of more and more observations is an important aim of "active realism" as I characterized it in Chap. 4. It is also an aim implicitly encouraged by constructive empiricism; although van Fraassen's focus was on the construction of empirically adequate theories, constructive empiricism in practice would be seriously crippled if it were not accompanied by an effort to increase the stock of observations against which to test the candidate theories. Now introduce incommensurability, and a pluralist position emerges: especially if it is the case that each system discourages or even precludes certain observations while facilitating others, then the pluralist recommendation is to *adopt all of those systems*. I think that was the path taken, for example, with the multiple operationalizations of the concept of "atom" (Chap. 3, Sect. 3.2.1)—by equivalent weights, combining weights, combining volumes, specific heat, and electrolysis. In such a way we can engage fully in the overall enterprise of learning about reality in all possible ways.

---

[28] I take these quotations from Lakatos (1970, 140), from the end of his discussion of the Proutian research program. The original source is Soddy, *The Interpretation of the Atom* (1932).

5.2.3 Benefits of Interaction

So far I have explored the benefits of a mere toleration of plurality, allowing various systems of practice each pursue its various aims in its own way. These benefits are very significant. However, the full force of pluralism can only be appreciated if we also consider the benefits arising from the interaction of different systems with each other. (The content of this section is programmatic, as it is a topic on which I still need to do much further work.[29]) Recall Peirce's metaphor quoted above in Sect. 5.1.1, which extols the strength of "a cable whose fibers may be ever so slender, *provided they are sufficiently numerous and intimately connected*". As Peirce pointed out, a chain is only as strong as its weakest link. Tolerant pluralism is like a bundle of strings, which is as strong as its strongest strand. But a real cable, a perfect image for interactive pluralism, is actually stronger than its strongest strand, due to the productive interaction between the strands. Metaphor aside, there are different types of benefits of interaction arising from different types of interaction, three of which I will outline below.

5.2.3.1 Integration

At least one key aspect of interactive pluralism has been articulated in some detail in the literature. There are situations in which not one of the available systems by itself, not even all of them additively, can achieve a certain aim. In such cases we may attempt to reach a better result by an *ad hoc* integration of different systems; this integration is by definition *ad hoc*, because if it weren't, then we would have a unified system rather than a pluralistic integration of different systems. Mitchell's (2003, esp. ch. 6) "integrative pluralism" spells out in convincing detail this kind of benefit: she especially considers biological systems, such as communities of social insects, whose great complexity precludes monistic explanations. This is similar to Otto Neurath's modest recipe for the unity of science, which recognizes the existing disunity in science but insists on "unity at the point of action" (see Cat et al. 1996, and references therein). A perfect contemporary example of such *ad hoc* integration at the point of action is the global positioning system (GPS), which I discussed briefly in Sect. 5.1.4.

In the history of atomic chemistry discussed in Chap. 3, there is also a good example of integration: the five systems of atomic chemistry that I described there were idealizations and most actual chemists seemed to mix-and-match between the different systems. That integrative aspect comes out most strongly in the table of atomic weights. Those engaged in an extensive enough practice of atomic chemistry, especially anyone writing a systematic textbook or treatise on the subject, had to work

---

[29] I would like to thank Sabina Leonelli and other colleagues in the "PPP" (Pluralism, Pragmatism and Phenomenology) reading group in London for pushing me to think more about the benefits of interaction.

out and present a whole set of atomic weights (or equivalents, as they may have called them). Each person made his own judgment about how best to fix all the atomic weights in the face of considerable uncertainty, and often helped himself to whatever seemed the most plausible elements of each idealized system, integrating them all into his own practice. Berzelius was perhaps the master artist of integration here: his analytical work was initially grounded in the weight-only system, but he made use of volumetric atom-counting in some crucial places (such as the fixing of the H_2O formula for water and the atomic weight of oxygen as 16); on top of that, electrochemical reasoning was of course an important ingredient in his practice; we have also seen how Berzelius adapted his system by accommodating findings from the substitution–type system by means of his new idea of copulas; he even used some geometric-structural thinking, as he was the early pioneer of the investigation of allotropy and isomerism.[30] So it seems that Berzelius integrated all five systems of the atomic chemistry of his day! The shape of Berzelius's practice is similar to the eclectic and extensive clue-hunting that Dmitri Mendeleyev did in constructing his periodic table.[31]

Galison's "intercalated" picture of physics (1988, 1997, ch. 9, esp. 799), briefly mentioned in the first section (Sect. 5.1), gives a similar message about integration, with an even more positive take on plurality. With Mitchell and Neurath, one gets the sense that they would prefer a fully and generally unified account, if one could possibly be had. Galison, on the other hand, argues that the strength of physics as a system of knowledge lies in its composition from independent strands (theory, experiment, and instrumentation), each of which develops with its own internal dynamics. At each point in time, the integration between the strands has to be worked out anew, as each strand develops independently. If there is a discontinuity in one of the strands, that does not disrupt the whole of physics because other strands continue, and interact with the discontinuous strand in such a way as to prevent it from either disintegrating or becoming dissociated from the rest of physics. Galison's observation on the strength of this type of structure can be extended to any other field of study comprised of distinct yet interacting systems of practice. Mitchell is concerned with integrating different instances of the same type of thing (theoretical models, mostly); Galison is talking about integrating different types of activities within physics; for Neurath, action brings entire fields of science together. Yet the same general lesson about the benefits of integration can be drawn in all cases.

5.2.3.2   Co-optation

Even when different systems of practice are not being pulled together to achieve a specific aim, one system can be helped in its development by the use of ideas and results taken from another. In order to help achieve one's own aims, one may co-opt

---

[30] On the last point, see Freund (1904), ch. 18. More generally and extensively on Berzelius's chemical system, see Melhado (1980).

[31] On Mendeleyev, see Scerri (2007) and Gordin (2004).

various types of elements from another system of knowledge: empirical results, theoretical ideas, mathematical techniques, instruments, materials, etc. Beneficial co-optation can happen without direction from the top, or even any meaningful two-way communication; even genuine incommensurability need not get in the way. Co-optation is a fact of life in science that is both pervasive and unpredictable (see Holton et al. 1996). The most salient case of co-optation I have treated in earlier chapters is the use of phlogistonist results by Lavoisier and his colleagues, discussed in Chap. 1. Lavoisier would not have arrived at his new chemistry without co-opting phlogistonist experimental results such as Priestley's discovery of oxygen and Cavendish's synthesis of water, as well as various experimental techniques of pneumatic chemistry developed by phlogistonists. Liebig (1851, 26) even claimed that Lavoisier had discovered *nothing* new:

> He discovered no new body—no new property—no natural phenomenon previously unknown.... all the facts established by him were the necessary consequences of the labours of those who had preceded him. His merit, his immortal glory consisted in this—that he infused into the body of the science a new spirit; but all the members of that body were already in existence, and rightly joined together.

Liebig's claim notwithstanding, the traffic of co-optation was not entirely one-way. Later on in the game, various phlogistonists did take up the Lavoisierian idea that oxygen combined with the combustible substances in combustion, while phlogiston left them; this allowed them to make more sophisticated versions of the phlogiston theory which were able to explain the weight-gain in combustion without difficulty. This is not to denigrate either of the two sides, or to claim that they could *never* have done what they did without the co-optations. Rather, I think we should simply acknowledge help where it was given or taken, and also recognize that without such help the developments would have been more difficult to make.

Still, it may be objected that useful elements of knowledge *can* and *should* be created within one's own system. Why should they need to be co-opted from another system? The general answer is that each system develops under certain constraints, which may prevent the production of elements that would actually help *its own progress*. Thinking back to the Lavoisier–Priestley dynamic discussed in Chap. 1, we may wonder why it is that the ingenious and diligent Lavoisier did not think of heating up a calx to release something from it. I do not have a clear answer to that question.[32] What is clearer is why Priestley did think of it: in his program of pneumatic chemistry, the input and withdrawal of phlogiston was one of the standard methods of producing new gases by transforming previously known materials, so he tried all sorts of experiments in that vein, and the red calx of mercury happened to be one of the many substances he tried out. One could say that Priestley's program was well suited for producing previously unknown substances, and Lavoisier's was not. But once the Lavoisierians co-opted important new substances such as oxygen, they knew better than Priestley how to use them in theoretical developments because

---

[32] As calxes were formed in the first place by heating metals, it may have been thought that further heating would not do anything interesting.

they had a stronger framework to drive the theoretical thinking. Something similar can be observed in early electrochemistry, too, as hinted in Chap. 2. Volta's battery was co-opted with spectacular success by those who did not accept his theory at all; however, it has to be admitted that what led Volta to the invention of the battery was his theoretical notion that contact between two different metals in contact with each other excited electrical action. The "chemical theorists" of the battery opposed Volta's theory with many cogent arguments; however, it is not likely that they would have invented the battery in the first place.

It needs stressing that co-optation is not such a straightforward process as using a result that someone else working in the same system was quicker to get but one would have come up with oneself anyway. Co-optation usually involves some sort of incommensurability, which has to be overcome before the alien element can be integrated into one's own system with benefit. For example, the chemical theorists of the battery had to rearrange Volta's battery physically and mentally before they could work with it sensibly, as Kuhn (2000, 23) points out. In Chap. 3 I discussed various cases of co-optation between the five major systems of atomic chemistry in the nineteenth century. Particularly notable is how the fruits of Berzelian radical theory were absorbed into the substitution–type system, and how the classificatory molecular types of the latter system became crucial devices in the physical volume–weight system. All this is analogous to the kind of adjustment and re-invention that cultures and economies have to make when importing objects and practices from each other. Co-optation is likely to be an interpretative and adaptive process, even when it is not drastic as turning dephlogisticated air into oxygen.

5.2.3.3 Competition

Even if there is neither integration nor co-optation, there can be a productive *competitive* relationship between co-existing systems of practice. Philosophers of science often speak about the competition between theories, but typically competition is seen merely as a tally of the accomplishments of each system, as if they were achieved in total isolation from each other. This limitation is apparent in Lakatos's (1970) account of the competition between research programs, which is probably still one of the best-developed accounts of scientific competition widely known within the philosophy of science. Real-life competition is an interactive process, in which competitors pay attention to each other and become influenced by each other's behavior. That is the case in science as much as it is in sports or the economy. The effect of competition is not necessarily positive, but what I would like to do here is to consider what the positive effects are, and how to facilitate them while moderating the negative effects.

In order to get a clearer insight into the effect of competition in any particular case, we need to ask what the accepted rules and customs of the competition are, and how they shape the behavior of the agents involved. In the economic realm, what capitalists hold sacrosanct is the positive effect of competition in increasing efficiency; this comes, for example, from firms wanting to cut prices in order to attract more customers. I am oversimplifying things, but I think even this crude level of thinking would offer an improvement on the usual philosophical discourse on competition in science. How does competition affect scientific behavior? There is no simple story here, because what scientists want is not simple: it may be fame and prestige, it may be better research facilities and bigger research teams, it may be a secure teaching job, it may be money, or it may actually be the satisfaction of problem-solving and understanding. In practice much of it comes through the production of useful results or attractive ideas that are impressive to one's peers, to the next generation of scholars, and sometimes to the general public.

It is interesting to note how having a plurality of systems affects what is considered impressive. In a monistic context such as Kuhnian normal science, attention and creativity will be focused on a smaller number of problems, since everyone will be agreed on what the cutting-edge problems are. There will also be a tendency to be agreed on the epistemic values, so that a narrow range of types of solutions will be prized. It will be easy to suppress or neglect those who want to pursue different types of solutions or pursue different problems altogether, because the choice for them will be either to give up their preferences or to leave science altogether. In a pluralistic regime, the presence of competing systems will make it much harder to make people toe the line in the same way. Even the practitioners of a dominant system will need to work harder at justifying their own approach and "selling" it to funders, students and potential collaborators. In the production of impressive results, pluralism will make it more important to impress people at large about the merits of one's system, than to impress people already within one's own system about how good one is. Some may mourn the passing of the time when scientists could just make the best intellectual case for their research projects, and governments and foundations just came through with the necessary support. But that was a very rare situation enjoyed by physics and allied sciences in mid-twentieth century America, only made possible by the combination of the atomic bomb and the Cold War.[33]

A genuine competition between multiple systems of practice within a scientific field can have a liberating and stimulating effect on the development of science. I think the co-existence of the five systems of atomic chemistry discussed in Chap. 3 served exactly this sort of role, and similarly for the competing systems within "compound-water electrochemistry" discussed in Chap. 2. And it is important not to be hasty and drastic in the removal of systems that are not doing very well. Consider the long-running competition between the wave theory and the particle theory of light. There were various moments at which one or the other theory seemed to have a clear upper hand, but both were kept around, and I think that had the positive effect of mutual stimulation. The benefits of competition can be maintained only

---

[33] Even in that situation scientists weren't really receiving no-strings-attached money that had no influence on the content of their research, as Paul Forman (1987) argues through the case of quantum electronics. One does not have to agree with every aspect of Forman's work in order to see the main point of it, namely that the outcome of the sponsorship was to take "pure science" away from the ideal that the scientists themselves would have wanted.

when there *is* competition. In the economic realm ruthless competition can lead to a monopoly by the winner, which defeats the whole purpose of competition; likewise for science. If we are going to eliminate losers, we also need to ensure that there is a mechanism by which new competitors can take their places to keep the competition going.

Caution against a hasty removal of competitors also arises in relation to what I call the "lacuna effect". Imagine a situation in which two systems of knowledge have different aims, and one system does well in achieving the aims laid down by itself, while the other one does not succeed in its own aims. Common sense would dictate that the system that can't even succeed in its own terms should be discarded. I want to argue the opposite: as long as it is generally agreed that the failing system's aims are worthwhile, then it should be kept around, because its failures will serve as a reminder of valuable aims that science as a whole should strive to satisfy. If we simply discarded the failing system, it would be easy to forget the unachieved aims; it would be like abandoning whole areas of medicine because quick success is not being achieved there. Part of my critique in Chap. 1 of the premature dismissal of the phlogistonist system was that it allowed chemists to forget too easily about the need to explain things like the common properties of metals and their relation to electricity; even if the phlogistonists had not provided very convincing or informative explanations, they at least would have kept the questions alive. Or consider attempts by electrochemists to explain the mechanism of electrolysis (Chap. 2), and by some atomic chemists to obtain a full physical picture of the atom (Chap. 3); these attempts were mostly unsuccessful until the end of the nineteenth century, but they kept the important questions alive. Just for another, quick example, I would argue that the dominance of special relativity with a ruthless stifling of dissent has caused many physicists to forget that it may be a useful thing to seek dynamic explanations of relativistic effects.[34]

5.2.4 Tasks for History and Philosophy of Science

Having made arguments in favor of normative epistemic pluralism, I now want to consider how to make it active, how to put it into practice. The main action point is to *proliferate*: to foster valuable alternative scientific systems of practice alongside the orthodox and the fashionable. I intend pluralism as a doctrine about knowledge-building, not just knowledge-evaluation. In a way, it is obvious that people who can best put pluralism into practice in science are practicing scientists. However, it is also likely that scientists are already being as pluralistic as their professional constraints allow, and at any rate it is unlikely that many scientists will be inclined to change the way they do science following some philosophical doctrine articulated

---

[34] Harvey Brown's award-winning work *Physical Relativity* (2005) is an important corrective to this neglect.

from outside their own field. So it may well fall to those who are not professional scientists to undertake active pluralist work, and there are some distinct lines of useful work that historians and philosophers of science can plausibly carry out. I would like to articulate some concrete ideas about what we philosophers and historians of science can do to help ensure that "many things go" in science, and how such attention to pluralism can also change our own scholarly practices. Successful pluralist work in history and philosophy of science may eventually inspire scientists themselves to become more pluralist. Even if it doesn't, it can still improve the quality of scientific knowledge; that is the mission of "complementary science", which I address further in Sect. 5.3.4.

5.2.4.1 Pluralist Historiography[35]

I begin by recalling the memorable opening sentence of Kuhn's *Structure of Scientific Revolutions* (1970, 1): "History, if viewed as a repository for more than anecdote or chronology, could provide a decisive transformation in the image of science by which we are now possessed." Pluralist historiography can serve as an important corrective to common views of how science has progressed. Kuhn's work did much to discredit the traditional historiography of science that was infused with a certain amount of triumphalism, which approached history from the viewpoint of the winners, especially at some prominent junctures. Triumphalism is a peculiar kind of historiographical monism. Unlike whiggism, which writes history as a progression toward the present, triumphalism takes the winner's viewpoint on any given occasion, focusing particularly on salient points of victory such as Lavoisier's over the phlogistonists (see Chang 2009b). These triumphal moments acquire their salience through a complex and contingent historical process of re-telling, re-packaging and selective commemoration, and their choice is rather haphazard if seen from a principled whiggish point of view.[36] Even in our postmodern condition, triumphalism in historiography is alive and well at a deep level, as illustrated by the standard accounts to which I have opposed the first three chapters of this book. The presence of triumphalism is most obviously exhibited in the choice of subject matter that historians of science choose to investigate: why is there still so much focus on Darwin, Newton and such heroes of science, even as many of the very historians who write on them strongly disavow hero-worship? The need to capture popular imagination by talking about household names is not a good enough excuse: to know the contrary, one only has to recall the massive best-sellers about John Harrison (*Longitude*, Sobel 1995) and Henrietta Lacks (Skloot 2010).

---

[35] This needs to be distinguished from pluralism *about* historiography, as defended by Theodore Arabatzis (2008), which I am also in favor of.

[36] See Bensaude-Vincent (1996) on the commemorations of Lavoisier, and Chang (2009b) on the difference between triumphalism and whiggism, especially in the accounts of the Chemical Revolution.

There is profound potential in a pluralist re-orientation of the historiography of science. Although a great deal has already been achieved in the last few decades, I think much more work is needed, along with better articulation of what the enterprise is. Pluralism can be framed as an explicit historiographical directive, in three steps.

1. Pay particular attention to losing sides in past scientific debates, and do your best to see if they can't be constructed and understood as sensible alternatives that *unfortunately* got dropped. Historiographical pluralism is founded on a commitment to put under examination the complacent triumphalist assumption that the winning side won because it was right. In Chaps. 1 and 2, a re-examination of the Chemical Revolution and Ritter's unorthodox electrochemistry in this light paid handsome dividends. I believe that the same critical pluralist spirit has been behind many recent classics in the history of science—Holton (1978) on Millikan and Ehrenhaft, Kuhn (1957) on the Copernican Revolution, Shapin and Shaffer (1985) on Boyle and Hobbes, Pickering (1984) on elementary particle physics, Collins (2004) on gravitational waves, Worboys (2000) on theories of communicable diseases, etc, etc.—though the authors of these works themselves have often not identified their orientation explicitly as pluralism.[37]

2. Turn away from the preoccupation with *closure* and its explanation. By making an effort to shift our historiographical focus away from consensus-formation, we can counter the retrospective tidying-up tendency displayed by many historians, most scientists, and even some sociologists. In Chaps. 2 and 3 much of my focus was on the middle of the nineteenth century, during which time electrochemistry and atomic chemistry each proceeded without an overall theoretical consensus within it. The narrative in Chap. 3 was admittedly driven toward a familiar consensus (the H_2O formula for water and the system of atomic weights to go with it), but I also made an effort to show that the consensus reached did not cover all aspects of atomic chemistry. Pluralist historiography would also seek out and celebrate rugged individualists and quirky sub-communities, especially those who lacked the ability and resolve of a Priestley to make themselves known down through the ages.

3. Bring out plurality as a normal feature of science. In highlighting the parts of the history of scientific development that did not display consensus, try to gain an *overview* of the multiple systems of practice operating in mutual interaction. We need to craft fresh historiographical frameworks for this, and I think my notion of "systems of practice" (see Chap. 1, Sect. 1.2.1.1) constitutes a good start. I made some attempts in this historiographical direction in Chap. 3 by showing the interlinked development of five systems of atomic chemistry on one canvass, and in Chap. 2 by displaying the plurality of theories within "compound-water" electrochemistry. In Chap. 1 I was only able to include a brief

---

[37] Also congenial to historiographical pluralism is Douglas Allchin's idea of "reverse whiggism", in which "one starts from a former theoretical position, now in disrepute, and follows it forward, rather than backward, through time" (Allchin 1992, 110).

glimpse of the plurality within the phlogistonist system and even the oxygenist system, but there is much more that could have been said there, too.

As I hope my work in the first three chapters of this book has illustrated, there is much to be gained from a pluralist re-telling of historical episodes, even those that are widely considered to have been "done to death" already. We can recover much historical material that is very interesting in itself. And after an experience of telling stories in which nobody really won yet much good science was done, it will also be easy to see that science is not only about winning, or even agreeing. Not surprisingly, pluralist historiography will bring out pluralist science in action, and the uncovering of new historical material will inevitably also help shape a new image of science, reminiscent of Ludwig Wittgenstein's attempt to cure the "philosophical disease" caused by a "one-sided diet" consisting only of a particular kind of examples (Wittgenstein 1958, 155e, §593).

5.2.4.2 Pluralist Philosophical Practice

If the pluralist historian's initial task is to reveal the plurality that has in fact existed in the past of science, the pluralist philosopher's initial task is to ferret out monist preconceptions that underlie the received views of science. Once identified, these assumptions can be subjected to critical scrutiny to see if they are warranted. If they seem to be unwarranted, we can replace them with appropriate pluralist assumptions to see what kinds of new questions and answers emerge.

An excellent place to start would be the philosophical discourse on theory-choice, which strongly reflects a philosophical ideal of monism held by many scientists. Even admitting that they do not know whether they are in possession of the ultimately true theory, scientists still tend to think that if one of the competing theories is clearly better than the others, then the latter need to be eliminated. This notion is shared by many philosophers, whose discourse reinforces the scientists' presumptions. Even among those who do not think science deals in "truth", there is a widespread idea that scientists ought to work with only one theory at a time. The emblematic example here is Kuhn, with his insistence that a paradigm does and should enjoy a monopoly within a given field of science in its "normal" phases. Extraordinary science, in which competing paradigms co-exist, is presented by Kuhn as a temporary and uncomfortable phase which inevitably settles into another period of normal science.

Lakatos is a key exception on this score, but only apparently. Against Kuhn he maintains that there should always be multiple research programs in a field of science; from that, one might imagine that he was a pluralist. On the contrary, Lakatos wants multiple programs only so that they can compete with each other in such a way that scientists choose the best one at the end and discard the rest. In Sect. 5.2.3.3 above I raised an objection to this kind of notion of competition. Lakatos does not explain why there should be a clean ending to a phase of scientific research defined by the competition between a set of research programs. He seems to take for granted that one and only one of those programs will in the end stand out in its progressiveness, but gives no reason as to why we should expect that to be the case. Perhaps a phase of science is simply *defined* by such an end-point if it comes to be, which means that Lakatos is not making an empirical prediction about whether and how soon a competition is likely to end. That would bring us back to the question of historiography: why are we so ready to privilege those moments of clear victory, rather than ongoing competition?

Why can't theory-choice simply be a matter of each scientist deciding to take a particular avenue of investigation, without implying that all the other avenues are inferior, and that the inferior avenues should be closed off? Scientific rationality should not have to consist in every individual making a monistic choice and all individuals agreeing in that choice. It is very often rational for scientists to refrain from making a monistic choice between diverse systems of knowledge in a given field of study, and we philosophers may help scientists see that point more clearly if we start talking about theory-evaluation in a pluralist way. A closely related issue is the question concerning the inevitability of certain scientific results (Hacking 2000; Soler 2008). It is useful here to pay some attention to the "able" in "inevitable", which gently suggests that the question is about our own abilities as much as anything else. Inevitability is unavoidability, and whether something is unavoidable is generally not something we can tell without having made an attempt to avoid it. We may benefit from applying methodological skepticism to apparently rock-solid items of scientific knowledge: do not grant inevitability to anything until we've tried and failed to avoid it—in other words, until plurality has been attempted. The study of water contained in this book is an object lesson in this.

Even if one does not entirely agree with pluralism, just thinking about it seriously will put one off certain monist ways of framing philosophical issues linked to theory-evaluation. For instance, the very idea of inference to the best explanation will start to seem quite futile. Even setting aside the difficulty of needing to know about *all* plausible alternatives before we can say which one is the best, we have to ask why we should be so concerned about making that monistic inference down to one best alternative. A similar sense of futility will set in about debates on such topics as eliminative induction or the Bayesian concern with convergence, as well as all one-dimensional wrangles about scientific progress. Meanwhile, other philosophical topics will receive fresh formulations that will make their discussion more productive and interesting; for example, reductionism can be treated as a question about the relationship between different scientific systems of practice impinging on the same domain of phenomena, rather than an issue of hierarchy between scientific theories or fields, or an issue of one theory/field eliminating another or making it redundant.

5.2.4.3   Complementary Proliferation

The basic re-orientation of philosophy and historiography suggested so far will enable us to do even more active work, which goes beyond philosophy or history as it is normally conceived. That further work is very much in line with my conception of history and philosophy of science as "complementary science" (Chang 2004, chapter 6).

Pluralism about competing systems of knowledge allows us to see that the judgment of inferiority does not and should not equal a death-sentence. If we look back at history from that perspective, we will begin to see that there is still life in many of the "false" and "outdated" systems. If a system of knowledge once was well-established on the basis of success in engaging with reality, it is difficult to see how it would suddenly become invalid or useless, unless there is some genuine ontological change in the very laws of nature. In fact scientists often do preserve and use systems of knowledge that are deemed to be invalid in an ultimate sense. I mentioned how the geocentric view of the world still underlies our everyday life, and quite justifiably so. There are many other similar cases. Newtonian mechanics is still in use in most practical applications, and taught in all systems of physics education because its concepts still do have a role to play in modern practices. Orbitals still form the basis of much work in chemistry, although they are not supposed to exist according to up-to-date quantum theory. Geometric optics still has its uses; classical wave optics even more so. It is of course acknowledged that the old theories do not apply well outside the domains in which they are well-established, but it is also acknowledged that they still function in their own right in practice, and the in-principle reductions to newer theories are often mere promissory notes. Scientists may pay lip-service to an overreaching monism, but their actual practices tend to be much more pluralistic, even in many areas of theoretical physics. Only Weinberg-style "dreams of a final theory" make our well-established systems of knowledge suddenly appear shabby and not worth keeping.

The active task of pluralist history-and-philosophy of science is the proliferation of scientific systems to complement the current orthodoxy. As mentioned in the first section (Sect. 5.1) already, the first step in such complementary proliferation is *conservation*, which can be compared with the conservation work people do with wildlife, or languages threatened with extinction. Conservation starts with an identification of what remains in place, and continues with the effort to preserve it. As well-informed and slightly removed observers of science, we historians and philosophers of science can appoint ourselves as guardians of worthwhile systems of knowledge threatened with extinction. When we survey the history of science, we may find that some supposedly rejected past knowledge actually lives on in some form, as in the various examples just given above. In such cases, we can highlight the survival of those systems, and attempt to provide better mental and physical space for their continued flourishing.

The next step in complementary proliferation is to revive systems of knowledge that were killed off without sufficient justification. And what we preserve and revive, we can also develop further. Douglas Allchin (1997) has even used the concept of phlogiston successfully in teaching modern-day students the chemistry of redox reactions. And as I argued in Chap. 1, Sect. 1.2.4.1, phlogiston was *in effect* revived in later chemistry, under different names: chemical potential energy as Odling pointed out, and electrons as Lewis recognized. These neo-phlogistonist concepts were certainly developed further with great benefit. May there be other such cases, actual and potential? In Chap. 4, Sect. 4.1, I hinted that the notion of elementary water and Berzelius's and Davy's outdated theories of electrochemistry may still have useful roles to serve, if we can learn to relax our notion of what "elementary" means. Other possibilities for a productive revival that I have been pursuing that are not presented in this book include the following: the one-fluid theory of electricity, the concept of the positive radiation of cold (see Chang 2002), and Volta's original theory of the battery (see Chang 2011c).

To sum up: the ultimate aim of the active normative epistemic pluralism that I advocate is to improve science by cultivating multiple systems of knowledge. The most active service that history-and-philosophy of science can perform in this connection, going beyond description and commentary, is to address *scientific* questions that are not being dealt with by scientists because they are restricted by monist traditions—sometimes due to the necessities of normal science, sometimes for lack of imagination. History gives us an effective starting point, if we approach it with sufficient philosophical acumen to discern elements of the past that became discarded or hidden without good reason. That same approach can be applied to current science, too. I have given the name of "complementary science" to my own brand of history-and-philosophy of science: using the intellectual tools and perspectives of history and philosophy to address scientific questions that are neglected by current specialist science. What I did not quite see when I initially put that idea forward was that the project of complementary science was the expression of a thorough-going pluralism.[38]

5.3   Further Notes on the Practice of Pluralism

5.3.1   *Pluralism vs. The Pluralist Stance*

Some may think that my brand of pluralism is too definite and decisive. I have tried to make pluralism more compelling by making it more exuberant and unapologetic, instead of trying to make it more palatable by watering it down. But especially as humility and prudence feature prominently among my motivations, wouldn't it make sense to adopt a more cautious approach to pluralism itself? Kellert, Longino and Waters (2006, xiii) advocate what they call "the pluralist stance", which is "empirically motivated" from the various studies contained in the volume they put together.[39] They define the pluralist stance as follows: "a commitment to avoid reliance on the monist assumptions in interpretation or evaluation coupled with an openness to the ineliminability of multiplicity in some scientific contexts."

---

[38] I will explore this connection further in Sect. 5.3.4.

[39] I would like to thank Ken Waters for pushing me to think about why I shouldn't retreat to the pluralist stance, and also John Norton for a similar discussion.

There are two aspects of the Kellert–Longino–Waters pluralist stance from which I diverge. The first is straightforward: they take pluralism (or the pluralist stance) as "an approach to interpreting the content and practices of scientific inquiry"; my active pluralism consciously goes beyond interpretation. My feeling is that merely interpreting science this way and that makes little difference, and Kellert, Longino and Waters anticipate this issue (p. xv): "What is the advantage of the pluralist interpretation? . . . it provides a means of avoiding senseless controversies that do not lead to progress. It also helps emphasize the partiality of scientific knowledge." While I agree with these aims, I think they are too passive. As explained in earlier parts of this chapter, I aspire to go beyond the removal of distractions and promote scientific progress directly (at least by learning to *live* with the recognized partiality of scientific knowledge).

The other point of divergence is more subtle: I think trying to choose between pluralism and the pluralist stance is going to be futile, because *in practice* the pluralist stance comes down to the same thing as full-fledged pluralism. Kellert, Longino and Waters (p. xiv) leave the question open as to whether pluralism or monism is the more productive approach in a given subject area: "We do not hold that for every phenomenon there will inevitably be multiple irreducible models or explanations. We hold that the task of identifying which situations require multiple approaches requires empirical investigation." Now, the only way we can answer that question is to try out both monist and pluralist approaches. But how would we try out pluralism? For that experiment, we have to make a genuine effort to create and cultivate a set of systems, and observe how they develop, each of them in itself and also through mutual interaction. We have to keep this going long enough to see whether any trends in successfulness that we detect are stable; if it turns out that the particular combination of systems that we try out really doesn't deliver the goods, we have to try some other combinations of systems before we give up on pluralism in general. By that point, we are up to our necks in pluralism with no clear end of the experiment in sight, so we might as well *be* pluralists! The empirical question can only really be answered post-commitment, and it is pointless to insist on treating a question as empirical if we are not going to try to answer it through real experience.

It is in principle possible that one would engage in this experiment seriously, and at long last conclude: monism consistently yields better results, so we should have been monist all along, and we should be monist in the future. I do not think this is a likely outcome, but one cannot rule out the possibility in principle. But who would be willing to engage in such an experiment? Even if operational difficulties could be overcome,[40] I do not see anyone with any non-negligible attachment to monism coming into this bargain. And who could blame them? Here is a very imperfect and colorful analogy, to illustrate the asymmetry between monism and pluralism in this regard. The pluralist stance is like trying to decide between monogamy and promiscuity by trying both out. This experiment will be no big deal for someone happy

---

[40] For one thing, there would be formidable difficulties in finding plausible methods of measuring the benefits of each scheme precisely enough.

with promiscuity, but for someone in a monogamous relationship, trying out promiscuity is a huge price to pay. By the time the experiment is sufficiently advanced, one *is* promiscuous, and will probably have been deserted by one's faithful partner! "Stepp'd in so far that, should I wade no more/Returning were as tedious as go o'er."[41]

So the experiment of the pluralistic stance would have to be undertaken by pluralists (or at least people without strong predisposition toward monism). It is imaginable for a community of pluralists to test out monism for a trial period. But we will find that there is no need for this monist trial. A great deal of scientific work in the last century or so has been experiments in monism! We have gained enough experience of monism, and we know what successes we have had with it. What we badly lack is any comparable data on the pluralist side, since pluralism has not been tried out on a large scale in recent science. So what we need to do now is get a great deal of pluralist science going, alongside the monist science that will no doubt be continuing. Some may say that in certain areas of biology and some newer fields of science, there is already a good deal of pluralism being practiced. If so, that is all to the good, but my sense is that a lot more needs to be happening if we are to generate a large enough body of experience on the pluralist side to match what we have got on the monist side. (I am not in favor of over-simplifying the field here into monism vs. pluralism, but I think my comments made in this section can easily be adapted to a continuous spectrum or even a multi-dimensional space for characterizing degrees and types of pluralism.)

5.3.2 Between Metaphysics and Epistemology

I have billed my position as epistemic pluralism. Should I not support a metaphysical pluralism as well? Doesn't epistemic pluralism actually require at least some degree of metaphysical pluralism? After all, I did premise my arguments on humility, which arises from a sense that reality is more abundant and complex than our minds can grasp through simple schemes. So I do not pretend that my epistemic position is entirely free from some basic metaphysical assumptions. However, I do want to stay clear of any *specific* metaphysical premises or conclusions, because I do not think it is possible to support them well enough. This is consistent with the bit of metaphysics that I do presume, which only amounts to saying that the true shape of reality (whatever that might really mean) is not directly accessible to us! There are a few particular ways in which I do not want to be metaphysical.

First, I make no assumptions of plurality in ontology. I think ontological pluralism is an unverifiable opinion just as much as ontological monism is, and it does not underpin my epistemic pluralism. If metaphysical pluralism means a denial of the existence of "the one world", I do not have much appetite for it, and talk of

---

[41] *Macbeth*, Act 3, Scene 4, lines 136–137.

possible worlds or multiverse does not contribute to the line of thought that I am trying to explore here. I would be happy to go with Israel Scheffler's (1999, 425) "plurealism", which affirms "one world of objects not of our making" (or, "upholds the existence of objects independent of our making and accessible to inquiry") but "denies that inquiry into such objects converges toward a unique world-version." This position is ontologically monistic to the extent that it assumes external reality to be one, but epistemically pluralistic in allowing that inquiry about the "one world" may coherently yield a plurality of "world-versions".

Such a position also involves no specific metaphysical commitment about what the presumed one world consists of. So it is fully compatible with ontological pluralism in the sense of granting reality to various types of entities. For example, John Dupré's "promiscuous realism" (1993, 7 and *passim*) accepts that many kinds of things are real, not just one (matter or mind), or two (matter and mind). Paul Feyerabend (1999, 3) puts the point more poetically: "The world we inhabit is abundant beyond our wildest imagination. There are trees, dreams, sunrises; there are thunderstorms, shadows, rivers; there are wars, flea bites, love affairs; there are lives of people, Gods, entire galaxies. . . . There is no limit to any phenomenon, however restricted." I think this pluralistic vision is esthetically wonderful, but epistemic pluralism does not require a commitment to it.

My version of epistemic pluralism is also not premised on a strong assumption about the complexity of nature. As mentioned in the first section (Sect. 5.1), I cannot convince myself that the world is a "dappled" place, as Nancy Cartwright argues. I do agree with her that the record of scientific practice very often shows the failure of schemes that presume unity and simplicity, and that real-world problems tend to be handled most effectively by particular and piecemeal approaches. Now, as William James put it, "pluralistic empiricism" suggests a picture of a world that is "a turbid, muddled, gothic sort of affair without a sweeping outline and with little pictorial nobility."[42] Again, if I had to make a metaphysical choice, I would assent to the picture of the world offered by James and Cartwright rather than the opposite. However, such a choice is neither necessary nor truly warranted. The evidence from the track record of our humble science from the last few centuries is not strong enough to indicate anything conclusive about the very shape of reality. Unifying schemes have also had some impressive success in science, although not an unmixed series of triumphs as some monists would have us think. And there might even be a mysterious kind of unity and simplicity in nature that I cannot comprehend and would therefore find very complicated if it were revealed to me!

What we know for reasonably sure is that the world is difficult for humans to fathom, not that it is either complex or simple in some absolute sense. Any known ontological complexity would tend to strengthen arguments for pluralism. However, it is also important to recognize that even the apparently simple things in nature can still be unfathomable and may exhibit an indefinite amount of complexity when probed in the right way. In this book I hope to have demonstrated that pluralism is a

---

[42] Quoted in Bernstein (1989, 10), from William James, *A Pluralistic Universe*.

cogent philosophy even in the "simplest" of physical domains, not just in obviously complex subjects such as the biology of social insects (Mitchell 2003), or the sciences of human behavior (Longino 2006). Again, the introduction by Kellert, Longino and Waters provides a friendly and articulate foil (2006, xi): "We think that some phenomena may be such (e.g., so complicated or nebulous) that there can never be a single, comprehensive representation of everything worth knowing, or even of everything causal (or fundamental), about the phenomenon." It may be that the basic constitution of water is such a complicated or nebulous thing, but in that case I submit that there is little hope of finding many scientific subjects that are much simpler.

In distinguishing my pluralism from a metaphysical position, it is also pertinent to stress that describing the nature of reality is not my ultimate aim in advocating pluralism. Rather, I unavoidably make some metaphysical assumptions with reluctance and discomfort, and I really only make negative assumptions, regarding what cannot be captured by our meager and simplistic schemes. All in all, in an attenuated version of Feyerabend's vision quoted above, I believe that human understanding is not capable of exhausting the abundance of nature. That is entirely consonant with both the humility I have advocated in this chapter and the striving of active realism that I expressed in Chap. 4.

5.3.3 Can Monists Help, Too?

It cannot be denied that there is a positive motivational side to monism. And perhaps the more fanatical the monism, the better—there are few things so motivating as a firm belief in the unique superiority of one's own system. In human history, many of the greatest achievements have been made through fanaticism. Think, for example, about the great medieval cathedrals of Europe. I am full of admiration for them—not only the supreme skill and organization that must have been required in their design and construction, but the dedication and conviction it must have taken to even think about erecting such formidable structures with so little by way of supporting technologies. With all due respect, I cannot avoid the conclusion that those who built the cathedrals must have been religious fanatics, and that fanaticism must have been essential to such great achievements. But at what cost, not only in material terms but in the lives of those who gave their sweat and blood to the enterprise? And we must consider what else the same fanaticism resulted in—everything from the burning of heretics to the hierarchical domination of society. Yes, the raising of the great cathedrals was an impressive achievement, which required fanaticism and human sacrifice at that stage of history. But our continuing admiration for that achievement should not stifle honest questions about whether it really was the best thing for medieval Europe to devote itself to. Even if we do not want to get into the business of judging the past, we can at least easily agree that today no fanaticism is required for the building of structures on a similar scale. Our society is sufficiently affluent and our technological systems sufficiently mature to build grand buildings without extreme human suffering or faith in any other-worldly purpose that they should serve.

The parallel to science is quite suggestive. Yes, it probably was necessary for a Newton, a Kepler or an Einstein to have a near-fanatical monistic faith in order to dedicate themselves sufficiently to their task, and this would have been the easiest if they thought in terms of truth, clearly linked with religious truth at least in the case of Newton and many of his contemporaries. (Is "monistic" really "monastic"? There is actually an etymological link, as "monastery" and "monk" both ultimately derive from the Greek *monos*.) And it was perhaps felt necessary that "everyone" should be "Newtonian" for some time, to muster and concentrate sufficient amounts of resources and manpower into the most promising direction, especially after the Cartesians and the Leibnizians had finally been defeated. But again, at what cost? Bridgman, for one, lamented the fact that physicists were so spellbound by Newtonian orthodoxy that they did not recognize for two centuries the operational meaninglessness of absolute space and time, and hoped that better awareness would "render unnecessary the services of the unborn Einsteins." (Bridgman 1927, 24) Nowadays much science seems to be done without fanatical monism. With greater available resources, massively enhanced computing power, and a greater number of researchers in the business, science can be more relaxed about trying out various different models instead of having to focus all its energy and resources in one direction in order to get anywhere. I think Kuhn was probably correct in detecting a necessity in paradigm-monopoly in some of the historical episodes that he was studying. But that was not always the case, as illustrated amply in Chaps. 1, 2, and 3 of this book. And even if Kuhn were right about history, that would still not imply that the present and the future of science will be, or should be, like its past. Perhaps we lose a certain type of spiritual value when we grow out of scientific monism, but we do gain greater maturity (see Chap. 4, Sect. 4.2.4). And there are other kinds of positive spiritual value in a pluralist regime, just as there is religious fulfillment offered by the more tolerant religious denominations.

The question about the value of monism can also be put in the form of a question about the reflexivity of pluralism, or pluralism at a meta-level. Shouldn't pluralism be applied to itself, too, so that we allow both monism and pluralism? That would also be consonant with the Kellert–Longino–Waters pluralistic stance: let monism and pluralism compete, and see which one does better. The first comment I have on this question is that there is no general obligation to jump to the meta-level. Pluralism as I advance it is a doctrine about science; it is not itself part of science; therefore, it does not apply to itself. It would be a different matter for a fully naturalistic philosophy of science, which regards itself as part of science. That much is simple. There is a subtler point, however, in relation to my notion of complementary science (see Sects. 5.2.4 and 5.3.4 for further discussion), which proposes that scientific questions that are neglected by scientists themselves should be taken up by philosophers of science. Doesn't that bring my brand of philosophy of science into the realm of science, so that the reflexivity question becomes applicable? Yes, but pluralism itself is not even part of complementary science, even though it provides a very important motivation and justification for it.

Still, we do need to deal with the question of the place of monism under a pluralistic regime of science, just as we need to deal with the political question about what to do with absolutists in a tolerant society. I propose to do this by straightforwardly considering the advantages and disadvantages of monism, without getting tangled up in reflexive questions. We can and should harness the energy that monism and only monism can generate, *in some people*. To enter into an analogy again: if some people will only do charitable work and lead an upright life by believing in a monotheistic god, then society should welcome this belief on the part of those people. And pluralism would recommend that various brands of monists should be encouraged to pursue their own enterprises, so that they each develop and bring benefits to society. However, there are two important qualifications we need to make to this pluralist appreciation of monism. First, if we only have monists, we can only reap the benefits of *tolerant* pluralism. To realize *interactive* pluralism, at least some of the people must be pluralists in order to allow genuine interactions. Even the benefits of competition will only be realized if people compete *with* each other according to the rules, rather than simply pursue their own systems paying no regard to others. Secondly, pluralism can only allow monists who respect pluralism sufficiently. Perhaps we can allow harmless bravado about pursuing the one unique truth, etc., but pluralism cannot allow intolerant monists who would exterminate the competition. There is an obvious analogy to the limits that a liberal democracy must place on totalitarian political movements.

5.3.4   Complementary Science Continued

In my previous book, *Inventing Temperature* (Chang 2004, 3), I defined complementary science as follows:

> [Complementary science] contributes to scientific knowledge through historical and philosophical investigations. [It] asks scientific questions that are excluded from current specialist science. It begins by re-examining the obvious, by asking why we accept the basic truths of science that have become educated common sense. Because many things are protected from questioning and criticism in specialist science, its demonstrated effectiveness is also unavoidably accompanied by a degree of dogmatism and a narrowness of focus that can actually result in a loss of knowledge. History and philosophy of science in its "complementary" mode can ameliorate this situation.[43]

The present book has been an attempt to continue the project of complementary science, and I would now like to give a brief assessment of how well it has served that aim. I will do that in two different ways. First, I will assess the fruits of this book according to the three main ways in which complementary science can advance scientific knowledge. Second, especially in the light of that assessment, I would like to clarify further the relation between pluralism and complementary science.

The three categories of complementary scientific contributions are recovery, critical awareness, and new developments (Chang 2004, 241–247). Has my work in

---

[43] The initial articulation of the idea was given in Chang (1999), which can still serve as a freestanding manifesto, although the statement in Chang (2004, ch. 6) is more extensive and further developed.

this book made much recovery of forgotten scientific knowledge? In the factual sense, I have not dug up such striking lost facts here as I have done in some of my previous works, such as the anomalous variations in the boiling point of water (Chang 2004, ch. 1; 2007b) and the reflection of cold (Chang 2002). However, I have presented here a brief glimpse from Priestley's electrochemical experiments that is the tip of an iceberg (Chap. 2, Sect. 2.3.2). In early electrochemistry there was a profusion of strange-sounding experimental results, which I have begun to reproduce and explore further (see Chang 2011c); this will be the subject of my next major project. In the current work, the main payoff of recovery has been in the rehabilitation of discarded ideas (rather than facts). Most significantly, I have been able to recognize the phlogistonist system of chemistry as a cogent and viable alternative to Lavoisierian chemistry. I hasten to add that this point is not truly original, as numerous historians of science have recognized it. However, I do think I have managed to make an unabashed and unfettered presentation of the point. I have also recalled past instances of the recovery of phlogiston by Odling and Lewis, themselves long-forgotten. Similarly, I have strengthened the rehabilitation of Ritter's interpretation of electrolysis, and connected it to the rehabilitation of phlogiston.

Perhaps the most serious complementary work contained in this book has been in the direction of critical awareness. The whole study has been driven by one overarching question: how do we know that water is H_2O? How did scientists come to agree on that? I have investigated in careful detail three of the most important developments in the shaping of the scientific view of water as H_2O. In each case I found very significant debates and disputes among scientists, and concluded in each case that the winning side did not have a complete and unequivocal superiority. The fact that the subject at hand was such a crucial item of scientific common sense as "Water is H_2O" gave a sense of urgency to the critical awareness. I believe that it is very important for us all to be aware that water is not *simply* H_2O, and to know the subtle and sophisticated reasons that scientists had for arriving at that belief. This sort of critical awareness enhances the quality of our scientific knowledge. And I submit that raising critical awareness is a key task of historians and philosophers, especially when we find past scientific decisions that are difficult to justify. I reject each of the following common assumptions that variously hamper critical awareness on the part of historians and philosophers of science: (i) scientists generally make the right decisions, and we should primarily look for good ways of rationalizing what they do; (ii) any lack of certainty and rationality in scientific theory-choice is only apparent, and must be dispelled by attention to neglected factors; (iii) we should judge the epistemic merits of past science according to the verdict of current science; (iv) or, we should not judge the epistemic merits of past science at all; (v) we should above all look for causal explanations of the scientific decisions that have been made.

In the direction of new developments, the immediate fruits of the current project have been meager. I think the revival of phlogiston sketched in Chap. 1 would have led to exciting new developments, but in the two long centuries after its premature death, chemists and physicists did eventually get at those developments through different routes, as explained in Sect. 1.2.4. It is perhaps not entirely trivial that my pondering about atoms, molecules and elements in Chaps. 3 and 4 has yielded some useful new insights about what it means to call something an "element". However, I think the farthest-reaching future consequence of my work on this book comes out of Chap. 2: namely, the opening-up of the complementary research project on basic electrochemistry, which is just under way. As shown in my very preliminary report (Chang 2011c), I have begun some research aimed at providing a renewed understanding of Volta's original battery (employing salt water as the electrolyte), going beyond purely historical work on the nineteenth-century debates on how it works. Theoretically, this research has so far laid down some groundwork for the incorporation of Volta's original concept of contact potential into the modern framework of electrochemistry, suggesting a modification of the latter at least as it is presented in elementary textbooks. Experimentally, I have produced some phenomena that are not readily familiar even to most research chemists of today, including the direct decomposition of H_2O at the cathode, the solution of gold at the anode, and a preferential production of Cu^+ ions over Cu^{++} ions, all in the electrolysis of NaCl solutions of sufficient concentration. I have also made observations of electrolyte-less cells employing de-ionized water, or a finger between two metals (reminiscent of De Luc's "dry pile"). There is much more to come.

I am happy to conclude that this project on the early history of water has continued the complementary science project in many exciting ways. As outlined above, ample affirmation has been given of the three main ways in which I envisaged complementary science contributing to the improvement of scientific knowledge. In addition, the articulation and defence of pluralism given in this chapter constitutes a further articulation and elaboration of the complementary science project, filling out a brief tentative statement made in my earlier exposition: "complementary science is inherently a pluralistic enterprise. . . . there is no unthinking dismissal of theoretical possibilities in complementary science. . . . When the complementary scientist picks up a rejected research program to explore its further potential, or suggests a novel research program, that is also not done with the crank's conviction that his particular heresy represents the only truth." (Chang 2004, 247) Complementary science is the expression of active normative epistemic pluralism in history and philosophy of science.

References

Allchin, Douglas. 1992. Phlogiston after oxygen. *Ambix* 39: 110–116.

Allchin, Douglas. 1997. Rekindling phlogiston: From classroom case study to interdisciplinary relationships. *Science and Education* 6: 473–509.

Anderson, Philip W. 1972. More is different: Broken symmetry and the nature of the hierarchical structure of science. *Science* 177(4047): 393–396.

Anderson, Philip W. 2001. Science: A 'dappled world' or a 'seamless web'? [Essay review of Nancy Cartwright, *The dappled world*]. *Studies in History and Philosophy of Modern Physics* 32: 487–494.

Arabatzis, Theodore. 2008. Causes and contingencies in the history of science: A plea for a pluralist historiography. *Centaurus* 50: 32–36.

Bensaude-Vincent, Bernadette. 1996. Between history and memory: Centennial and bicentennial images of Lavoisier. *Isis* 87: 481–499.

Bernstein, Richard J. 1989. Pragmatism, pluralism and the healing of wounds. *Proceedings and Addresses of the American Philosophical Association* 3(63): 5–18.

Birks, J.B., ed. 1962. *Rutherford at Manchester*. London: Heywood.

Bloor, David. 2007. Epistemic grace: Antirelativism as theology in disguise. *Common Knowledge* 13: 250–280.

Bridgman, Percy Williams. 1927. *The logic of modern physics*. New York: Macmillan.

Brown, Harvey. 2005. *Physical relativity: Space-time structure from a dynamical perspective*. Oxford: Oxford University Press.

Cartwright, Nancy. 1999. *The dappled world: A study of the boundaries of science*. Cambridge: Cambridge University Press.

Cat, Jordi, Nancy Cartwright, and Hasok Chang. 1996. Otto Neurath: Politics and the unity of science. In *The disunity of science*, ed. Peter Galison and David Stump, 347–369. Stanford: Stanford University Press.

Chang, Hasok. 1995. The quantum counter-revolution: Internal conflicts in scientific change. *Studies in History and Philosophy of Science* 26: 121–136.

Chang, Hasok. 1997. Can Planck's constant be measured with classical mechanics? *International Studies in the Philosophy of Science* 11: 223–243.

Chang, Hasok. 1999. History and philosophy of science as a continuation of science by other means. *Science and Education* 8: 413–425.

Chang, Hasok. 2002. Rumford and the reflection of radiant cold: Historical reflections and metaphysical reflexes. *Physics in Perspective* 4: 127–169.

Chang, Hasok. 2004. *Inventing temperature: Measurement and scientific progress*. New York: Oxford University Press.

Chang, Hasok. 2007b. The myth of the boiling point. http://www.cam.ac.uk/hps/chang/boiling. First posted on 18 Oct 2007.

Chang, Hasok. 2009b. We have never been whiggish (about phlogiston). *Centaurus* 51: 239–264.

Chang, Hasok. 2011c. How historical experiments can improve scientific knowledge and science education: The cases of boiling water and electrochemistry. *Science and Education* 20: 317–341.

Collins, Harry M. 2004. *Gravity's shadow: The search for gravitational waves*. Chicago: University of Chicago Press.

De Regt, Henk, Sabina Leonelli, and Kai Eigner, eds. 2009. *Scientific understanding: Philosophical perspectives*. Pittsburgh: University of Pittsburgh Press.

Duhem, Pierre. 1962. *The aim and structure of physical theory*. New York: Atheneum.

Evans, James, and Alan S. Thorndike, eds. 2007. *Quantum mechanics at the crossroads: New perspectives from history, philosophy and physics*. Berlin: Springer.

Faraday, Michael. 1834. Experimental researches in electricity, seventh series. *Philosophical Transactions of the Royal Society* 124: 77–122.

Feyerabend, Paul. 1975. *Against method*. London: New Left Books.

Feyerabend, Paul. 1999. *The conquest of abundance: A tale of abstraction vs. the richness of being*. Chicago: University of Chicago Press.

Forman, Paul. 1987. Behind quantum electronics: National security as basis for physical research in the United States, 1940–1960. *Historical Studies in the Physical and Biological Sciences* 18(1): 149–229.

Fortun, Mike, and Herbert J. Bernstein. 1998. *Muddling through: Pursuing science and truths in the 21st century*. Berkeley: Counterpoint.

Freund, Ida. 1904. *The study of chemical composition*. Cambridge: Cambridge University Press.

Fuller, Steve. 2008b. *Dissent over descent: Intelligent design's challenge to Darwinism*. Cambridge: Icon Books.

Galison, Peter. 1988. History, philosophy, and the central metaphor. *Science in Context* 2: 197–212.

Galison, Peter. 1997. *Image and logic: A material culture of microphysics*. Chicago: University of Chicago Press.

Gillies, Donald A. 2008. *How should research be organised?* London: College Publications.
Gordin, Michael. 2004. *A well-ordered thing: Dmitrii Mendeleev and the shadow of the periodic table*. New York: Basic Books.
Hacking, Ian. 1983. *Representing and intervening*. Cambridge: Cambridge University Press.
Hacking, Ian. 2000. How inevitable are the results of successful science? *Philosophy of Science* 67: 58–71.
Hartley, Harold. 1971. *Studies in the history of chemistry*. Oxford: Clarendon Press.
Hertz, Heinrich. 1899. *The principles of mechanics* (trans: Jones, D.E., and Walley, J.T.). London: Macmillan.
Holton, Gerald. 1978. Subelectrons, presuppositions, and the Millikan–Ehrenhaft dispute. *Historical Studies in the Physical Sciences* 9: 161–224.
Holton, Gerald, Hasok Chang, and Edward Jurkowitz. 1996. How a scientific discovery is made: A case history. *American Scientist* 84: 364–375.
Kellert, Stephen H., Helen E. Longino, and C. Kenneth Waters, eds. 2006. *Scientific pluralism*. Minneapolis: University of Minnesota Press.
Kim, Kiheung. 2006. *The social construction of disease: From scrapie to prion*. London: Routledge.
Kitcher, Philip. 1993. *The advancement of science: Science without legend, objectivity without illusions*. New York/Oxford: Oxford University Press.
Kitcher, Philip. 2011. *Science in a democratic society*. Amherst, NY: Prometheus Books.
Knight, David. 1967. *Atoms and elements*. London: Hutchinson.
Kuhn, Thomas S. 1957. *The Copernican Revolution: Planetary astronomy in the development of Western thought*. Cambridge, MA: Harvard University Press.
Kuhn, Thomas S. 1970. *The structure of scientific revolutions*, 2nd ed. Chicago: University of Chicago Press.
Kuhn, Thomas S. 1977. Objectivity, value judgment, and theory choice. In *The essential tension: Selected studies in scientific tradition and theory change*, 320–339. Chicago: University of Chicago Press.
Kuhn, Thomas S. 2000. *The road since Structure: Philosophical essays, 1970–1993, with an autobiographical interview*. Chicago: University of Chicago Press.
Lakatos, Imre. 1970. Falsification and the methodology of scientific research programmes. In *Criticism and the growth of knowledge*, ed. Imre Lakatos and Alan Musgrave, 91–196. Cambridge: Cambridge University Press.
Laudan, Larry. 1981. A confutation of convergent realism. *Philosophy of Science* 48: 19–49.
Liebig, Justus. 1851. *Familiar letters on chemistry, in its relations to physiology, dietetics, agriculture, commerce, and political economy*, 3rd ed. London: Taylor, Walton, & Maberly.
Longino, Helen. 2006. Theoretical pluralism and the scientific study of behavior. In *Scientific pluralism*, ed. Stephen H. Kellert, Helen Longino, and C. Kenneth Waters, 102–131. Minneapolis: University of Minnesota Press.
Melhado, Evan M. 1980. *Jacob Berzelius: The emergence of his chemical system*. Stockholm: Almqvist & Wiksell International.
Miller, Arthur I. 1998. *Albert Einstein's special theory of relativity: Emergence (1905) and early interpretation (1905–1911)*. New York: Springer.
Mitchell, Sandra D. 2003. *Biological complexity and integrative pluralism*. Cambridge: Cambridge University Press.
Mitchell, Sandra D. 2009. *Unsimple truths: Science, complexity, and policy*. Chicago: University of Chicago Press.
Pickering, Andrew. 1984. *Constructing quarks: A sociological history of particle physics*. Chicago: University of Chicago Press.
Popper, Karl. 1970. Normal science and its dangers. In *Criticism and the growth of knowledge*, ed. Imre Lakatos and Alan Musgrave, 51–58. Cambridge: Cambridge University Press.
Popper, Karl. 1981. The rationality of scientific revolutions. In *Scientific revolutions*, ed. Ian Hacking, 80–106. Oxford: Oxford University Press.
Preston-Thomas, H. 1990. The international temperature scale of 1990 (ITS-90). *Metrologia* 27: 3–10.

Priestley, Joseph. 1790. *Experiments and observations on different kinds of air, and other branches of natural philosophy, connected with the subject*, vol. 3, 2nd ed. Birmingham: Thomas Pearson.

Putnam, Hilary. 1995. *Pragmatism: An open question*. Oxford: Blackwell.

Ratzsch, Del. 1996. *The battle of beginnings: Why neither side is winning the creation–evolution debate*. Downers Grove: InterVarsity Press.

Ruse, Michael. 2005. *The evolution–creation struggle*. Cambridge: Harvard University Press.

Scerri, Eric R. 2007. *The periodic table: Its story and significance*. New York: Oxford University Press.

Scheffler, Israel. 1999. A plea for pluralism. *Transactions of the Charles S. Peirce Society* 35: 425–436.

Shapin, Steven, and Simon Schaffer. 1985. *Leviathan and the air-pump: Hobbes, Boyle, and the experimental life*. Princeton: Princeton University Press.

Shea, William. 1987. The quest for scientific rationality: Some historical considerations. In *Rational changes in science: Essays on scientific reasoning*, ed. Joseph C. Pitt and Marcello Pera, 155–176. Dordrecht: Reidel.

Skloot, Rebecca. 2010. *The immortal life of Henrietta lacks*. New York/London: Crown Publishers.

Sobel, Dava. 1995. *Longitude: The true story of a lone genius who solved the greatest scientific problem of his time*. New York: Walker.

Soler, Léna. 2008. Revealing the analytical structure and some intrinsic major difficulties of the contingentist/inevitabilist issue. *Studies in History and Philosophy of Science* 39: 230–241.

Stanford, P. Kyle. 2006. *Exceeding our grasp: Science, history and the problem of unconceived alternatives*. New York: Oxford University Press.

Thagard, Paul. 1978. Why astrology is a pseudoscience. *PSA 1978: Proceedings of the Biennial Meeting of the Philosophy of Science Association*, vol. 1, 223–234.

Van Fraassen, Bas. 1980. *The scientific image*. Oxford: Clarendon Press.

Watkins, John. 1970. Against normal science. In *Criticism and the growth of knowledge*, ed. Imre Lakatos and Alan Musgrave, 25–37. Cambridge: Cambridge University Press.

Weinberg, Steven. 1992. *Dreams of a final theory*. New York: Random House.

Wimsatt, William C. 2007. *Re-engineering philosophy for limited beings: Piecewise approximations to reality*. Cambridge: Harvard University Press.

Wittgenstein, Ludwig. 1958. *Philosophical investigations* (trans: Anscombe, G.E.M.). New York: Macmillan.

Worboys, Michael. 2000. *Spreading germs: Diseases, theories, and medical practice in Britain, 1865–1900*. Cambridge: Cambridge University Press.

Wylie, Alison, ed. 2006. *Epistemic diversity and dissent, Part 1*. Special issue of *Episteme: A Journal of Social Epistemology, vol. 3, issue 1/2*.

Index

A

Abduction hypothesis. *See* Invisible transport
Ability, 215–216
Absolute space and time, 270, 277, 295
Absolutism, 261
Académie des Sciences, 76
Accuracy, 36, 137, 139, 164–165, 207, 230, 273
Acetic acid, 175–178
Acid, xvii, 6–9, 11, 20–21, 28, 32–33, 44, 48, 54, 56, 60, 61, 75, 83, 86, 93, 96, 102–104, 109, 115, 116, 118, 121, 123, 140–142, 148, 155–156, 162, 166–167, 215, 226
Acidity, theory of. *See* Acid; Lavoisier, theory of acidity
Action (of human agents), xix, 15, 17, 51, 64, 253–298
Action at a distance, 109
Active realism. *See* Realism, active
Activity. *See* Epistemic activity
Ad hoc, xviii, 11, 24, 54, 93, 97, 145, 279
Aepinus, F., 107, 237
Affinity, 39–40, 43, 59–62, 102, 108, 109, 122, 125, 138, 145, 161, 186, 187
Affirmation, 214, 298
Age, 57, 78, 94, 99, 238
Aim (of a system of practice), 16
Aim (of an epistemic activity), 15–17
Aim (of science), 112, 219, 260, 270, 273–275
Aim, epistemic, 112, 151
Air, 1–11, 20, 23, 26, 30, 40–45, 52–55, 59–61, 74, 80, 81, 91, 118, 125, 140, 282
Alcohol. *See* Ethyl alcohol
Alkali, 75, 102–103

Alkali metal, 87, 103, 209
Alkaline earth, 103
Allchin, D., viii, 25, 31, 32, 44, 45, 65, 286, 289
Allotropy, 280
Almeder, R., 196, 197
Amateur science, 2, 134
Ammonia, 24, 91, 96, 102, 142, 148, 150, 171, 179, 182–183
Ammonia type, 171, 182–183
Ampère, A.M., 107, 144, 168
Anderson, P.W., 259
Annales de chimie, 58, 98
Anode, 83, 102, 109, 115–121, 211–212, 298
dissolution of, 118–120
Anti-anti-phlogistonist, 30, 31, 34, 58
Antiphlogistic (or antiphlogistian) chemistry, 15, 30
Anti-realism, 192, 215, 223, 245–246
Aqua fortis. *See* Nitric acid
Arabatzis, T., viii, 285
Aristotle, 1, 270
Arrhenius, S., xviii, 78, 79, 87, 101, 109–114, 185, 194, 226
Arsenal (Paris), 24, 276
Asbestos, 76, 98–99, 105
Atom
    chemical, xv, 147, 152–163
    counting, 133, 158–160, 167, 170, 182, 183, 186, 190–191, 211, 280
    fixing power, 179–182
    physical, xix, 139, 149, 152, 153, 156, 163, 178, 195
    reality of, 194–195
Atom-binding power. *See* Atom, fixing power

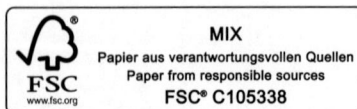

Printed by Books on Demand, Germany